THE PETROLEUM DICTIONARY

THE PETROLEUM DICTIONARY

DAVID F. TVER
RICHARD W. BERRY, PH.D.

VNR VAN NOSTRAND REINHOLD COMPANY
NEW YORK   CINCINNATI   TORONTO   LONDON   MELBOURNE

Copyright © 1980 by Van Nostrand Reinhold Company Inc.

Library of Congress Catalog Card Number: 79-19346
ISBN: 0-442-24046-5
ISBN: 0-442-28529-9 pbk.

Manufactured in the United States of America

Published by Van Nostrand Reinhold Company Inc.
135 West 50th Street, New York, N.Y. 10020

Van Nostrand Reinhold Publishing
1410 Birchmount Road
Scarborough, Ontario M1P 2E7, Canada

Van Nostrand Reinhold
480 Latrobe Street
Melbourne, Victoria 3000, Australia

Van Nostrand Reinhold Company Limited
Molly Millars Lane
Wokingham, Berkshire, England

15 14 13 12 11 10 9 8 7 6 5 4 3 2

Library of Congress Cataloging in Publication Data

Tver, David F.
    The petroleum dictionary.

    1. Petroleum—Dictionaries. I. Berry, Richard W., 1933-   joint author. II. Title.
TN865.T83   553'.282'03   79-19346
ISBN 0-442-24046-5
ISBN 0-442-28529-9 pbk.

PREFACE

The Petroleum Industry is varied and complex and is dependent upon a vast body of scientific personnel and technology. Each area of specialization is almost an industry within itself. The principle functions are exploration, geophysics, drilling, production, gas processing, refining, pipeline, marketing and supporting research functions. Each area is further broken down into separate supporting functions.

In addition, whole industries have been built from particular segments. The drilling industry, pipeline, chemical, muds, offshore drilling and support facilities, seismic contracting, engineering, construction and supporting equipment are but a few. The petrochemical industry is related, but in a sense is a separate function.

In exploration there are many sciences that blend together to complement each other and are broken down into exploration, exploitation, and production geology, seismology and geophysics.

Production includes the drilling of oil wells both onshore and offshore, the design and construction of surface well equipment, the building of pipelines to carry the crude to the refineries, the construction and maintenance of gas plant facilities, workover and remedial work. Also allied is reservoir engineering, with secondary and enhanced recovery.

The basic principle of oil refining is to separate crude oil into its major chemical components simply by heating and distilling off the various fractions in a tower. However, the processes to achieve this end are very complex and require complicated equipment and procedures to obtain petroleum products such as fuels, lubricants, asphalts, and waxes from the various types of crudes processed through the refinery.

All these areas are covered in the Petroleum Dictionary. In addition, other related subjects have been covered because of their direct and indirect relationships, such as soils, oceanography, vulcanism, stream erosion, glaciation, and shore processes.

A dictionary could be prepared on each segment of the industry because of the complex terminology and individual technical aspects. There are dictionaries on the market covering lightly various aspects of the petroleum industry, or in detail, on specific areas of specialization. It is believed that this Petroleum Dictionary is the first comprehensive combination dictionary-handbook that covers virtually all aspects of the petroleum industry, with comprehensive coverage of geology, geophysics, seismology, drilling, gas processing, production, a detailed analysis of various refining operations and processes, offshore technology, and description of various materials and supporting techniques used in drilling and production of oil and gas.

This Petroleum Dictionary should be of interest and of use to everyone involved in the petroleum industry and supporting activities, the student, and a reference source for every library.

David F. Tver
Richard W. Berry

THE PETROLEUM DICTIONARY

A

abandon (prod).   Cease efforts to produce oil and gas from a well; plug depleted formation and salvage all material and equipment.

able closed tester (refin).   Instrument for determining flash and fire points of kerosene and other similar products.

abrasion.   Wearing away or rounding of surfaces by friction or rubbing as from action of glaciers, wind, and waterborne sand on rock fragments and minerals.

abrasion platform.   Surface of marine denudation formed by wave erosion, still in its original position at or near wave base with marine forces still operating on it.

absolute age.   Time past, stated in years; referring to geologic events, generally based on measurement of radioactive decay rate, isotopic ratios, and products of minerals or rock substances.

absolute alcohol.   100 percent alcohol, usually applied to ethyl or methyl alcohol.

absolute date.   Date expressed in terms of years and related to present by a reliable system of reckoning.

absolute pressure.   Pressure measure with zero equal to a perfect vacuum; total pressure exerted by a gas or liquid. The absolute pressure of a vacuum is zero pounds per square inch and of the atmosphere, about 14.7 pounds per square inch.

absolute system of units.   System of units in which a small number of units are chosen as fundamental and all other units are derived from them.

absolute viscosity.   The force that will move one square centimeter of plane surface with speed of one centimeter per second relative to another parallel plane surface from which it is separated by a layer of the liquid one centimeter thick. This viscosity is expressed in dynes per square centimeter; its units being the poise equal to one dyne-second per square centimeter.

absolute vorticity.   Vorticity of a fluid particle expressed with respect to an absolute coordinate system.

absolute zero.   Temperature at which bodies would possess no heat whatever. Since heat is the kinetic energy of motions of molecules of a substance then, at absolute zero, there is no motion of molecules with respect to one another; equals -273°C.

1

absorber (refin). Tank or tower employed in manufacture of natural gasoline; also called absorption tower, bubble tower or fractionating tower.

absorptance. Ratio of radiant flux absorbed by a body to which it is incident.

absorption. (1) Process in which incident radiant energy is retained by a substance. (2) Penetration or apparent disappearance of one substance into another. Particles absorbed are of molecular size. (3) Action by which absorption oil collects gasoline during absorption process.

absorption (refin). Light oil used in absorption process to absorb gasoline in wet gas during manufacture of natural gasoline. Oil is returned to absorption tower for reuse after the gasoline has been distilled from the absorption oil.

absorption gasoline. Gasoline extracted from natural gas or refinery gas, as contacting the gas with an oil and subsequently distilling the gasoline from the heavier oil.

absorption plant (refin). Plant for recovering condensable portion of natural or plant gas by absorbing these heavier hydrocarbons in an absorption oil, often under pressure, followed by separation and fractionation of absorbed material.

absorption process (gasoline). Method of separating the condensable gasoline present in wet gas to make natural gasoline.

absorption system (refin). In gasoline manufacturing, a system of manufacturing casing head gasoline in which light vapors of casing head gas are absorbed in heavy oil by being passed upwards through continuous spray of oil in a tower-like apparatus.

absorption tower (refin). Tower or column effecting contact between a rising gas and a falling liquid so that part of the gas may be taken up by the liquid; employed in manufacturing natural gasoline by absorptive process.

abyssal (marine). Deep within the hydrosphere, signifying depth in miles; beyond depth of light penetration in oceans or lakes, referring to 6,000 feet or more in oceans and 1,000 feet or more in lakes.

abyssal plains (marine). Flat or nearly level areas in deepest portions of ocean basin.

abyssopelagic (marine). Pertaining to that portion of ocean which lies below depths of 2,000 fathoms (3,700 meters), and to ocean-derived sediments found at that depth.

accretion. Gradual building of land over a long period of time, by deposition of water or wind borne sediments on a beach, in a stream channel or dune field. (*See* also continental accretion).

acretionary limestone. Limestone formed in situ by slow accumulation of organic remains such as coral or shells.

accumulator (refin). Vessel for temporary storage of gas or liquid; usually used for collecting sufficient material for continuous charge to some refining process.

accumulator still (refin). Shell still used for the purpose of removing moisture and light products from feed to pipe still; also acts as a feed reservoir.

acetone. Inflammable liquid composed of carbon, hydrogen and oxygen. Under normal conditions, a colorless liquid having a mint-like odor; used as a solvent for fats, resins, and as an absorbent for acetylene gas.

acetylene gas. Colorless, inflammable gas composed of carbon and hydrogen; found sparingly in natural state and produced synthetically by action of water on calcium carbide in acetylene generation.

acicular (min). Slender or needle-shaped.

acid. Any chemical compound containing hydrogen, capable of being replaced by positive elements or radicals to form salts. In terms of the dissociation theory, a compound which on dissociation in solution, yields excess hydrogen ions. Acids have a pH below 7. Some acids or acidic substances are hydrochloric acid, tannic acid, sodium acid pyrophosphate.

acid blowcase (refin). Small tank constructed of material to withstand corrosion and pressure; sulfuric acid is blown by compressed air to agitator in treatment process. When tank is of cast iron, is usually termed an "egg."

acid conductor (refin). Vessel in which hydrolyzed acid is refortified by heating and evaporation of water or, in some places, by distillation of water under partial vacuum.

acid free. Uncombined acid content of a body or portion of acid whose active hydrogen atom has not been replaced in mixing acid with a second substance.

acid heat test (refin). Test, under controlled conditions, measuring temperature rise resulting from addition of commercial sulfuric acid to a petroleum distillate.

acid number (refin). Number of milligrams of potassium hydroxide required to neutralize the total acidity in one gram of fat, oil, wax, free fatty acids, etc.

acid recovery plant (refin). Auxiliary department of certain refineries where sludge acid is separated into acid oil, tar, and weak sulfuric acid, with provision for reconcentration later.

acid refined oils (refin).   Term commonly referring to a class of linseed oils from which mucilaginous "break" has been removed by treating raw oil with sulfuric acid.

acid rock (petrology, old).   Igneous rock containing high proportion of silica, contrasting with basic rock in two-division classification of rocks; contains more than 66 percent silicon dioxide.

acid sludge (refin).   Black viscous residue left after treating petroleum oil with sulfuric acid for removal of impurities; containing spent acid and impurities.

acid treatment (refin).   Oil refining process in which unfinished petroleum products as gasoline, kerosene, diesel fuel, and lubricating stocks are contacted with sulfuric acids to improve color, odor, and other properties.

acidizing.   Acid treatment of oil-bearing limestone or carbonate formations increasing production. Hydrochloric acid or another type of acid is injected, under great pressure, into formation bringing about an enlargement of the pore spaces and passages through which the reservoir fluid flows. Acid is held under pressure for a period of time and then pumped out. (See Fig. p. 5).

acidity.   The relative acid strength of liquids measured by pH.

activated clay (refin).   Used in petroleum refining and re-refining as an absorptive filtering medium; removes solids and certain liquid compounds by both contact and percolation methods of filtration.

activity coefficient.   Measure of deviation from ideal to non-ideal conditions of a solute, gas or liquid, in a solvent (stationary phase).

acylation.   Process introducing an acyl group (RCO-) into a molecule.

additive (refin).   Additive agents are used for imparting new, or for improving existing characteristics of mineral lubricating oils.

adhesion.   Force of attraction between molecules of two different substances as water to skin, paint to wood, etc.

adiabatic temperature changes.   Compression of a fluid without gain or loss of heat to the surroundings, is work performed in the system producing a rise or fall of temperature.

adiabatic vaporization.   Vaporization of a liquid with practically no heat ex-

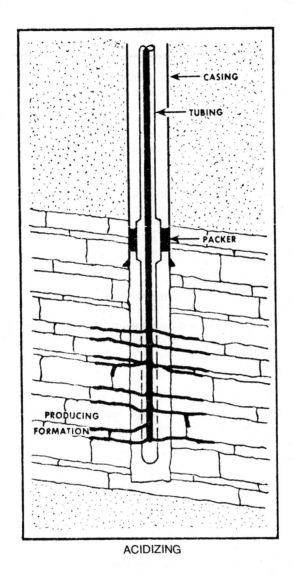

ACIDIZING

change between liquid and its surroundings while normally vaporization of a liquid is accompanied by absorption of heat by that liquid.

adiabatic wall temperature. Temperature assumed by a wall in a moving fluid
stream when there is no heat transfer between wall and stream.

adit. Nearly horizontal passageway from the surface into a mine, as in the side
of a hill.

adobe (min).   Very fine grained rock as coherent as many shales yet crumbles rather readily; has the harsh feeling of loess; color range from pinkish white to deep buff, grayish brown or even chocolate brown; consists of fine grained products of weathering in an arid environment; found in place of origin, residual soil or transported and deposited by fluvial processes. Clay is the principal mineral constituent.

adsorption.   A surface phenomenon exhibited by a solid (adsorbent) to hold or concentrate gases, liquids, or dissolved substances (adsorptive) upon its surface; property due to adhesion.

aeolian (geol).   Designates rocks and soil constituents carried and laid down by atmospheric currents (wind); also applying to erosion and other geologic effects accomplished by wind.

aerify.   Changing a liquid into a gaseous vapor by infusion or forced introduction of air.

aerometer.   Instrument for ascertaining weight or density of air and other gases.

affinity.   Attractive force exerted in different degrees between atoms, causing them to enter into, and remain in combination.

aftershock.   Earthquake which follows a larger one, originating near the focus of the larger one.

age.   Period of earth history of unspecified duration characterized by dominant or important life forms or particular event i.e., the Age of Fishes, the Ice Age; may also refer to the position of anything in the geologic time scale possibly expressed in years.

age dating.   Calculation of absolute age of a material by such means as fossil record or radioactive determination of the number of atoms of a stable radiogenic end product relative to number of atoms of its radioactive parent.

agglomerate (geol).   Accumulation of coarse, angular or subangular pyroclastics; coarse volcanic breccia.

aggradation (geol).   Upward building of a surface by deposition of sediment as natural filling of the head of a watercourse at any point of weakening of the current.

aggrade (geol).   Deposition of sediments on a stream bed or valley floor.

aggregate (nonbituminous).   Inert material as sand, gravel, shell, slag, or broken stones, or combination thereof, mixed with cementing material to form a mortar or concrete; divided into coarse, fine, graded, dense grades, and macadam aggregate.

agitator (refin). Apparatus for shaking and mixing; in refining of petroleum, a structure in which a body of oil is kept in constant motion while being treated with sulfuric acid or other chemicals either by mechanical stirring or by forcing a number of jets of compressed air into mixture.

agitator batch (refin). Agitator piped and fitted for batch operation.

agonic line. Line through all points on Earth's surface at which magnetic declination is zero; i.e., the locus at which magnetic north and true north coincide.

A-horizon. Oxidized soil zone immediately below surface, from which soluble material has been leached downward by water seeping into the soil. Varying amounts of organic matter give A-horizon a gray to black color.

air blown asphalt (refin). Asphalt produced by blowing air through residual oils or similar mineral oil products at moderately elevated temperatures.

air condenser (refin). Condenser in which air is used as cooling medium.

air sweetening (refin). Use of air or oxygen to oxidize lead mercaptides to disulfide instead of using elemental sulfur.

alcogas (refin). Composite fuel secured by blending alcohol and gasoline.

alcohol. Group of colorless organic alkyl compounds containing hydroxyl (OH) group, often made from petroleum hydrocarbons. Principal petroleum alcohols are amyl, butyl, methyl, propyl, and in some instances, their isomers.

alcohol, butyl. Series of alcohols having the chemical formula C_4H_9OH and boiling points between 182° and 243°F.

alcohol, ethyl (enthanol). Colorless volatile inflammable liquid of formula C_2H_5OH; fermented and distilled liquors.

alcohol, isopropyl. Alcohol made from petroleum gas of formula C_3H_7OH. Heavier than ordinary grain alcohol; used as a solvent and for manufacture of other chemicals.

alcohol, methyl (methanol). Poisonous liquid of formula CH_3OH; lowest member of the alcohol series. Methanol is also known as wood alcohol, since originally its principal source was by destructive distillation of wood.

alcoholysis. Intermediate between alcohols and organic acids differing from alcohols in having two fewer hydrogen atoms in molecule; any one of a class of compounds typified by acetaldehyde.

algal. Pertaining to algae as algal limestone and reefs, largely formed by algae; or algal structures generally laminated crusts, ooids and pisoids.

algal limestone. Limestone composed largely of calcium-carbonate precipitated by algae.

algorism. Art or system of calculating with any species of notation, as in arithmetic with nine figures and a zero; a computational procedure as a computer program; also called algorithm.

alicycle hydrocarbons. Hydrocarbons containing a ring of carbon atoms but not belonging to the aromatic series.

aliphatic compound. Major class of organic compounds in which carbon atoms are in long chains.

aliphatic hydrocarbons. Organic hydrocarbon compounds as ethane, butane, octane, acetylene, in which carbon atoms are in open chains, as opposed to ring structures of aromatics and naphthenic compounds.

aliphatic petrochemicals (refin). Group of organic compounds produced from methane, butane, ethane, and propane, the four paraffin saturated open members of the gas family, together with five companion olein unsaturated members of gas family, ethylene, propylene, butene-1, butene-2, and isobutene.

alkali. In chemistry, any substance having marked basic properties. In its restricted sense, term applying to hydroxides of potassium, sodium, lithium, and ammonium; soluble in water; have power of neutralizing acids and forming salts and turn red litmus, blue. In a more general sense, term also applying to hydroxides of so-called alkaline earth element, barium, strontium, magnesium and calcium.

alkali (grease). Refers to a class of products combined with fats to form soaps (sometimes called sponges) used in grease manufacture; among these, calcium, used to make lime base grease, and caustic soda, used to make sodium base grease.

alkali test (refin). Test to determine presence or absence of free alkali in finished oils after chemical purification.

alkali test (kerosene). Treatment of kerosene with caustic soda to determine presence of certain undesirable compounds, as minerals.

alkaline. Having properties of an alkali, opposed to acidic.

alkaline wash (refin).   Process through which kerosene is treated with a solution of caustic soda to chemically purify it and render it more suitable for illuminating purposes.

alkalinity. Combining power of a base measured by the maximum of equivalents of an acid with which it can react to form a salt. In water analysis, represents the carbonates, bicarbonates, hydroxides, and occasionally the borates, silicates, and phosphates in the water, determined by titration.

alkaloid.   An organic nitrogenous base occurring naturally in the animal and vegetable kingdoms and manufactured synthetically. Many alkaloids are of great medical importance as morphine, strychnine, atropine, cocaine, quinine.

alkylate.   Product obtained in the alkylation process.

alkylation.   Involves a combination of an isoparaffin with an olefin generally ethylene, propylene, butylene, and anylene or of aromatic hydrocarbon benzene, with ethylene to form ethyl benzene, one of the basic materials used in the manufacture of styrene and synthetic rubber. Formation of complex saturated molecules by direct union of a saturated and an unsaturated molecule. Process in which an alkyl group, R-, is introduced into a molecule.

alkylation (refin).   Process combining an isobutane with an olefin to produce a liquid of superior stability and anti-knock quality suitable for blending aviation gasoline or motor fuel. When butylene is the olefin feedstock, the major component in the alkylate product is 2,2,4-trimethyl pentane, commonly called isooctane, the material designated as the 100 octane number reference for anti-knock ratings of all other carbons.

allochem.   Marine sediment formed by chemical or biochemical precipitation includes intraclasts, fossils, and pellets.

allochthon.   Large displaced body of rocks; as a rock mass moved a considerable distance by thrust faulting or land slide; opposite of autochthon.

allochthonous.   Applied to rocks of which dominant constituents have not been formed in situ.

allogenic.   Applied to rock or sediment constituents originating at a different place and time to the rock of which they are now part, as pebbles in a conglomerate.

alluvial fan.   Land counterpart of a delta; assemblage of sediments marking place where stream moves from steep gradient to gentler gradient and suddenly loses its transporting power. Typical, but not confined to, arid and semiarid climates, they are low, cone-shaped heaps, steepest near the mouth of the valley and sloping gently outward with ever decreasing gradient.

alluvium. General term for all detrital deposits resulting from operation of modern rivers, includes sediments laid down in riverbeds, floodplains, lakes, fans, at foot of mountain slopes, and estuaries; unconsolidated detrital deposits ranging from clay to gravel sizes, generally poorly sorted, typically fluviatile in origin.

alpha decay. Radioactive transformation of nuclide by alpha particle emission.

alpha emitter. Radioactive element, natural or artificial, changing into another element by alpha decay.

alpha particle. Positively charged particle from an atomic nucleus; has identical properties to nucleus of a helium atom, two protons and two neutrons. Alpha particles have great ionizing power; are dangerous to living tissue, but can easily be shielded out since they have a low penetrating power.

alpha ray. Ray or stream of alpha particles thrown off by some disintegrating atoms.

alteration. Change in mineralogical composition of rock by action of hydrothermal solutions.

aluminum grease. Grease made by thickening petroleum oil with aluminum stearate or other aluminum soap.

aluminum naphthenate. Aluminum salt of refined naphthenic acids.

aluminum oleate. Soap-like compound of aluminum oil oleic acid, which is introduced into lubricating oils and greases to increase viscosity.

aluminum stearate. Metallic soap used in paints, varnishes, and lacquers imparting special properties as thickening oils and organic solvents; aids in suspension and deflocculation of pigments; imparts water proofness and flatness to organic finishes.

Amagat law. Volume of a mixture of gases is equal to the sum of the volumes of the component gases, each taken at the pressure and the temperature of the mixtures. The Amagat law is exact when applied only to a mixture of ideal gases.

Amagat volume unit. Unit of volume; the volume of one gram-mole of gas at 32°F (0°C) and at one standard atmospheric pressure; value determined by gas considered.

amber. Fossilized resin, generally of coniferous trees.

amino. Members of the -NH₂ group as amino acids, carboxylic acid bearing an amino group.

amino acid. An organic compound containing an amino group (-NH₂) and a carboxyl group (-COOH); may be linked together to form the peptide chains of protein molecules. Twenty different amino acids are polymerized in various combinations by living organisms to make proteins.

ammonia oil. Lubricating oil having a low pour test and suitable for use in an ammonia compressor; used in connection with mechanical refrigeration.

ammonolysis. Cleavage of a molecule by ammonia.

amorphous. Rocks and minerals having no definite crystalline form or orderly arrangement of atoms.

amorphous graphite. One natural form of graphite, named for its earthy appearance.

amphoteric. Possessing both base (alkaline) and acid properties.

amplitude. Magnitude of displacement of a wave from a mean value; for a simple harmonic wave, maximum displacement from mean, for more complex wave motion, usually taken as one-half mean distance (or difference) between maximums and minimums.

amu. Atomic mass unit; an expression of the atomic mass of a substance.

amygdale. Vesicle or vapor cavity in volcanic rock filled with secondary minerals; diminutive form is amygdule.

amygdaloid. Extrusive igneous rock containing many gas-formed vesicles more or less filled by secondary minerals as quartz, calcite, zeolite, etc.

amygdaloidal basalts. Porous, blackish lava whose holes or "blisters" (amygdules) are filled with other minerals, the commonest of these, quartz, calcite, and zeolite.

amyl acetate. Liquid having a banana-like odor; used as solvent in preparation of photographic film, flavoring compounds, and perfuming; also called banana oil.

amylene. Product found in petroleum having boiling point within gasoline boiling point range.

anaerobic sediment. Highly organic sediment rich in hydrogen sulfide formed in absence of free oxygen; characteristic of some fjords and marine basins where little or no circulation or mixing of bottom water occurs.

analysis, core (geol). Laboratory examination of geological samples taken from the well bore; used to determine capacity of formation to contain oil and gas; possibility of oil and gas passing through formation; degree of saturation of formation with oil, gas and water, and for other purposes.

analysis, mud (prod).   Examination and testing of drilling mud to determine its physical and chemical qualities.

andesite line (old).   Postulated geographic and petrographic boundary between the andesite-dacite-rhyolite rock association of the margin of the Pacific Ocean and the olivine-basalt-trachyte rock association of the Pacific Ocean basin and its included islands.

angle of incidence (seis).   Angle at which a ray of energy or an object impinges upon a surface, measured between direction of propagation of the energy or object and a perpendicular to the surface at point of impingement or incidence.

angle of reflection (seis).   Angle at which a reflected ray of energy leaves a reflecting surface, measured between direction of the outgoing ray and a perpendicular to the surface at point of reflection.

angle of refraction (seis).   Angle at which a refracted ray of energy leaves the interface at which refraction occurred, measured between direction of the refracted ray and a perpendicular to interface at the point.

angular unconformity.   (1) Erosional loss of geologic record between two series of rock layers; rocks of lower series meet rocks of upper series at an angle therefore the two series are not parallel. (2) Unconformity in which the older strata dip at different angles (generally steeper) than the younger strata.

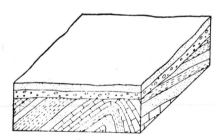

aniline.   Colorless, poisonous oil fluid, largely made by processing nitrobenzene secured from coal tar.

aniline point.   Lowest temperature at which equal volumes of freshly distilled aniline and an oil, which are being tested, are completely miscible; test gives an indication character paraffinic, naphthenic, asphaltic, or aromatic character of the oil.

anilol.   Mixture of aniline and alcohol; used as a blending compound.

anion.   A negatively charged atom or radical as CL^-, OH^-, $SO_4^=$, etc., in solution of an electrolyte. Anions move toward the anode (positive electrode) under the influence of an electrical potential.

annular space (drill).   Space surrounding pipe suspended in the well bore; outer wall of annular space may be an open hole or string of large pipe.

annular velocity.   Circular movement, usually expressed as number of revolutions per minute or per second.

anomaly.   Departure from the usual or expected in surface or subsurface geologic structure or earth energy fields; may indicate conditions favorable to accumulation of oil or natural gas.

antecedent platform theory (old).   Theory of coral atoll and barrier reef formation postulated as a submarine platform, 50 meters, or more below sea level from which barrier reefs and atolls grew upward to water surface without changes in sea level.

antecedent stream.   Stream that maintains, during and after uplift, the course it had established prior to uplift.

anthracene oil.   Oil of high specific gravity.

anthracite (hard coal).   A compact, dense, brittle rock whose color is steely or jet black; has a glassy or almost metallic luster, an uneven or conchoidal fracture and a hardness of two or more. Joints, well developed and bedding observed in all, except small pieces. Anthracite contains 86 to 99 percent carbon, one to fourteen percent volatile material, and a small amount of clay and other ashy impurities; burns slowly with a pale blue flame and with much less smoke than from soft coal. Hard coal occurs in regions where rocks have been folded, heated, and squeezed.

anti-acid additive.   Prevents or retards formation of certain acids, as an anti-corrosive additive retards formation of corrosive acids in crankcase oils during use.

anticlinal (struct. geol).   Pertaining or related to an anticline.

anticlinal theory.   Theory that water, petroleum, and natural gas accumulates in uparched strata in a specific order (water, lowest), provided structure contains reservoir rocks in proper relation to source beds, and capped by an impervious barrier.

anticline. A configuration of folded, stratified rocks which dip in two directions away from a crest, as in the principal rafters of a common gable roof dip away from the ridgepole; the reverse of syncline.

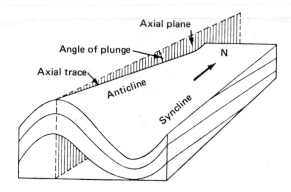

anticlinorium. Broad arch of strata folded into a series of anticlines and synclines.

anti-foam agent. Additive used to control aeration in petroleum lubricating oils and gear lubricants.

anti-knock compound. (1) Chemical such as tetraethyl lead to improve anti-knock quality of a fuel. (2) Resistance to detonation of "singing" in spark ignited engines.

anti-knock agents. Chemical compounds, when added in small amounts to fuel charge of an internal-combustion engine, have property of suppressing, or at least of strongly depressing, knocking.

anti-oxidants. Chemical added to gasoline, lubricating oils, waxes, and other products inhibiting oxidation.

apparent viscosity. The viscosity of a fluid appears to have on a given instrument at a stated rate of shear; a function of the plastic velocity and the yield point.

aqueous. Pertaining to, or containing water; as an aqueous solution.

aqueous emulsion. Tiny particles of oil, or other substances, surrounded with water, as oil-in-water emulsion.

aqueous soluble oil. Cutting fluid or oil soluble in water.

aquiclude. Formation which, although porous and capable of absorbing water slowly, will not transmit it fast enough to furnish an appreciable supply for a well or spring.

aquifer. (1) Permeable rock formation of subsoil through which ground water moves more or less freely. (2) Formation, group of formations, or part of a formation that is water bearing.

Archeozoic Era (old). Early Precambrian time when oldest system of rocks were formed.

archipelago. Group of islands more or less adjacent to each other, arranged in groups and dispersed across portions of the sea.

arc shooting (seis). A method of refraction, seismic prospecting in which the variation of travel time with the azimuth from a shot is used to infer geologic structure.

arenaceous. Pertaining to rocks containing sand; sandy.

argentiferous. Producing or containing silver.

argillaceous. Pertaining to rocks containing considerable clay; shaly or clay bearing, non-productive formation.

argillite (petrology). Nonfissile sedimentary rock that is much harder and more dense than shale resembling it in origin, mineral, and general appearance. Generally cementing material is silica or one of the compounds of iron and silica; since most argillites are slightly metamorphosed, some of their mineral grains are recrystallized and enlarged. Tiny flakes of mica are developed in addition to those originally in the rocks.

argon. One of the gases in the atmosphere remaining unchanged during combustion of fuel vapor in an engine. Symbol Ar; atomic number 18; atomic weight 39.948; boiling point -303°F (-186°C).

argon-potassium method (or Argon-40 method). A method of radiometric dating based on measuring the decay of radioactive potassium to argon.

arkose. Sandstone whose mineral composition is similar to and originating from granite rock; may consist of little except feldspar and quartz; when firmly cemented it may look very much like granite. The decided angularity of the grains shows they were deposited soon after weathering loosened them from granites and were not carried far. When grains are large, arkose grades into breccia.

arm. Any deep and comparatively narrow branch of the sea extending inland; as opposed to gulfs and firths.

aromatic. Group of cyclic hydrocarbons, the principle one being benzene; so called because many of their derivatives have a sweet or aromatic odor; have relatively high specific gravity, possessing good solvent properties, and are the source of industrial aromatic hydrocarbons.

aromatic (benzene) series. Aromatic (or benzene) series of hydrocarbons, so named because many of its members have a strong or aromatic odor; an unsaturated, closed ring (carbocyclic) series having the general formula CnH_2n- . Benzene (C_6H_6), member of the series found in petroleum, other members include toluene (methylbenzene $-C_6H_5CH_3$) and xylene (dimethylbenezene-C_6H_4-CH_3CH_3). While aromatics are present in all petroleum, the percentage is generally small; benzene, and its derivatives, also occur extensively in light oil fractions of tars obtained from dry distillation of coals at temperatures above 1,000°C.

aromatization catalytic process (refin). Process producing toluene from ring-shaped members of the coal family when used in combination with dehydrogenation process to remove excess hydrogen.

aromatic compounds. Compounds derived from benzene, with one or more benzene rings of carbon atoms; distinct from aliphatic or alicyclic charger.

aromatic hydrocarbons. Hydrocarbons, as benzene, toluene, xylene, having a high specific gravity, good solvent properties and aromatic or sweet odors.

aromex (refin). Extraction of aromatics for petroleum hydrocarbons; producing high-octane motor fuel blending components, high purity aromatics (benzene, toluene, xylenes, and heavier aromatics), or dearomatized raffinate for use in specialty solvents, jet fuels, and high smoke point kerosenes.

aromizing (refin). Upgrading low-octane naphthas; producing C_6-C_8 aromatics for petroleum use, and producing LPG. A high severity process using continuous catalyst regeneration for the purpose of producing high purity aromatics; uses a wide range of naphtha cuts. The main advantages of aromizing are: increased yields in aromatics with respect to catalytic reforming; substantial conversion of the C_6 paraffins; production of pure aromatics by distillation (extractive distillation for benzene cut only).

arosorb process (refin). Process which recovers aromatic members as benzene and toluene from residue in hydroformer. ARO- stands for aromatic, and -SORB, for absorption, accomplished by silica gel.

arostat (refin). Saturating aromatics by catalytic hydrogenation. The process is exemplified by high hydrogen utilization and flexibility while requiring low capital investment and operating costs; producing high quality jet fuels and low aromatic content solvents; meeting regulations on permissible aromatic concentrations; also producing high-purity cyclohexane from benzene.

artesian.   Pertaining to subsurface water under sufficient pressure, making it rise above the level of the aquifer, as artesian basin, spring, well, etc.

artesian basin.   Subsurface geologic structural feature in which water is confined under hydrostatic pressure.

artesian water.   Water, under pressure when tapped by a well and able to rise; it may or may not flow out at ground level. Diagram illustrates the structural conditions favorable to artesian wells.

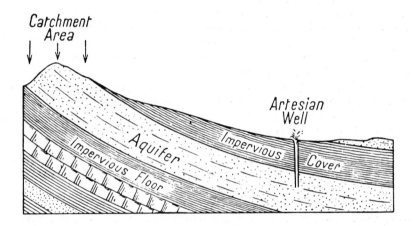

artificial element.   Chemical element that does not exist naturally in the earth but can be manufactured in an atomic furnace by forcing extra atomic parts into the nucleous of a natural element as atomic fuel plutonium.

artificial graphite.   Made by Acheson process; baking powdered coke in an electric furnace at extremely high temperature.

aschistic rocks.   Generally occurs in dikes having the same chemical composition and, insofar as they are crystallized, the same mineral composition as the parent bodies. Dikes are apophyseal to a larger plutonic or subjacent mass and may be regarded as chilled samples of the source rock. Generally aschistic rocks, intrusive texturally, are fine grained or fine-grained porphyry.

aseismic.   Not subject to earthquakes, as an aseismic region.

ash.   Fine unconsolidated pyroclastic material, commonly a by-product of an explosive erupting of viscous (acidic) lava.

ash fall.   Rain of airborne volcanic ash falling from an eruption cloud; characteristic of volcanic eruptions. A deposit of volcanic ash resulting from such a fall.

asphalt.   Brown to black, solid or semisolid bituminous substance, occurring in nature but also obtained as a residue from refining of certain hydrocarbons. Comparatively nonvolatile, bitumen are of variable hardness, related in origin to petroleum.

asphalt base course.   Foundation consisting of mineral fragments bound together with asphaltic material.

asphalt base crudes.   Crude oils which, upon processing, yield relatively large amounts of asphaltic residues and possesses naphthenic characteristics.

asphalt blowing (refin).   Process obtaining various types of bitumen with minimum thermal susceptibility and a higher softening point for a given degree of penetration than does distillation; the charge is vacuum asphaltic residue.

asphalt cement.   Flux or unflux asphalt especially prepared as to quality and consistency for direct use in manufacture of bituminous pavement.

asphalt content.   Percent of weight of 100- penetration asphalt contained in an asphaltic product.

asphalt flux.   Frequently refers to an oil made from a naphthene base crude, used as diluent to reduce consistency of asphalt to specified viscosity.

asphalt pyrobitumens.   Harder and more infusible than asphaltites; specific gravity does not range above 1.25 or melt, rather swells and decomposes before melting temperature is reached; insoluble in petroleum naphthas and only slightly soluble in carbon disulphide. Wurtzilite, elaterite and abbertite are examples.

asphaltene.   Product of oxidation found in some crude oil residue, in asphalt and in high temperature sludge; soluble in aromatic solvents but not soluble in naphtha; components of bitumen in petroleum, petroleum products, asphalt cements and solid native bitumens.

asphaltic.   Essentially composed of, or similar to asphalt.

asphaltic concrete.   Surface made by mixing proportional quantities of graded aggregate and asphalt at a central plant, conveying completed mix to point of placement, and spreading by mechanical finisher; must be completed when in a heated condition.

asphaltites.   Harder solid hydrocarbons, as gilsonite and glance pitch, with melting points between 250° and 600°F; specific gravity is less than 1.20; solubility in petroleum naphtha ranges between 0 and 60 percent, and in certain carbon disulphide, between 60 and 90 percent. Gilsonite and glance pitch are asphaltites.

assay. An examination, not a complete analysis of a mineral or ore to determine the amount of certain of its constituents.

assimilation. Incorporation of material, originally present in wall rock, into magma.

asthenosphere. Layer of materials in earth's interior extending from the bottom of lithosphere to a depth of possibly several hundred miles.

ASTM distillation. Distillation test made on such products as gasoline and kerosene to determine initial and final boiling points and boiling range; any distillation made in accordance with ASTM distillation procedure.

ASTM gum test. An analytical method of determining amount of existing gum in a gasoline by evaporating a sample from a glass dish on an elevated temperature bath with aid of circulating air.

atoll. Roughly circular, elliptical, or horseshoe-shaped island or ring of islands of reef origin, composed of coral, algal rock, shellfish and calcareous sand; enclosing an open lagoon.

atollom. Large reef ring in the Maldine Islands consisting of smaller reef rings; word atoll derives from this name.

atoll reef. Ring-shaped coral and limestone reef often carrying low-sand islands, enclosing a body of water.

atom. Smallest piece of an element that can exist alone and still be that element. There are over 100 elements, each with its own unique atom, but all made of the same subatomic parts: electrons, protons, and neutrons in various combinations.

atom, excited. Some, or all electrons of an atom raised to a higher than usual energy state; excitation may occur by chemical reaction, absorption of heat, or photons.

atom smasher. A machine causing changes in atoms by bombarding them with parts of other atoms; bombardment either knocks subatomic particles off the target atom or forces additions to it.

atomic energy. Power released when atoms change from one kind into another; change can occur through splitting (fission) in which one heavy atom breaks into two medium size atoms, or can occur through joining (fusion) in which two light atoms unite into one medium size atom.

atomic fission. Breaking down of a large, heavy atom with liberation of energy in form of heat, alpha, beta or gamma radiation, and creation of new elements from the larger fragments.

atomic heat.   Thermal capacity of a gram atom of an element; obtained by multiplying its specific heat by its atomic weight, expressed in grams.

atomic nucleus.   Small dense core of an atom containing all of the positive charge and most of its mass.

atomic particle.   One of the particle constituents of an atom as electron, neutron, or positively charged nuclear particle.

atomic pile.   Controlled and self-sustained nuclear fission reactor releasing a great amount of energy.

atomic radiation.   Radiation or radioactivity resulting from decomposition of a nuclear fissionable, radioactive material, or from the fusion of atomic nuclei.

atomic radius.   Radius of an atom; average distance from the center to the outer most electron of the neutral atom, commonly expressed in angstrom units (10^{-8} centimeters).

atomic reactor.   Device designed to maintain a controlled nuclear chain reaction.

atomic spectrum.   A series of discrete wavelengths of energy emitted by the atoms of an element when excited above ground state.

atomic theory.   States that matter cannot be divided into smaller and smaller pieces indefinitely and still remain matter.

atomic weight.   Weight of an atom compared to weight of an oxygen atom, taken as exactly 16; approximately equal to the number of protons and neutrons in an atom's nucleus (the mass number).

atomizer.   Nozzle device used to break up fuel oil into fine spray, bringing the oil into more intimate contact with air in the combustion chamber.

Atterberg grade scale.   Logarithmic (geometric) decimal grade scale for sedimentary particle size; uses two millimeters as a reference point.

Atterberg limits.   Indices (LL, PL) of water content of a sediment at the boundaries between semiliquid and plastic state (liquid limit), and plastic and the semisolid state (plastic state).

auger stem (drill).   A section between the bit and rope socket in a string of cable tools supplying needed weight.

aureole. Zone of contact metamorphism.

authigenic. Products of chemical and biochemical action originating in sediments at the time of or after deposition, and before burial and consolidation.

autochthon. Large body of rocks in situ, opposite of allochthon.

autofining (refin). Desulfurizing distillate stocks, e.g., feed to synthetic natural gas (SNG) plants, without requiring an external source of hydrogen. Charge is straight-run distillates up to 465°F true boiling point (TBP) cut point; almost complete desulfurization obtained.

average boiling point. Unless otherwise indicated, the sum of ASTM distillation; 10 percent boiling temperatures are added at the 10 percent point and ending with the 90 percent (inclusive), divided by nine.

Avogadro Law. Equal volumes of gases at the same pressure contain the same number of molecules.

axial plane. A plane through a rock fold that includes the axis and divides the fold as symmetrically as possible. Diagram of an upright symmetrical fold.

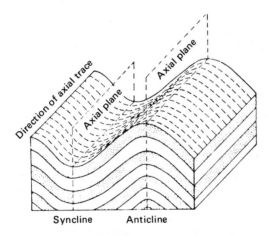

Syncline   Anticline

axial trace. Line created when an axial plane intersects a bedding plane of a fold; linear trend of the crest of an anticline or the trough of a syncline.

azeotrope. Liquid mixture which shows a maximum or a minimum boiling point; boiling at a constant temperature either higher or lower than that of the components.

azimuthal projection. Map projection in which directions of all lines radiating from a central point, or pole are the same as the direction of the corresponding lines on the sphere; when centered on one of the poles, sometimes called a *polar projection.*

Azoic era (old).   Time in earth's history meaning "without life."; after earth's development as a separate planet.

B

Babcock tube. Special flask with a long graduated neck, used in testing spray oils for unsulfonated residue.

Babo's Law. Addition of a nonvolatile solid to a liquid in which it is soluble, lowers the vapor pressure to the solvent in proportion to the amount of substance dissolved.

back arc. Region on the opposite side of a volcanic arc from the subduction zone; may contain a continental shelf, a basin, or a fold and thrust belt if the arc is along a continental margin. In mid-ocean, region may contain a basin or zone (center) of spreading.

back off (drill). Term meaning to unscrew pipe by reversing direction of rotary table.

back pressure (prod). Pressure resulting from restriction of full natural flow of oil and gas.

back pressure valve (prod). A valve designed to control flow rates so upstream pressure remains constant; may be operated by diaphragm, spring, or weighted lever.

back reef. Shallow sea or lagoon shelf area behind a reef or between it and land, commonly characterized by nonfossiliferous deposits of limestones, dolostones, and evaporites as in western Texas area of the Permian reefs.

backscattering (seis). Part of reflected energy returning to the transducer.

backshore. Storm washover or aeolian zone of beach; shoreward of non-storm, high tide level; also called back beach. (*See* sketch of beach profile)

backslope. Gentle sloping side of a ridge; in contrast with escarpment, a steeper slope.

back sweetening (refin). Deliberate but controlled addition of commercial grade mercaptans to a stock having excess free sulfur; reducing free sulphur by forming a disulfide, thus improving the copper-strip corrosion.

backup (drill). The act of "backing up" or holding one section of pipe while another is screwed out of it, or into it.

backup tong (drill). Heavy tong used to break tool joints of a drill pipe.

backup scatterance (seis).   Ratio of radiant flux scattered through quadrant 90 to 180 degrees, from a beam to incident flux.

badlands.   An arid terrain deeply, closely gullied, and canyoned by streams; difficult to cross on foot, horseback, or by motor vehicle, hence, "bad land" to cross.

baffle plate (refin).   Partial restriction, generally a plate located to change direction, guide flow, or promote mixing within equipment.

baffler (prod).   Device supplying and controlling flow of lubricating oil from pumps.

bail (drill).   In drilling rig place to hang goose neck of swivel, may weigh over two tons on the travelling block.

bailer (drill).   Large piece of pipe used to bail out drillings in form of mud from cable drilled wells. Lower end has a check valve allowing mud to enter bailer while going down, and closing when it starts moving upwards.

bailing (drill).   Operation of cleaning mud cuttings, and other material from bottom of well bore with a bailer.

bajada.   A gradually sloping surface of coalescing alluvial fans formed of detritus, extending along base of mountain range and sloping toward adjacent basin or bolson; usually restricted to arid and semiarid climates.

ball-and-ring method.   Method of testing melting and softening temperatures of asphalts, waxes, and paraffins. A ring is filled with substances to be tested and a steel ball is placed upon substance in ring; standard melting point is temperature at which substance softens sufficiently to allow ball to fall through ring.

ball and seat (prod).   Parts of check valve preventing return of pumped fluids.

band wheel (drill).   Large wheel used in oil drilling operations; by means of a crank, wheel transmits power from engine to walking beam.

bank.   Elevation of ocean bottom, above which water is relatively shallow but sufficient for navigation.

bank reef.   Reef lying within outer margins of rimless shoals, in contrast to barrier and atoll reefs rising directly from deep water.

bar.   A submerged or emerged embankment of sand, gravel, or mud built on river or lake bottoms, in seas or oceans, in shallow water by waves and currents. [See Fig. (top) p. 25].

barometric condenser (refin).   Device condensing steam, producing partial vacuum in a piece of refinery equipment, as a vacuum pipe still.

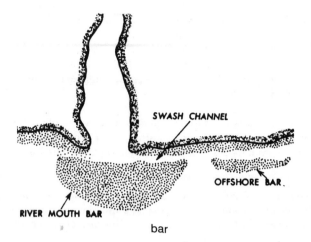

bar

barred beach sequence.   A sequence of longshore bars, barrier beaches, and lagoons occurring under low energy conditions where waves transverse a broad continental shelf before impinging on a shoreline abundantly supplied with sand sized sediments.

barrel (USA).   A standard barrel contains 42 U.S. gallons, or approximately 35 Imperial gallons. In petroleum industry, both the 42 gallon barrel (231 cubic inches) and the 50 gallon barrel are in common use.

Barrett jack (drill).   Double-acting travelling jack on a semi-circular track; used for making (screwing) and breaking out (unscrewing) cable tool joints.

barrier beach.   (1) A bar extending a long distance essentially parallel to shore; bar crest is usually above high water line by a few meters along its entire length or composed of a series of islands, separated from shore by a lagoon (also *barrier island*). (2) Occasionally synonym for bay barrier where elevated bar connects headlands of bay, transforming it into a lagoon.

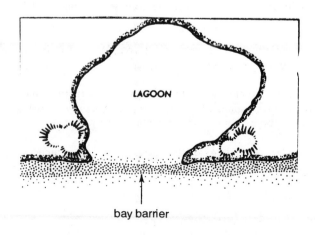

bay barrier

barrier chain. Series of barrier islands, barrier spits, and barrier beaches extending along a considerable length of coast.

barrier reef. Reef, parallel to, and separated from coast, by a lagoon that is too deep for growth of reef builders as coral and other sedimentary organisms, having calcareous shells or skeletons.

barytes (drill). A mineral used as a weighting material in drilling fluids (muds) to increase the efficiency of removing cuttings from around the drill bit; has a specific gravity of about 4.2, i.e., about 4.2 times heavier than water. (Also called tiff or barites)

basal conglomerate. Coarse, lithologically homogenous sedimentary deposit found just above an erosional break; initial stratigraphic unit overlying an unconformity; conglomerate formed over an old erosion surface, as near shore deposit of a slowly advancing sea.

basalt. A dark, heavy, fine-grained, extrusive igneous rock consisting almost equally of plagioclase feldspars and ferro-magnesium minerals; most common variety of volcanic rock.

base. Metal or metal-like group with hydrogen and oxygen in proportion to form an −OH radical ionizing in aqueous solution to yield excess hydroxyl ions; forms when metallic oxides react with water; increases the pH.

base constituents (lubricating oils). Organic and inorganic bases, amino compounds, salts of weak acids (soaps), basic salts of polyacidic bases, salts of heavy metals, and additional agents as inhibitors and detergents.

base element. Easily oxidized element, opposed to a noble element.

base level. Limits downward erosion by streams; inclined slightly toward sea level; below which land surface cannot be reduced by running water.

base map. Map drawn with considerable accuracy serving as a base for geological work.

basement. Igneous or metamorphic rock complex underlying sedimentary or volcanic rock; commonly of pre-Cambrian age.

basement complex. Rocks, generally with complex structure and distribution of individual rock types, commonly composed largely of crystalline igneous, metamorphic rocks, or both, that underlies a sedimentary sequence.

basement, seismic. A complex of rock below a layered sequence of sediment or sedimentary rock; returns no useful information or recognizable seismic signature. Somewhat arbitrary term relative to ability of seismic instrumentation, skill of operator and interpreter, also applies to depth of seismic penetration, not necessarily to rock type or structure.

base, petroleum.   Principal classes of petroleums, paraffin base oils, carrying a predominating amount of paraffin in the residue, and asphaltic base oils, with a greater percentage of asphalt than paraffin.

base stock, gasoline.   Type or kind of gasoline making up bulk of commercial gasoline; straight run and cracked gasoline are examples; smaller quantities of other gasolines, blending agents or blending components, are added to base stocks.

basic (old).   Pertaining to igneous rocks having a silica content less than two-thirds (generally less than one-half) of total constituents; as basalt, peridotite.

basicity.   Ability to neutralize, accept protons or hydrogen ions from acids; pH value above 7.

basic rock (old).   Igneous rock with dominant minerals comparatively low in silica and rich in metallic bases, as ferro-magnesium minerals.

basic sediment (prod).   Residue settling in oil kept in a tank; generally found covering tank bottom when cleaned.

basin.   Term used by petroleum geologists; (1) rounded structural depression of great extent, or a large area with no particular shape containing very thick sediments; (2) large structural depressions within a geosyncline or larger structural basin.

basin plain.   See *abyssal plain*.

batch.   Definite amount of oil, mud, acid, or other liquid in a tank or pipe line.

batch coke still (refin).   Apparatus in which the charged oil, usually reduced crude oil, is run through a coke bottom; all volatile matter being driven off and only hard, dry coke remaining.

batch distillation (refin).   Fractional distillation of a batch of oil in a batch still; remaining residue cleaned out, still refitted with another batch, and process repeated.

batch oil (refin).   Light body oil used principally to prevent sticking in molds, or as a lubricant in the manufacture of ropes and cords.

batch still (refin).   Petroleum still; distillation carried out in "batches"; the entire still charge introduced before fires are lighted, and distillation completed without introduction of an additional charge.

batch system (refin).   Distillation made by heating a charge of oil and removing the various fractions as the temperature increases.

batholith.   Huge intrusive body of igneous rock originating deep in the earth, flaring outward, usually with an undefined bottom; commonly form the cores of mountain ranges. Huge masses or rock intruded into formations that were already folded and uplifted; may cut across folded formations, and may also be the cause of folding and faulting of preexisting sediments; consist of granite and other coarsely crystalline rocks. Earth surface exposure exceeds forty square miles by arbitrary agreement. Block diagram illustrates the characteristic features of a batholith.

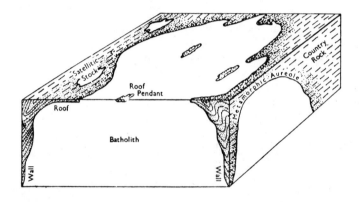

bathyl zone.   Depth zone of the marine (benthonic) environment; refers to the continental slope and water depths between 200 and 2000 meters. Sediments are richer in organic carbon than in the shallower (neritic) zone.

battery (prod).   An area where storage tanks are installed to receive produced fluids; may include several tanks, separation and treating equipment; also called tank battery. In a refinery, a series of stills, boilers, or cracking coils operated as a unit or under a single supervisor or crew.

Baume gravity.   An arbitrary scale on a hydrometer stem graduated in Baume degrees determines specific gravity of liquids.

BB fraction (refin).   Boiling points of butane and butene are so close, they are collected as one fraction, called the BB cut, or BB fraction.

BCPMM (prod).   Barrels condensate per million; i.e., barrels of condensed liquid per million cubic feet of gas.

beach cusps. Small points projecting seaward from a beach, usually evenly spaced with an interval that varies from a few feet along beaches with coarse sediment and/or small waves to several hundred feet along beaches with fine sediment and/or large waves.

beach profile. Common features of a beach viewed in cross section at right angle to the shore. Details vary with availability of sediment, grain size of sediment, energy of waves, and resistance to erosion at the headlands.

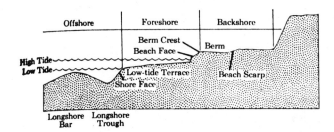

beam (prod). Walking beam of pumping jack unit.

beam well (prod). Well using a pumping jack, or unit and rod, to lift fluid.

bean (prod). Choke used to regulate flow of liquid from a well. Different sizes of beans are used for different producing rates.

bean back (prod). Using a smaller size bean or choke to make the amount of production smaller.

bed. Tabular body of rock lying in a position essentially parallel to the surface, or surfaces on or against which it was formed, whether these be surfaces of weathering and erosion, planes of stratification, or inclined fractures; usually restricted to tabular bodies of sediments of widely varying thickness deposited in a nearly horizontal orientation.

bedded. Applying to sedimentary rocks exhibiting planes of separation, designated as bedding planes.

bedding. Beds or layers of sedimentary rocks separated by parallel bedding planes along which rocks tend to separate or break. The varying thickness of layers in a given sedimentary rock reflects the changing conditions that prevailed when each deposit was laid down. In general, each bedding plane marks termination of one deposit and beginning of another. Diagram: common types of bedding: (a) regular bedding (b) cross or current bedding (c) graded bedding (d) slump structure. (See Fig. p. 30)

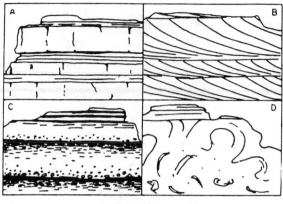

bedding

bedding cleavage.   Cleavage or break parallel to bedding.

bedding, cross.   Complicated structure produced when layers of sediment are deposited on an inclined surface, as the frontal slope of a small delta, the lee slopes of sand dunes and subaqueous ripple marks. Depositional laminae laid down on the sloping surface are usually lenticular and gently curved. Slight erosion by the current, flowing over them carries away their thin, upper parts leaving a surface of erosion truncating the inclined laminae. New laminae are then laid down above this surface in the same pattern as those below, so that a contrast is produced between the asymptotic curvature at the base of each group of laminae and the sharp erosional plane of truncation at their top. (Older terms: current bedding and false bedding)

bedding, graded.   When a turbulent mixture of particles of different sizes, shapes, and densities is brought to the site of sedimentation, the coarser, heavier, and more nearly spherical grains tend to settle more rapidly than the others. As the smaller, lighter, and less spherical ones follow in a more or less progressive series, the bed of sediment finally accumulated tends to show a progression of the particles from coarse or heavy at the bottom, to fine or light particles at the top.

bedding joint.   Joint parallel to the bedding.

bedding plane.   Division planes in sedimentary or stratified rocks, separating individual layers, beds, or strata.

bed form.   Structure created by currents on stream, channel beds, or on ocean bottom sediments, particularly shelf environment. Small scale ripple marks are normal but strong tidal or fluvial currents can develop dune structures at a scale measured in meters. When properly interpreted much can be learned about current direction, strength, and general depositional environment.

bed load.   Sediment moved by a current in a rolling or bouncing manner; usually applying to fluvial transport but may be used in conjunction with any current including wind and waves, which moves sediment along the sediment surface; opposite to *suspended load.*

bedrock.   More or less solid, undisturbed rock either at the surface or beneath superficial deposits of gravel, sand, or soil.

beehive coke (refin).   Coke manufacturing in a bee-hive, rectangular, or similar form oven chamber where heat for coking process is by combustion.

bell cap (refin).   Hemispherical or triangular casting placed over riser in a tower to direct vapors through liquid layer in a tray.

bell hole (drill).   Bell-shaped hole dug beneath a pipe line to provide room for use of tools.

bell nipple (drill).   Bell-mouthed nipple inserted into top of casing in an oil well permitting easy entry of drilling tools and protecting top of casing while drilling.

bench.   Outcropping bed forming a conspicuous, relatively horizontal ledge; location for a permanent reference mark or support given to a known elevation; used in engineering surveys when running levels.

bench mark.   Marker in ground indicating an exact location and elevation above sea level.

Bender sweetening (refin).   Sweetens distillates; particularly applicable in converting mercaptan sulfur, in the range of 0.004 to 0.08 weight percent to disulfides. Charge is distillate streams as kerosene, jet fuel and No. 2 fuel oil, especially effective in the kerosene, No. 2 fuel oil range. A sweet, bright, non-corrosive distillate.

benthonic.   Pertaining to the ocean bottom regardless of depth; as an example, benthonic organisms dwell on the ocean bottom whereas planktonic organisms dwell near the air to sea interface.

bentonite.   Rock composed of clay minerals and derived from alteration of volcanic tuff or ash. Color range of fresh material is from white, to light green or light blue; on exposure, may darken to yellow, red or brown. Highly plastic and expands, when wet; small amounts can be mixed with kaolin clays to increase their plasticity; absorbs water and swells accordingly. A major constituent of drilling mud. A natural clay used in petroleum products to improve color.

bentonite grease.   Consistency of No. 2 cup grease and containing a small percentage of water and bentonite as a substitute.

benzene. Member of the gasoline family having six carbons and six hydrogen atoms; liquid hydrocarbon, at normal temperatures having a ring-shaped structure; secured from distillation of coal tar and from aromatic hydrocarbons in crude oil.

benzol. Hydrocarbon compound obtained from coal; has low specific heat value; high freezing point; slow burning rate and may be compressed to a high degree; an active solvent.

benzol-acetone process. Solvent dewaxing process; chills mixture of oil, solvent, and wax, then removes wax by means of a filter.

berm. Nearby horizontal portion of a beach or backshore having an abrupt fall and formed by deposition of material by wave action; marks limit of ordinary high tides. (*See* **beach profile**)

Bernoulli's Principle. At a constant level, the sum of the kinetic and potential energies of a fluid is constant, i.e., if its velocity increases, its pressure decreases and vice versa.

beta decay. Process by which certain radioactive atoms disintegrate; atomic core loses a neutron, gains a proton and at the same time expels an electron from the atom; involved in conversion or ordinary uranium into atomic fuel plutonium.

beta emitter. Radioactive element, natural or artificial, changing into another element by beta decay.

beta particle. A negative electron or a positive electron (positron) emitted from a nucleus during beta decay; the symbols β, β^-, and β^+ are reserved for electrons of nuclear origin.

beta radiation. Emission of either an electron or a positron by an atomic nucleus.

beta ray. A stream of electrons. Name beta ray was applied before scientists knew what it was. Still commonly used to describe an electron expelled from a disintegrating atom.

B horizon. An oxidized soil zone lying below the A horizon in the soil profile; colloids, soluble salts, and fine mineral particles from near the surface, accumulate in this zone.

bight. Bay formed by natural bend of the coastline.

bioclastic rock. Sedimentary rocks consisting of broken shell and skeletal material, usually calcareous and of fragmental organic remains.

bioherm.   A reef or mound built principally by sedimentary organisms such as corals, mollusks or other sedimentary organisms having calcareous shells or skeletons.

biostrome (old).   Bedded structures consisting of skeletal shell and remains of organisms which built them.

bioturbation.   Mixing of sediment by the action of browsing and burrowing organisms. Primary sedimentary structures are destroyed (bedding, ripples, etc.). Sediment may be homogenized to depths approaching one meter below sediment to sea interface.

biozone.   Rock distinguished by fossil remains of specific kinds of organisms (species or genus), from lowest to highest and including laterally farthest.

bit.   Drilling tool that actually cuts the hole in rock. Grouped into three broad categories: rotary bits, percussion bits, and combination rotary percussion bits. Cabletool and rotary tool bits are of various designs, depending on the kind of rock being drilled.

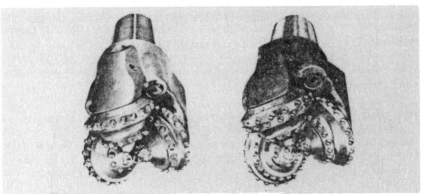

Rotary drill bits

bitumen.   Any of the inflammable, viscid, liquid or solid hydrocarbon mixtures soluble in carbon disulphide; often used interchangeably with hydrocarbons. Solid or semi-solid organic material, dark brown to black in color and usually a residue from distillation of crude petroleum.

bituminous.   Containing oil, or yielding oil on distillation as in Kerogen shale and asphaltic sand.

bituminous (soft coal).   Compact brittle rock ranging from grayish black to deep black in color, distinctly banded or laminated; some layers shiny, while others dull; shiny layers, called glance or vitrain, consists of flattened tree trunks and limbs whose cell walls are preserved in a crumpled state, though all soft parts of the tree have vanished; dull layers, called matt, dull coal or durain, consist of tiny fragments of plant material mixed with resistant fragments and clay. Bituminous coal is well jointed and breaks into angular blocks. Common impurities include clay grains, sulphur, and pyrite; hardness generally is between 1 and 2.

bituminous shale. Shale containing hydrocarbons or bituminous material; when rich in such substance, yields oil or gas on distillation; also called *"oil shale."*

black acids. Sulfonates found in acid sludge; insoluble in naphtha, benzene, carbon tetrachloride and in 30 percent sulfuric acid but very soluble in water. In dry, oil-free state, sodium soaps are black powders.

black grease. Applies to grease, black due to use of asphalt, either as such, or as normal constituents of fuel oils or residues used in the manufacture of the greases.

black oil (fuel). Residue remaining in the still after distillation; not a product of distillation. Consists of several grades of viscous low gravity oils.

black sands. Local deposits of heavy minerals concentrated by wave action and current action on beaches, consisting largely of magnetite, ilmenite, and hematite, associated with other minerals as garnet, rutile, zircon, chromite, amphiboles, and pyroxenes.

blanket sand. Body of sand or sandstone covering a considerable two-dimensional area; often called sheet sand.

blank off (prod). Close off, as with a blank flange or bull plug.

bleeder. Connection located at a low place in a gas line or vessel by means of which condensate, water, and oil can be drained from the line or vessel without discharge of the gas.

bleed into. Cause a gas or liquid to mingle slowly with another gas or liquid, usually by pressure.

bleed off (or bleed down). Reducing pressure by letting oil or gas escape at a low rate.

blending. Process of mixing two or more oils having different properties to obtain an oil of intermediate properties. Certain classes of lubricating oils are blended to viscosity, whereas naphthas may be blended to meet a distillation specification as end point.

blending compound (refin). Non-organic material used to improve the antiknock quality of gasoline, as in tetraethyl lead.

blending naphtha (refin). Refinery term for a distillate used to thin heavy stocks to facilitate process as thinning lubricating oil with gasoline to prevent the oil from congealing during the dewaxing process.

blending stocks (refin). Blends with commercial gasoline, including natural gasoline, straight run gasoline, cracked gasoline, polymerized gasoline, alkylate, toluene, tetraethyl lead, anilol, etc.

block faulting. Breaking of large rock masses into blocks bound by normal faults with or without appreciable tilting.

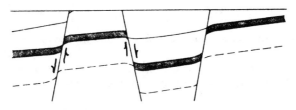

Block-faulting

bloom. Color of oil by reflected light differing from its color by transmitted light.

blotter press (refin). Plate and frame press with blotter paper for the filtering medium.

blowing. Agitating a liquid by introduction of compressed air near the bottom of a tank or container. In "blowing bright," air assists in carrying off moisture; in acid treating, air used for agitation.

blown asphalt. Asphalt treated or heated with air or steam blown through it, in a blowing still at relatively high temperature.

blown oil. Oxidizing oil by blowing air through it while hot. Procedure increases the viscosity of certain animal and vegetable oils as rapeseed, castor and similar oils; subjecting oil to a stream of heated air or steam to remove relatively low boiling point products at temperatures below their actual boiling point and thus increases the viscosity and retards decomposition of remaining stock or residue.

blowout. An uncontrolled escape of drilling fluid, gas, oil, or water from a well, caused by the formation pressure being greater than the hydrostatic head of the fluid in the hole.

blow-out preventer (drill). Rams which can quickly and effectively seal the annular space between drill pipe and string of casing to which attaches preventer; located under the derrick floor and rotary table. Operates manually or hydraulically.

blue mud (old).   Variety of deep sea mud; consisting mainly of land derived fine silt and clay; turns red when oxidized.

blue oil.   Produced from heavy oil or paraffin secured from shale or ozokerite; cooled and pressed to separate the hard paraffin scale and then refined and fractioned into lubricating oil.

body.   Referring to consistency or viscosity; heavy or high body synonymous to high viscosity.

body waves.   Earthquake waves that travel through the body of the earth; primary and secondary waves.

bog.   Swamp or tract of wet land; covered in many cases with peat.

boiling point.   Temperature at which vapor pressure of a liquid becomes equal to pressure exerted on the liquid by surrounding atmosphere.

boiling range.   Temperature range usually determined at atmospheric pressure in standard laboratory apparatus, over which boiling or distillation of an oil commences, or finishes.

boiling water reactor.   Apparatus in which water is allowed to boil directly inside reactor, thus producing steam without intermediate step of transferring heat from a coolant to a boiler where steam is made.

bolson.   A Southwestern United States term denoting a large basin that receives drainage from surrounding highlands but has no outlet.

bomb.   Steel cylinder used as a testing device for conducting chemical tests under high pressure; used for tests as gum in gasoline, sulfur, and vapor pressure; also used to synthesize igneous and metamorphic rocks and minerals.

bombs, volcanic.   Ellipsoidal or spindle-shaped masses of viscous lava ejected from a volcano but solidifying before striking the ground; ranges in size from a fraction of an inch to several feet.

bond, covalent.   Linkage between two atoms in a molecule, with no difference in electric charge in the two atoms; linkage formed by sharing of electron pairs.

bond, ionic.   Linkage between two atoms, with separation of electric charge of the two atoms; linkage formed by the transfer or shift of electrons from one atom to another.

bonnet.   Part of a valve packing off and enclosing the valve stem.

borderland.   Continental regions of the world where continental margins and active plate boundaries coincide; sites of current or recent tectonic and volcanic activity.

borehole.   Hole drilled into the earth to obtain samples of, and to measure physical properties of rock and sediments penetrated.

bottled gas.   Ordinarily butane, propane, or butane-propane liquefied mixtures, bottled under pressure for domestic gas use.

bottomed (drill).   Elevation or depth below surface at which a well is completed.

bottom hole contract (drill).   Payment of money or other considerations provided by contract upon completion of a well to a specified depth.

bottom hole pressure.   Well bore pressure at a depth of the producing interval; usually recorded by a gage run on wire line.

bottoms.   Liquid collecting in the bottom of a vessel, either during fractioning process or while in storage.

bottomset beds.   Layers of fine alluvial sediments carried out and deposited on the bottom of the sea in front of a delta.

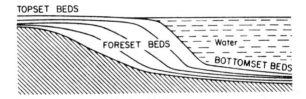

bottom settling (BS).   Residue of substances settling from crude oil in storage; or where used oil is stored in a settling tank.

bottom settling and water (BS&W).   Results of a test determining the presence of solids and water.

bottom steam (refin).   Steam introduced into oil during distillation from perforated pipes on bottom of still.

bottom water.   (1) Water occurring in producing formation below the oil or gas. (2) In oceanography, referring to older, colder, heavier water found at great depths in the earth's oceans.

Bouguer anomaly (geoph).   Difference between theoretical value of gravity and observed value corrected to sea level, taking into account altitude of gravity, station, and rock between station and sea level.

boulder clay.   Unsorted glacial drift with rock fragments of boulder size embedded in clay matrix; till.

Bouma Sequence.   Sequence of lithologies and structures typical of turbidite deposition. Five subdivisions from bottom to top: **(a)** Massive, graded sand to granule-sized particles, **(b)** Plane parallel laminae of sand, **(c)** Ripples, wavy or convoluted laminae in sand and silt, **(d)** Upper parallel laminae in sand and silt, **(e)** Interturbidite (generally shale).

Boyle's Law.   At constant temperature, the volume of a given quantity of gas varies inversely with the pressure.

breaking out (drill).   Unscrewing drill pipe when coming out of drilled hole to change drilling bit, or for any one of many other reasons.

breakout (drill).   Refers to the act of unscrewing one section of pipe from another section, especially in case of drill pipe while it is being withdrawn from well bore.

breakout block (drill).   Heavy plate fitting in rotary table and holding drill bit while it is being unscrewed from drill collar.

breakpoint bar.   An elongated sand bar parallel to a beach, deposited as a result of breaking waves. (*See* longshore bar)

breathing.   Movement of gas, oil vapors or air, in and out of the vent lines of storage tanks due to alternating heating and cooling.

breccia.   Fragmental rock resembling conglomerate but having angular, instead of rounded pieces; cemented mass of rock fragments, most of which are sharp-edged or angular, characteristic of not being carried far before deposited; some are cemented talus.

breccia, fault.   Breccias produced in a fault zone where rocks grind against each other, breaking off angular chips.

breccia, intraformational.   A stratum or formation consisting of angular fragments of itself; may form if bed is involved in mass movement (slumping) before consolidation is complete.

breccia, reef.   Angular fragments of coral and limy algae broken from reefs and deposited on their slopes.

breccia, residual.   Fragments produced by weathering and cemented by iron compounds or calcite.

breeder.   An atomic furnace manufacturing new fuel at the same time as it consumes fuel to generate energy. A true breeder makes more new fuel than it uses up; besides creating heat, splitting fuel atoms, throw off fragments that can convert certain other materials into atomic fuels.

bridge (drill).   Usually formed by the caving of the wall bore or by the intrusion of a large boulder.

bright.   Term applying to lubricating oils, meaning clear or free from moisture.

bright stock (refin).   An oil in the SAE 60 or 70 grade, clay filtered, dewaxed, or otherwise processed to improve color, to make it bright; used as a heavy lubricator oil and as a blending stock with neutral oil to secure various grades of blending oil.

brine.   Water saturated with, or containing a high concentration of salt; any strong saline solution containing sodium chloride or other salts as calcium chloride, zinc chloride, calcium nitrate, etc.

Brinell hardness.   A measure of the hardness of material in terms of a specified indentation procedure. The indenter is a 10 millimeter hardened steel ball.

British thermal unit (Btu).   Amount of heat required to raise the temperature of one pound of water, one degree Fahrenheit.

bromine.   Elemental halogen, similar in properties to chlorine; dark brown to reddish brown; easily liquified gas; soluble in water and most liquid solvents.

bromine number.   Degree of unsaturation of petroleum products expressed as the number of grams of bromine consumed by 100 grams of sample under standard conditions.

bromine test.   Unsaturated hydrocarbons present in the test sample are absorbed by the bromine determining the base or source of crude oil. The lower the rate of bromine absorption, the more paraffinic the test sample.

bromine value.   The number of centigrams of bromine absorbed by one gram of oil under certain conditions; test for the degree of unsaturation of a given oil.

brown acids.   Oil soluble petroleum sulfonates found in acid sludges and extracted by use of naphtha solvent.

Brown-Beovere test.   Apparatus designed to determine the resistance of an oil to the action of oxidation.

brown petroleum.   Solid or semisolid products produced by action of air in asphalt.

brownstone. Ferruginous sandstone where grains are coated and cemented with iron oxide.

Brunton pocket transit. An instrument measuring dip, strike, etc. in rock outcrops; compass especially adapted to geologic mapping.

BS doctor (refin). Black material, mainly lead sulfide, formed in treatment of "sour oil" with doctor solution and found in the interface between the oil and doctor.

bubble cap (refin). Inverted cap with a notched or slotted periphery dispersing vapor in small bubbles beneath surface of the liquid in the bubble plate of a distillation column.

bubble chamber (refin). Device containing superheated liquid through which charged subatomic particles are shot, initiating boiling, giving rise to bubbles that reveal existence and behavior of protons, electrons, mesons and other particles.

bubble point (refin). Temperature at which first incipient vaporization of a liquid occurs in a liquid mixture; corresponds to the equilibrium point of zero percent vaporization or 100 percent condensation; pressure should be specified, if not one temperature.

bubble pulse (seis). Pulsation attributable to bubble produced by a seismic charge, explosion or spark, fired in water. Bubble pulsates several times with a period proportional to the cubic root of weight of charge; each oscillation produces an identical, unwanted seismic effect.

bubble tower or column (refin). Fractioning tower so constructed that vapors rising, pass up through the layers of condensate in a series of plates. The vapors pass from one plate to the next, above the bubbling, under one or more caps, and out through the liquid plate. The less volatile portions of vapor condense in bubbling through the liquid in the plate and overflow to the next lower plate, and ultimately back into the boiler, thereby affecting fractionation.

bubble tray (refin). Horizontal tray fitted to the inside of a fractioning tower; used to secure intimate contact between rising vapors and falling liquid in the tower.

buffer. Any substance or combination of substances when dissolved in water, producing a solution that resists change in its hydrogen ion concentration (pH) upon the addition of acid or base.

bull wheel (drill). Rope driven hoist for lowering and raising cable tools.

Bump down (prod). Too long a length of rod between the pumping jack and the pump seat causing the pump to hit down, or the bottom, in the downstroke.

bumper (drill).   Supporting stay between the main foundation sill and the engine block; a fishing tool for loosening stuck cable tools.

bumping (refin).   Knocking against the wall of a still during the violent boiling of a petroleum product containing water; usually ceases before the water is entirely evaporated.

bunker oil.   Heavy and inexpensive grade fuel used for bunkering ships.

buradiene.   Member of the gas family having four carbon and six hydrogen atoms; used extensively for making buna S and butyl types of synthetic rubber.

burning point.   Lowest temperature at which volatile oil in an open vessel continues to burn when ignited by a flame held close to its surface; temperature determines the degree of safety at which kerosene and other illuminates may be employed.

burrows.   Openings created by burrowing organisms, generally preserved by being filled with sediment upon the demise of the organism or its migration.

Butamer process.   Economically converting normal butane to isobutane catalytically in the presence of hydrogen, using mild processing conditions. The initial product approximates equilibrium concentrations of isobutane; the recyclable operation product is high-purity isobutane. The Butamer process is a fixed-bed vapor process promoted by the injection of trace amounts of organic chloride and the addition of a minor amount of hydrogen to suppress polymerization of the traces of olefins formed during the isomerization reaction.

butane.   Member of the gas family, like propane is preserved in natural gas and often making up the bulk of the product sold commercially as LP-gas, bottle gas, or liquified gas.

butte.   A flat-topped, steep-walled hill, usually a remnant of horizontal beds, smaller and narrower than a mesa; characteristic of arid or semi-arid regions.

butyl alcohol.   Water white liquid made from petroleum; does not attack natural rubber; used as a shock absorber and hydraulic brake fluid; also for thinning lacquers.

butyl rubber.   A type of synthetic rubber made from a member of the gas family found in refinery gas. Petroleum products are combined to produce butyl rubber crumbs, which are then washed, dried, and mixed to make butyl rubber.

by-pass valve.   Small pilot valve used in connection with a larger valve to equalize pressure on both sides of the disc of a larger valve before it is opened.

C

cable-tool drilling (drill). Method of drilling. A steel bit is fastened to a drill stem and jars, suspended from a wire line. Given an up-and-down motion, the bit cuts the rock by impact. Drill cuttings are removed by bailing after tools have been raised from the hole. Diagram of a cable tool rig.

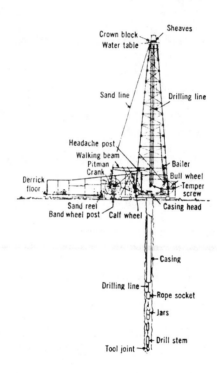

calcarenite. Limestone composed of sand-sized grains, consisting of more or less worn shell fragments or pieces of an older limestone; a clastic sediment or limestone.

calcirudite. A limestone conglomerate or sediment, composed of fragments of coral shells or limestone, cemented or mixed with calcite and calcareous sand or mud.

calcium. Element of the alkaline-earth family; a crystalline metal soluble in water with formation of hydroxide and readily dissolved in acids with formation of salts; used in the manufacture of calcium base grease; component of lime, gypsum, limestone, etc.

calcium base grease.   Soap used the manufacture of a cooked type grease using calcium, slack lime, as the alkali.

calcium carbide.   Made by combining carbon and calcium; used for generating acetylene gas, water plus calcium carbide produces acetylene gas.

calcium carbonate (drill).   An insoluble salt ($CaCO_3$), as in limestone, shells, etc.; sometimes used as a weighting material in specialized drilling fluids.

calcium chloride (drill).   A very soluble calcium salt ($CaCl_2$); sometimes an addition to drilling fluids to impart special properties, but primarily increases the density of the fluid phase.

calcium hydrate.   Active ingredient of slacked lime. Used in the manufacture of calcium base grease.

calcium hydroxite (drill).   Active ingredient of slacked lime ($CaCOH_2$); the main constituent in cement when wet; referred to as "lime" in oil field terminology.

caldera.   Large basin-shaped volcanic depression, diameter of which is many times greater than that of the included volcanic vent or vents; classified into three major types; explosion, collapse, and erosion.

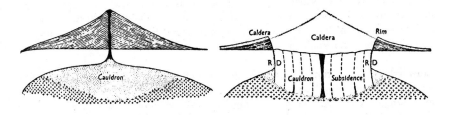

collapse caldera

caliche.   Precipitant of calcium carbonate or calcite in the B-horizon of arid and semi-arid soils; cements residual silt, sand, and gravel into a hard impervious layer (See *hardpan*).

caliper log (drill).   Ascertains the diameter of the wellbore, or the internal diameter of tubular goods; indicates undue enlargement of the wellbore due to caving, wash-out, or other causes. In tubular goods, reveals internal corrosion, scaling, or pitting.

calorie.   Measurement of heat; a great or kilo calorie is the amount of heat required to increase the temperature of one kilogram of water by one degree celsius. Small calorie is the heat required to raise one gram of water, one degree celsius at, or near maximum density.

calorific power.   Amount of heat, in calories, evolved in complete combustion of one gram of any material, more particularly fuel.

calorific value.   Potential heat contained in a substance and measured in terms of Btu's per pound; fuels and burning oils range between 18,000 and 19,000 Btu's per pound.

calorimeter bomb.   Apparatus determining the heat content of fuels and other petroleum products.

Cambrian Period.   Earliest division of the Paleozoic Era defined a century ago in Wales (Latin name, Cambria). During this period, large amounts of the North American continent was submerged under shallow seas. Most of the land was low and flat, apparently devoid of land, plants or animals. The climate was probably warm. Sedimentary rock of this period contains large numbers of invertebrates, as jellyfish, sponges, brachiopods, snails, trilobites, and cephalopods.

canadol.   Petroleum product having a specific gravity slightly higher than petroleum ether; used as a solvent and as an anesthetic.

canal coal.   Variety of sub-bituminous coal or bituminous coal formed in special parts of coal swamps. Very dense, deep black, lusterless rock, looking much like dull tar; containing bitumen and breaking with a conchoidal fracture; contains a large proportion of hydrogen, burning with a bright blue flame that describes its name meaning canal coal.

canyon delta.   A variation and more specific definition of a type of fan describing a sloping, cone-shaped accumulation of sediments at the mouth of a canyon having a single, deep sea channel and high natural levees on its upper portion with multiple shallow distributary sea channels in its lower portion; also called deep sea fan and fan-delta.

capillary action.   Phenomenon causing surface of a liquid to be elevated, or depressed, when in contact with a solid; attractive force between two unlike molecules, as evidenced by wetting of a solid surface by a liquid.

capillary fringe.   Ground water zone of variable thickness, rising above the water table. Water is drawn up by capillary action from the zone of saturation into fine pores and interstices in the lowest portion of the zone of aeration.

capillary viscosimeter.   Apparatus determining viscosity of light oils using capillary tubes.

caprock.   (1) Impervious rock layer overlying a reservoir bed, deformed into a structural trap as an anticline, homocline, or salt dome. Prevents upward migration of oil. (2) Disk-like mass of rock immediately overlying the salt of a salt dome.

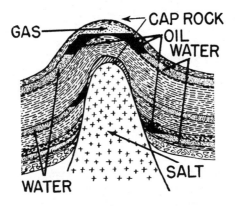

carbene.   Components of bitumen, soluble in carbon disulfide, but insoluble in carbon tetrachloride; with exception of solubility, similar to asphaltenes.

carbon.   Chemical element of atomic number 6 and atomic weight of 12; each nuclei of a common isotope of carbon, contains 6 neutrons and 6 protons. In form of graphite, very useful as a neutron moderator, slowing neutrons without absorbing them appreciably; in form of diamonds, embedded in drill bits designed to drill quickly through hard rock. A nonmetallic element existing in form of diamonds, graphite, and is amorphorous; combined as carbon dioxide, carbonates, and exists in all living things.

carbon 14.   A radioactive isotope of carbon with atomic mass of 14; useful in determining age of carbonaceous material.

carbonaceous.   Containing carbon. In geology, containing coal in well-defined beds, or small disseminated particles of carbon mingled with inorganic constituents; applying especially to black shale or other sedimentary rock containing particles of carbon distributed evenly throughout the mass.

carbonaceous organic rocks.   When plant tissue accumulates under a protective layer of water preventing too rapid oxidation, new accessions may overcompensate for loss by decay so extensive deposits result. Anaerobic bacteria, heat, and pressure expel impurities from the deposit in form of hydrogen sulfide, ammonia, and water, leaving nearly pure carbon behind. Process may be so gentle that details of trunks, twigs, and leaves are preserved. Chemical and physical changes produce a series of products—peat, lignite, sub-bituminous coal, bituminous coal, semi-bituminous coal, semi-anthracite, anthracite, and graphite.

carbonate.   Any compound formed when carbon dioxide contained in water combines with oxides as calcium, magnesium, potassium, sodium, and iron; common carbonate minerals are dolomite, siderite, and calcite.

carbonation.   Chemical weathering; minerals are altered to carbonate by carbon dioxide dissolved in ground water (carbonic acid).

carbon black.   A deposit of very finely divided carbon resulting from oil and gas burnt in an insufficient supply of air.

carbon color test.   Determines the carbon contents of a liquid by comparing its color to a standard solution.

Carboniferous.   Geologic Age of Coal and Amphibians comprising both the sub-periods, Mississippian and Pennsylvanian.

carbonization.   Process of converting an organic compound as an oil into carbon by means of heat; a substance into a residue of carbon by removing other ingredients, as in charring of wood, natural formation of anthracite, fossilization of leaves and other plant organs.

carbon residue.   Result of evaporation and pyrolysis of a petroleum product; not entirely composed of carbon, but a coke which can be further changed by pyrolysis. Tests for the determination of carbon residue provide some indications of the relative coke forming propensities of an oil.

carbon tetrachloride.   A non-inflammable product combining methane and chlorine; used for cleaning, as a solvent, fire extinguisher, etc.

carbureted water gas.   Blue gas enriched with cracked or vaporized oil to increase its heat content from approximately 300 Btu to between 500 and 560 Btu per cubic feet.

carried interest.   Working interest participation in production property; operator is reimbursed for his investment out of an oil before recipient receives a percentage share of net income.

casing cementing (drill).   Practice of filling area between casing and hole with cement to prevent fluid migration between permeable zones and to support the casing.

casing centralizer (drill).   A device secured around the casing to centralize the pipe in the hole and thus provide a uniform cement sheath around the pipe.

casing cutter (drill).   A heavy cylindrical body fitted with a set of knives used to free a section of pipe in the well. Apparatus is run on a string of tubing or drill pipe; cutting performed by the action of the knives on the inner walls of the pipe through rotation.

casing design chart (drill). A graphic means for selecting the most economical, safest grades and weights of pipe for the casing string. Safety factors, relating to such stresses as collapse, burst, and tension are incorporated in the chart.

casing elevator (drill). Tool for lowering or raising the casing during drilling operations.

casing head (drill). A heavy steel fitting connecting the first string of casing and providing a housing for the slips and packing assemblies by which subsequent strings of casing are suspended and the annulus sealed off.

casing head gas. Natural gas flowing from the casing head. Associated and dissolved gas produced along with crude oil from oil completions.

casing head gasoline. Liquid hydrocarbon extracted from casing head gas by compression, absorption, or refrigeration.

casing pressure. Gas pressure built up between the casing and tubing.

casing protectors (drill). Short, threaded nipples screwed into the open end of the coupling and over the threaded end to protect the threads from damage and dirt accumulation; also called *thread protectors.*

casing shoe (drill). A short, heavy, hollow, cylindrical steel section, beveled on the bottom edge, placed on the end of the casing string to serve as a reinforcement shoe and to aid in cutting off minor projections from the borehole wall as the casing is being lowered.

casing string (drill). Total footage of casing run in a well. In deep wells, high tensil strength is required in the top casing joints to carry the load, while high collapse resistance and high internal yield strength is needed in the bottom joints. In the middle of the string, average tensile strength, average collapse, and burst resistance are sufficient. By using the appropriate combination of kinds and weights of pipe in a long string, a combination string can be designed to best withstand the conditions encountered during the life of the well and also allow for efficient production at a minimum of cost. Diagram of casing strings and pipe used in an oil well. (See Fig. p. 48)

castings (fos). Indigestible remnants of meals swallowed by burrowing invertebrates. Lugworms swallow sand in order to eat small organisms. Once this food has been extracted, the sediment is regurgitated in the form of contorted castings.

catalyst (refin). Substances initiating or accelerating a chemical reaction but tak-

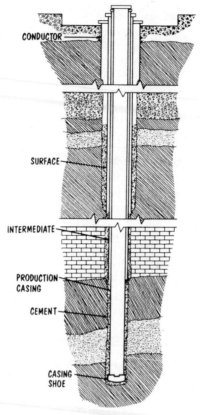

CONDUCTOR

SURFACE

INTERMEDIATE

PRODUCTION
CASING

CEMENT

CASING
SHOE

CASING STRING

ing no part in the reaction; they remain in the same composition at the completion of the reaction as at the beginning. Platinum is a catalyst used in refining of gasoline.

catalytic agent (refin). Substance modifying the velocity of a reaction by its presence, and recovered unaltered at the end of the reaction.

catalytic cracking (refin). Basically the same as thermal cracking since heat is used, but differs in its use of a catalyst to direct the cracking reaction to produce more of the higher octane hydrocarbons; provides a motor spirit of 10 to 15 octane numbers higher than that of the thermally cracked product and is more effective in producing isoparaffins and aromatics, all of which are of high anti-knock qualities or value. Modern cracking units are one of two types, the "fluid process" uses a finely powdered catalyst which is moved and circulated through the system by a "fluidized-solid" technique, and the other, moving bed process, in which pellet catalysts are circulated by elevators or gas-lift method.

catalytic cracking, fluid (refin). Fluid catalytic cracking process is characterized by the use of large quantities of a finely powdered catalyst moved through a processing unit. Catalyst particles are of such a size that when aerated or "fluffed up" with air or hydrocarbon vapor, the catalyst behaves like a liquid and can be moved through pipes and controlled by valves.

catalytic cracking, Houdriflow process (refin). In the Houdriflow process, either liquid or vaporized cracking stock enters the top of a reactor where it contacts a hot, falling bed of catalyst pellets. The cracked material is separated from the catalyst below the reactor and is conducted to the fractioning section of the unit. The catalyst falls into a kiln or regenerator located below the reactor. The coke is burned from it and a stream of flue gas and steam lifts it from the bottom of the kiln to a hopper above the reactor. The regenerated catalyst drops continuously from the hopper into the reactor to renew the cycle. The cracking termperature is in the range of 850°F to 1925°F and pressure of about 10 pounds per square inch.

catalytic cracking, Orthoflow process. A process using the fluidized solids technique. A gas oil cracking stock is fed into the bottom of the reactor. At this point, it contacts the heated catalyst delivered to the reactor through standpipes from the regenerators above; cracked vapors leave the top of the reactor; spent catalyst drops to the bottom of the catalyst bed and is lifted by air up a pipe in the center of the reactor to the regenerator above; more air burns coke from the catalyst, then drops back into the reactor; cyclones within the regenerator separate catalyst dust from the flue gas as the gas leaves the top of the regenerator. The Orthoflow process operates with a cracking temperature of 885° to 950°F, under a pressure of about 15 pounds per square inch.

catalytic reforming (refin). Increases the anti-knock quality of naphtha to get blending stocks for motor fuels. Although a number of reactions take place during reforming, the predominant one is the dehydrogenation of naphthenes to form aromatics. Some of these aromatics are isolated to become petrochemical feedstocks, but most become motor fuels blending stocks of high anti-knock quality. Most feedstocks for reforming are hydrotreated first to remove arsenic, sulfur, and nitrogen compounds, otherwise they would poison the reforming catalyst. Hydrogen is a by-product of catalytic reforming. Some of this hydrogen is recycled to sustain reformer reactor pressure and to suppress coke formation. The major portion of the hydrogen is available for such processes as: hydrotreating, hydrocracking, isomerization, and the manufacture of petrochemical.

catalytic reforming, Platform process (refin). Initial step is the preparation of the naphtha feed; heated to 850°F to 1000°F and passed into a series of three catalyst cases under a pressure of 200 to 1,000 pounds per square inch. Further heat is added to the naphtha between each of the catalyst cases in the series. The material from the final case is reacted with hydrogen rich gas and reformate.

catalytic reforming, Powerforming (refin). Process based on frequent regeneration, carbon burn-off permitting a continuous operation. Reforming takes place in four or five reactors, while regeneration is carried out in a fifth or swing reactor.

The cyclic process assures a continuous supply of hydrogen gas for hydrofining operations and tends to produce greater yields of higher octane reformate.

catastrophism. Hypothesis accounting for difference in organic remains found in successive rock layers by assuming destruction of old organisms by a catastrophe and creation of new ones later to replace them; antithesis of evolution.

catching samples (drill). Geological information obtained by studying samples of the formations penetrated by the drill. Members of the drilling crew obtain these samples from the drilling fluid as it emerges from the wellbore or from the bailer in cable tool drilling. Cuttings so obtained are carefully washed until free of foreign matter, then dried and labeled to show the depth at which they were found.

cat head (drill). A spool-shaped attachment on a winch around which rope is wound for hoisting and pulling. The breakout cat head is the rotating spool located on the side of the draw works used as a power source for breaking out, unscrewing the drill pipe. The make-up cat head is used as a power source for making up, screwing, joints of pipe.

catline (drill). A hoisting or pulling line powered by the cat head; used to lift heavy equipment around the rig.

catwalk (drill). Ramp to the side of the drilling rig where pipe is laid out to be lifted to the derrick floor by the catline.

caustic. Corrosive; a strong alkaline material; caustic soda or sodium hydroxide; used in many refining processes.

caustic lime. Used in a hydrated form in the manufacture of calcium base greases.

caustic soda. Same as ordinary household lye. Chemically it is sodium hydroxide and used in the refining of petroleum (1) to remove mercaptans and naphthenic acids, (2) to neutralize sulfuric acid in chemical refining, and (3) as the alkali in the manufacture of sodium base grease.

caving (drill). Collapse of the walls of the wellbore; falling in of the material surrounding the borehole; sloughing.

cavitation. Turbulent formation, generally mechanically induced; includes growth and collapse of bubbles in a fluid, occurring when static pressure at any point in fluid flow is less than fluid vapor pressure.

cc, cubic centimeter. A metric system unit for the measure of volume; essentially equal to the milliliter and commonly used interchangeably. One cubic centimeter of water at room temperature weighs approximately one gram.

cellar (drill).   A hole dug, usually before drilling of a well, to allow working space for the casing head equipment.

Celsius scale.   A temperature scale with fixed points of freezing and boiling of water at 0° and 100°, respectively.

cement (drill).   A mixture of calcium aluminates and silicates made by combining lime and clay while heating. Slaked cement contains about 62.5 percent calcium hydroxide, a major source of trouble when cement contaminates the mud.

cementation.   Process of precipitation of a binding material as quartz, calcite, or dolomite around grains or minerals in rocks.

cement copper.   Copper precipitated by iron from copper sulfate solution.

cementing (drill).   Application of a liquid slurry of cement and water to various points in an oil well, inside or outside the casing. Primary cementing refers to the cementing operation taking place immediately after the casing has been run into the hole, and which provides a protective sheath around the casing, segregates the producing formation, and prevents the migration of undesirable fluids. After the primary cementing operation has been completed, any subsequent cementing operation is generally referred to as secondary cementing. Among the most useful of secondary cementing methods, squeeze cementing, in which the slurry is squeezed out through perforations in the casing, by the application of great pressure. Squeeze cementing is used to isolate a producing formation, to seal off water, to repair casing leaks, etc. Another secondary cementing method, plug back job, a plug of cement is positioned at the desired point and allowed to set. Wells are plugged back to shut off bottom water or to reduce the depth of the well.

cementing materials (drill).   Cement slurry normally composed of portland cement and water, although a number of admixes are frequently used. Mainly designed to affect either the density of the mixture or its setting time. Portland cement used may be of the high early strength type, common, or standard cement, or slow setting. Additives include accelerators, as calcium chloride; retarders as gypsum; weighting materials, as barium sulfate; light-weight additives, as bentonite or pozzolans; and a variety of lost circulation materials, as mica flakes, cellophane flakes, etc.

Cenozoic Era.   Latest of the great eras of earth history. Also called the Age of Mammals. Sometimes divided into two periods, the Tertiary and the Quaternary but some European geologists prefer calling them Paleogene and Neogene, respectively.

centipoise, Cp.   Unit of viscosity equal to 0.01 poise. A poise equals one gram per meter-second, and a centipoise is one gram per centimeter-second. Viscosity of water at 20°C is 1.005 Cp (1 Cp = 0.000672 lb/ft-sec).

center irons (drill).   Saddle supporting the walking beam on the Samson post.

centrifugal separator.   Machine utilizing centrifugal force to separate two phases of differing gravity, as wax from oil, or acid sludge from treated oil.

centrifuge.   An instrument for separating substances having different specific gravities by centrifugal or rotating force. In the petroleum laboratory used for precipitating solids from a liquid sample measuring the amount of solids by the graduations on the glass tube, centrifuge tube.

centrifuge refining.   A process in which lubricating oil is heated, then mechanically mixed with sulfuric acid. After the necessary contact time between the oil charge and acid, the mixture is centrifuged, separating the oil from the acid sludge formed from the action of the acid.

centrifuge test.   A carefully measured sample of oil is diluted with naphtha in a graduated centrifuge tube, revolved at high speed in a centrifuge forcing the solids into the cone shaped end of the tube. Diluting the used oil with petroleum ether or some other volatile liquid, the fine fuel soot present is readily collected with other larger and coarser solid particles. The amount of solids is then calculated from the graduations of the tube and reported as the precipitation number.

cetane number.   Measure of ignition quality of diesel fuels by matching the fuel, under test, with a mixture of n-cetane and x-methylnaphthalene; the percentage of the fuel is the cetane number. Routine tests employ reference fuels of 18 to 70.5 cetane numbers.

Cgs system.   System of physical measurements in which the fundamental units of length, mass, and time are centimeter, gram, and second, respectively.

chain compound.   Characterizing atoms of a compound bound together in chain-like formation.

chain isomerism.   Isomerism arising from difference in the types of carbon chains in the isomers.

chain reaction.   Process propagating fission from nucleus to nucleus in a fissionable isotope, as U_{235} or Pu_{23}, by means of the neutrons emitted during fission. Reaction increases rapidly in intensity, as in the A-bombs, unless controlled, as in a nuclear reactor.

chain tongs (prod).   Apparatus used for turning pipe or fittings when diameter is larger than ordinary pipe wrenches.

chalk.   A soft bioclastic limestone made up of microscopic shells of one celled coccoliths and planktonic Foraminifera. Ordinarily nearly pure white in color, but iron oxide, carbon, and other minerals may stain it buff, light gray or flesh color.

changing rams (drill). Process changing the size of the blowout preventer rams on a rotary drilling rig when putting into service drill pipe of a different size from that previously used.

channel (drill). Cavity behind casing in a faulty cement job.

channel-fill deposit. Accumulation of sediment in a stream channel when a stream is losing energy and unable to remove sediment from a given locality as fast as it is being carried in.

channel-lag deposit. Relatively coarse-grained sediment left behind in a stream channel when finer sediments were carried away down stream.

channel process (refin). In the production of carbon black, a system of iron channel beams used as its depository surface. These channel beams are turned with the flat side downward over the horizontal rows of stationary burners and are given a slow reciprocating motion. From these beams, the black is scraped off and removed by spiral conveyors.

channel sands. Sedimentary deposits in the bed of a stream channel. Deposits become sandstone upon lithification. Channel sands may have markedly better permeability than surrounding rock in which case they may become a reservoir bed if associated with a source bed. Sometimes called shoestring sand.

channel test. Gear lubricant test showing lowest temperature at which lubricant can be used without channelling.

charged particle. An ion, an elementary particle carrying a positive or negative electric charge.

charging line (refin). A pipe line transporting fresh charge or charging stock of crude oil, gas, oil, etc. to the still.

charging stock (refin). A product charging a still; may be any product recovered through previous distillation as gas oil, fuel oil, or any product selected for further distillation or refining.

Charles law (gases). Volumes assumed by a given mass of gas at different temperatures with constant pressure are, within moderate ranges of temperature, directly proportional to the corresponding absolute temperature.

chase threads (prod). Straightening and cleaning threads of any kind.

cheese (refin). Mixture of wax and oil taken from wax presses to be later refined by the sweating process or by separation of oil remaining in the wax.

chemical acid. One of the acids accounting for activity in oil. Originates from chemical or chemicals used during chemical refining of the oil, as sulfuric acid.

chemical additive (grease). Increases the ability of grease to (1) resist oxidation, (2) prevent corrosion to metal surfaces, and (3) impart extreme pressure characteristics.

chemical bond. Binding or linking force holding atoms together; involves only electrons in the outermost shell of an atom.

chemical desalting (refin). Removes inorganic salts from crude oil that cause plugging of exchangers, coking of furnaces and corrosion; process also removes arsenic and other trace metals acting as poisons to catalytic cracking catalysts; product is crude, containing five to 10 pounds, or less of salts per thousand barrels.

chemical limestone. Inorganically precipitated nonclastic limestone, micrite.

chemical octane number. Anti-knock quality or octane number built into fuel at the refinery or by addition of anti-knock compounds.

chemical refining. Refining raw distillate by chemical action.

chemically neutral oil. An oil free of substances or elements causing corrosion when used under conditions for which it was made.

chenier. Beach ridge showing typical dune and beach features; shell debris abundant and grain size increases upward. Usually deposited on a swamp or marsh beds.

chernozem. A dark brown or black soil rich in humus and lime; type of pedalfer soil having a very thick and dark A horizon.

cherry picker (drill). Fishing tool used to take a friction hold on a section of lost tubing or pipe.

chert. Extremely fine grained crystalline chalcedony (quartz) occurring as nodules in limestone and occasionally in beds; difficult to drill through.

chert clause (drill). Clause in a drilling contract stipulating that when chert is encountered in a drilling well, footage rates are no longer applicable and day-work rates become effective.

choke (prod).   A type of orifice installed in a line to restrict the flow and control the rate of production. Surface chokes are a part of the Christmas tree assembly and contain a choke nipple or bean with a small diameter bore restricting the flow. Positive chokes replace the choke nipple with one of a different size changing the rate of production. Adjustable chokes utilize a conical needle and seal, allowing variations in chokesettings.

C-horizon   Lowermost soil horizon; found immediately below the b-horizon and consisting of partially oxidized bedrock. Original bedrock structure and texture are somewhat altered but still recognizable. C-horizon material is usually weaker and more friable than unweathered bedrock.

Christiansen effect.   Finely powdered substances as glass or quartz, immersed in a liquid of the same index of refraction obtain complete transparency only with monochromatic light. With white light, the transmitted color corresponds to the particular wave-length for which the substances, solid and liquid have exactly the same index of refraction. The indices of refraction match only a narrow band of the spectrum due to differences in dispersion.

Christmas tree (prod).   Assemblage of valves, pipes, and fittings at the top of a well controlling flow of oil and gas.

chromate (drill).   A derivative of chromium with a valence of 6, as in sodium bi-chromate; added to drilling fluids either directly or as a constituent of chrome lignites or chrome lignosulfonates; sometimes widely used as an anodic corrosion inhibitor, often in conjunction with lime.

chromatography.   Means of separating gaseous liquid, or solid mixture into identifiable components by passing the mixture through a solvent and then an absorbent.

chromometer.   Instrument determining color of oils, using some fixed scale as the Saybolt chromometer.

cinder cone.   Built when solid or nearly solid volcanic ejecta or cinders fall to earth in the immediate vicinity of a volcanic vent. Cinders may be loose and easily displaced from the cone if solid when they struck the ground. Nearly solid cinders may weld to each other resulting in a solid, cohesive cone. Cinder cones may consist of silica poor lava (basalt) or silica rich lava (rhyolite or anderite). Silica rich lava creates more numerous and larger cinder cones.

cinders.   Volcanic glass, pumice, or scoriaceous fragments of gravel and coarse sand size ejected to form a volcano in a solid or nearly solid state.

circulate (drill).   Revolving a drilling fluid through drill pipe and well bore while drilling operations are temporarily suspended; conditions drilling fluid and well bore before hoisting drill pipe and obtains cuttings from bottom of well bore before drilling proceeds.

circulation (drill).   Movement of drilling fluid from the suction pit through pump, drill pipe, bit, annular space in the hole, and back again to the suction pit; circulation time refers to process time.

circulation head (drill).   Device attached to the top of casing during cable drilling enabling the operator to drill against heavy gas pressure and circulating mud without danger of a blowout.

circulation, lost (drill).   Resulting from drilling fluid escaping into the formation by way of crevices or porous media.

circulation rate (drill).   Volume flow rate of the circulating drilling fluid, usually expressed in gallons or barrels per minute.

cirque.   Basin from which a mountain glacier flows, the focal point for the glacier's nourishment. After a glacier has disappeared and all its ice has melted away, the cirque is revealed as a great amphitheater or bowl, with one side partially cut away. The back wall rises a few hundred feet to over 3,000 feet above the floor, often as an almost vertical cliff. The floor of a cirque lies below the level of the low ridge separating it from the valley of the glacier's descent. A lake forming on the bedrock basin of the cirque floor is called a tarn.

cis-transisomerism.   Isomerism arising from different arrangement of atoms or groups around a double bond or other rigid structural unit.

clarifier.   Apparatus or device for removing color or cloudiness in an oil or water by separating foreign material by mechanical or chemical means.

clastic.   Texture of rock composed principally of detritus transported mechanically to its final place of deposition; common clastics are sandstones and shales. Diagram details a classification of common clastic rocks based upon percent occurrence of members of cobbles, pebbles, sand, silt, and clay. (See Fig. p. 57)

clay.   (1) A particle size term denoting natural material with diameter less than 5 microns (.005 millimeters—between silt size and colloidal. (2) A naturally occurring soft substance becoming plastic when mixed with water. (3) A group of mica-like minerals commonly occurring as weathering products in soil. Drilling "mud" consists primarily of bentonite; others include kaolinite, illite, and montmorillonite.

clay recovery (refin).   A method for recovering clay as fuller's earth; used as a filtering medium or in clay wash operations.

clay refining (gasoline).   Use of clay to neutralize and to improve the color of gas-

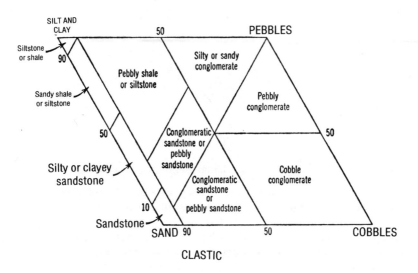

oline treated with acid in the chemical refining process; used alone or in conjunction with a neutralizing solution as caustic soda.

clay, residual. Result of both physical and chemical weathering of rocks, the latter being the more important. Clay is mixed with silt particles, weathered crystals and pebbles, or even large fragments remaining from the original rock; none of this material is sorted. The colors of residual clays vary according to oxide impurities, particularly iron oxide, and the climates of the regions in which they develop.

clay-revivifying system (refin). Method by which fuller's earth or other decolorizing clays are cleaned for reuse; most common process by "roasting" or heating the clay in some form of rotating cylinders, mechanically stirring and mixing the clay.

clay wash (refin). A light oil, as kerosene or naphtha, used to clean fuller's earth use as a filter.

cleavage. A property of crystalline material expressed as easy breakage along directions of weakness in the crystalline structure.

Cleveland open cup tester. An apparatus used for the determination of the flash fire points of all petroleum products above 175°F, with the exception of fuel oils.

clinograph (drill). An instrument for determining and recording the deviation from vertical of a well during drilling.

clinometer. During oil exploration, instrument measuring angles of surface structures from the plane of horizon; also determines the dip of sedimentary beds. (See Fig. p. 58)

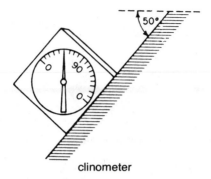

clinometer

closed steam (refin).   Steam used for heating oil, etc., in a still using a closed coil with no direct contact between steam and the other phases taking place.

closed-ring hydrocarbon.   Members having one or more closed rings with the prefix cyclo, meaning a closed ring structure as cyclohexane.

cloud and pour points.   Tests useful in estimating the relative amount of wax in an oil. As all oils will solidify if cooled to a low enough temperature, these tests do not indicate the actual amount of wax or solid material in oil. Tests indicate that most of the wax melting above the pour point has been removed.

cloud point.   Crystallizing or separating temperature of paraffin or other solid substances from the solution, imparting a cloud appearance.

coal.   Rock composed of combustible matter derived from partial decomposition of plants; considered a biochemically formed sedimentary rock.

coalescence.   Combination of globules in an emulsion caused by molecular attraction of the surfaces. Act of combining or uniting to form one body, either through chemical affinity, simple mixture, or concentration.

coal oil.   Mixture of hydrocarbon ring compounds of the aromatic series; obtained by destructive distillation of bituminous coal.

coastal terms.

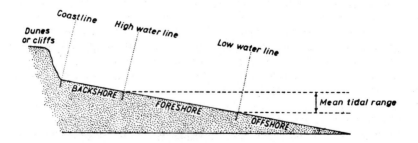

cobble.   Rock fragment larger than a pebble but smaller than a boulder, diameter between 64 and 256 millimeters

coefficient of cubic (volumetric) expansion.   Change in volume of any material with a temperature change of one degree. Coefficient of cubic expansion is three times the coefficient of linear expansion, using the same temperature scale.

cohesion.   Attractive force between constituent molecules, i.e., force holding the molecules of a substance together.

cohesiveness.   Characteristic of an oil enabling it to retain its body.

coils (refin).   Number of turns of piping or series of connected pipes in rows or layers for purposes of radiating or absorbing heat.

coke.   Generic term applied to infusible, cellular, coherent solid material obtained from coal, pitch, petroleum, petroleum residues, and some other carbonaceous materials, resulting from destructive distillation; has a characteristic physical structure resulting from the decomposition and hardening of a fused or semi-liquid mass.

coke breeze (refin).   Fine screening from crushed coke or from coke ovens.

coke drums (refin).   Vessel in which coke is formed and which can be cut off from the process for cleaning.

coke knocker (refin).   Mechanical device for breaking loose coke formations within a tower or drum.

coke scrubber (refin).   An apparatus filled with coke moistened with oil to purify gases. Moistened coke retains any vapors dissolved in the gas.

coke still (refin).   Apparatus where distillation is continued until only coke remains.

coking (refin).   Coking processes are severe forms of thermal cracking. Their feedstocks are residuals resisting cracking by other means. For many years, coke was a by-product of the processes since the primary purpose of coking was to get more lighter stocks from a barrel of crude. Coke can be a valuable product. When its sulfur and metal content are low enough, the coke is suitable for the manufacture of electrodes and aerospace components. Coke with intermediate sulfur content makes a reasonably good fuel for generating electricity.

cold press (refin).   Apparatus used for pressing chilled wax distillate to separate oil from wax. The wax remains between plates, canvas covered or perforated and the oil flowing from the press is pressed distillate.

cold set grease. Calcium-base grease made by the cold mixed or cold set procedures.

cold settling (refin). Process for removal of petroleum wax from cylindrical stock and high viscosity distillates. A naphtha solution of the oil is chilled and the wax crystallizes out, then allowed to stand until the wax settles to the bottom, leaving a clear, nearly wax-free oil-naphtha mixture at the top stripped of naphtha after percolation through clay to produce bright stock.

cold test. Temperature at which oil becomes solid, generally considered to be 5°F lower than the pour point.

collar (drill). A coupling device used to join two lengths of pipe. A combination collar is a coupling with left hand threads at one end and right hand threads at the other.

collar bound (drill). Casing stuck in a drilled hole, with mud collecting at the collar joint.

colligative property. One of the four characteristic properties of solutions, interdependent changes in vapor pressure, freezing point, boiling point, and osmotic pressure, with a change in amount of dissolved matter.

colloid. A state of subdivision of matter consisting either of single large molecules or of aggregations of smaller molecules physically dispersed to such a degree that the surface forces become an important factor in determination of its properties. Particle size is less than 0.1 micron; size and electrical charge of the particles determines the different phenomena observed with colloids, as in Brownian movement. A colloid is intermediate between a physical suspension and a chemical solution, showing some of the properties of each.

color (petroleum). Color of petroleum by transmitted light varies from light yellow to red; some very dark or black oils are opaque; higher the specific gravity or lower the API gravity, the darker the oil. The cause of the color is not known but thought to be related to the aromatic series of compounds. By reflected light, crude oil is usually green because of its fluorescence. Special refining produces nearly colorless oils. Color is commonly determined with the Saybolt colorimeter.

colorimeter (oils). An optical instrument to determine the color of oils by comparison with standard colored fluids in bottles or with standard color discs; may be done electronically with a spectrometer.

colorless asphalt. Produced by treating asphalt with chemicals as sulfuric acid to remove dark colored bodies. Although dark colored bodies are removed, the asphalt is by no means colorless. In the process, the softening point is lowered.

color stability. Resistance of oil to discoloration due to light, aging, etc.

color test.   Test showing the extent to which a petroleum product will transmit or reflect light.

columnar jointing.   Cleavage of rock producing vertical columns, generally hexagonal in section, as in many basaltic flows; frequently caused by thermal contraction of a cooling lava flow or igneous dike. (*See* columnar structure)

columnar section.   Graphic representation of rock succession in a vertical column with lithology shown by more or less standard symbols and thickness plotted to scale.

columnar structure (igneous rocks).   Special kind of jointing dividing igneous rocks into long blocks or columns with flat sides that meet at angles. Columns develop when fluid rock cools, hardens and contracts. Instead of dividing the rocks into blocks, joints start at the top or sides and work inward. Columnar structure appears best in hardened lavas cooled and shrunk very quickly, also found in dikes, sills, and some laccoliths.

combination cracking (refin).   Process involving crude distillation, viscosity breaking, naphtha reforming, and gas oil cracking; in the extreme case might involve gas recovery and coking.

combination tower (refin).   Tower designed for both flashing and fractionating operations.

combined water (grease).   Calcium base grease contains water acting as a stabilizer. This water will not cause corrosion and is reffered to, as combined water.

combustion.   Act or process of burning. Chemically, a process of rapid oxidation caused by the union of oxygen in the air, a supporter of combustion, with any material capable of oxidation.

compaction.   Reduction in volume or thickness of a sediment under load through closer crowding of constituent particles, accompanied by decrease in porosity, increase in density, and expulsion of interstitial water. Load pressure results from the weight of new sediment deposited on older layers; increase with depth in the sedimentary deposit.

competence of a stream.   Diameter of the largest particle, a stream can move in suspension; directly proportional to stream turbulence or turbulent velocity.

component distillation (azeotropic).   Process separating a part or fraction not obtainable by normal distillation; employs a third component maintaining a low constant boiling mixture.

composite cone.   Large volcanic cone built of alternating layers of lava and cinders.

compound.   A material consisting of a union of two or more elements in a fixed ratio by weight, as water, consists of hydrogen and oxygen in a two-to-one ratio, symbolized as H_2O; similar with carbon dioxide CO_2.

compound shoreline.   Shoreline whose essential features combine elements of at least two of the other shoreline classifications, as emergent, submergent, high, or low energy.

concentrate.   Separation of useful mineral from gangue in a rock by washing or mechanical methods of concentration; increase in percentage value. The amount of a substance in weight, moles, or equivalents contained in unit volume. The process of increasing the strength of a solution by the evaporation of liquid or volatile ingredients.

conchoidal fracture.   Curved, shell-like form of a surface produced by the fracture of brittle minerals and rocks as in quartz and volcanic glass.

concordant.   Having parallel structures.

concretions.   Lumps or masses of mineral matter generally harder and more resistant than the sediments around them ranging in size from a fraction of an inch to several feet in maximum dimension. Most are shaped like simple spheres or disks, although some have fantastic and complex forms.

condensate.   Hydrocarbons, in gaseous state under reservoir conditions but becoming liquid either in passage up the hole or at the surface.

condensation.   Process of changing vapor to a liquid or a lighter one to another denser form, by depression in temperature or increase in pressure.

condensation nuclei.   Microscopic chemical particles in which water vapor condenses forming cloud droplets.

condensed deposit.   Thin but not interrupted sedimentary material accumulated very slowly.

condenser (refin).   Unit in which vapors from a still are condensed into liquid. Liquids from the condenser are condensates, and the process is condensation.

condenser reflux (refin).   Condenser constantly condensing vapors and returning liquid to the original distilling flasks or to lower levels of a fractioning tower.

conductivity.   A measure of the quantity of electricity transferred across a unit area per unit potential gradient per unit time; reciprocal of resistivity.

conductor pipe (drill).   Short string of casing of large diameter whose principal function is to keep the top of well bore open and to provide means of conveying upflowing drilling fluid from well bore to slush pit.

cone of depression.   A conical depression in the water table developing around a well as it is being pumped.

confining pressure.   An equal, all sided pressure. In the crust of the earth, lithostatic pressure resulting from the load of overlying rocks. In experimental work, hydrostatic pressure, generally produced by liquids.

confluence.   Point where two streams flow together to form one stream.

conformable.   Superimposed strata deposited without interruption in accumulation of sediment; parallel beds.

conglomerate.   Cemented mixture of rounded pebbles and larger rocks, often with some fine material. Large pieces may consist of various kinds of rocks, though hard, resistant kinds as quartzite and granite are specially common; pebbles may be grains of minerals as quartz and feldspar. Cementing material varies greatly may consist of sand, clay, limestone, iron oxide or a mixture of these substance.

coning.   Process by which reservoir water invades the oil column and enters well due to uncontrolled production.

conjugated double bonds.   Two or more double bonds joined together by a single bond.

connate water.   Water trapped in pore spaces of a sedimentary rock when the sediment was deposited.

Conrad discontinuity.   Seismic discontinuity in the earth's crust where velocity increases from 6.1 to 6.4 or 6.7 kilometers per second; occurs at various depths and is supposed to mark contact of "granitic" and "basaltic" layers.

consequent.   Having a course or direction dependent on, or controlled by the geologic structure or by the form and slope of the surface, related mainly to streams and drainage.

consequent stream.   A river whose course was determined by the original slope and irregularities of the surface on which it developed.

conservation gasoline (refin).   Gasoline made in a conservation plant. Charging stocks consist of vapors collected from distillation and cracking stills, from storage tanks and at other points where condensation gasoline vapors may be lost.

consistency. Viscosity of a non-reversible fluid in poise, for a certain time interval, at a given pressure and temperature.

consolidated sediments. Sediments converted into rocks by compaction, by deposition of cement in pore spaces, and/or by physical and chemical changes in the constituents.

consolidation. In geology, any and all of the processes whereby loose, soft, or liquid earth materials become firm and coherent.

contact. Bonding surface between different rock units or between mineral bodies and enclosing rocks. The surface, in many cases, irregular, constitutes the junction of two bodies of rock.

contact filtration (refin). Method by which oil comes in contact with clay in an agitator or a still during distillation, after which oil and filtering medium are separated by a suitable filter.

contact-metamorphic ore deposit. Ore deposits occurring at or near the contacts of intrusive rocks with sedimentary beds and carrying minerals characteristic of contact metamorphism, as garnet, pyroxene, and epidote.

contact metamorphism. Metamorphism generally relating to the intrusion or extrusion of magmas and taking place in rocks at or near their contact with a body of igneous rock.

continental accretion. Concept of continental growth from an ancient nucleous. Growth takes place along continental margins as a result of mountain building.

continental basalt. Basalt erupting onto continents, frequently in large masses as in Columbia and Snake river basalts. These are several thousand feet thick and extend over thousands of square miles. Lava originated from the lower crust or upper mantle and exhibits minor but significant chemical differences from oceanic basalt.

continental crust. Thickened crust underlying the continents; thickness is variable, measured in tens of kilometers; consists of an upper "granitic" or silica rich portion and a lower "basaltic" or silica poor layer. Moho or "M" discontinuity marks the boundary of the crust and mantle.

continental glacier. Massive low altitude glacier thousands of feet thick, covering huge areas as in Greenland and Antarctica.

continental margin. Zone separating the emergent continent from the deep sea bottom; generally consisting of the continental shelf, slope, and rise.

continental nucleus.   Stable, usually centrally located core of a continent, consisting of ancient crystalline rocks; younger rocks are deposited against and over core to enlarge the mass of the continent (*See* continental accretion).

continental rise.   Part of the submerged margin of the continent that lies at the foot of the continental slope and curves gently down to the ocean basin.

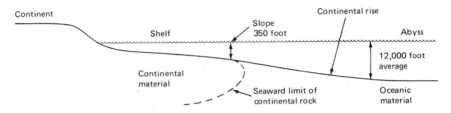

continental shield.   Stable interior portion of a continent underlaid by ancient rock eroded to a surface of low relief; often coincides with continental nucleus.

continental slope.   Relatively steep slope leading from the outer edge of the continental shelf to the deep ocean basin.

continuous contact filtration (refin).   Refinement principally of lubricating oil stocks, but also applies successfully to the finishing of waxes and specialty oils. The charge is petroleum hydrocarbons in the lubricating oil or specialty oil range solvent extracted and acid treated. Solvent extracted stocks are dewaxed, usually before charging. The product is finished lube oil or specialty oil base stocks ready for blending and compounding.

continuous distillation (refin).   A method separating a volatile product or fraction while the charging stock is in a state of continuous flow through the system. Distillation by continuous operation where the oil is successively heated, fractionated, and condensed.

continuous treaters (refin).   Series of closed drums with proper connecting lines and pumps in which acid and caustic treating of oil is done continuously and counter-currently.

contour line.   A line connecting points of equal elevation above and below a datum plane, as sea level.

contract depth (drill).   Depth of the well bore at which the drilling contract is fulfilled.

contraction, adiabatic.   Process of contracting a substance without loss or gain of heat.

contraction, heat.   Diminishing volume of gases, liquids, and solids due to cooling, resulting in contraction, as when hot iron shrinks when cooled.

controlled filter.   A filter with a metering device regulating the amount of oil flow through the filter.

control rod.   A metal rod, often containing boron, used to control the intensity of the chain reaction in a nuclear reactor by means of absorption of neutrons.

convection.   Process of mass movement of portions of any fluid medium, liquid or gas, in a gravitational field as a consequence of different temperatures in the medium and therefore different densities; moves both the medium and the heat, term convection signifying either or both.

convection current.   Transmission of heat by the mass movement of heated particles of water or magma.

convection section (refin).   Portion of the furnace in which tubes receive heat by convection from the flue gases.

convergent plate margin.   Boundary between two crustal plates moving toward each other. Margin is marked by subduction and associated seismic activity.

conversion.   Process in which a fissionable material is made by neutron capture while another is being consumed inside the reactor, as plutonium in a uranium-fueled reactor. Term is usually applied when the amount of fissionable material made is less than the amount consumed.

conversion per pass (refin).   Ratio expressed as percent of gasoline boiling below 400°F to the total oil pumped through the coils.

cooker (refin).   Lead-lined vat or tank in which acid sludge resulting from chemical treatment of lubricating oils is subjected to agitation for recovery of acid. These vats are also used for mixing the acid sludge from the light distillate with water.

coolant.   Material used in a nuclear reactor to remove the heat energy resulting from fission. In a power reactor, this heat energy is used to generate electrical power.

coolers (refin).   Any type of structure cooling water by evaporation.

copolymer.   A substance formed when two or more substances polymerize at the same time to yield a product, not a mixture of separate polymers but a complex having properties different from either polymer, as polyvinyl acetate-maleic anhydride copolymer (clay extender and selective flocculant), acrylamide-carboxylic acid, and copolymer (total flocculant), etc.

coprolite (dung stones).   Applied to feces preserved by petrifaction or as molds or casts.

coral reefs.   A ridge or mass of limestone built up of detrital material deposited around a framework of the skeletal remains of mollusks, colonial coral, and massive calcareous algae. Coral reefs occur in three forms, fringing reefs, barrier reefs, and atolls. Fringing reefs are attached directly to the rocky shore; barrier reefs stand on the sea floor and are separated from the shore by a belt of calm shallow water; atolls are small, irregular ring-shaped reefs unattached to any visible land, occurring especially in the Pacific Ocean.

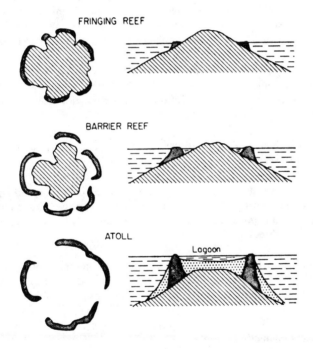

core (drill).   Cylindrical piece of material cut and brought to the surface by special types of rock-cutting bits during the process of drilling.

core analysis (drill).   A laboratory study of the core to determine the following properties of the formation: porosity, permeability, fluid content, angle of dip, geological age, lithology, and probable productivity.

core barrel (drill).   A device used in rotary drilling to cut cores, varies in length from 25 to 60 feet, runs at the bottom of the drill pipe in place of a bit, or in conjunction with a special type of bit.

core, earth's.   Innermost zone of the earth beginning at a depth of 2,900 kilometers and extending to the center of the earth. Measurements of the earth's mass, rotational characteristics, magnectic field, and deep penetrating seismic waves, indicate that the core consists of a nickle-iron alloy and has some of the properties of a liquid in its outer portion but is solid in the center.

coring.   Obtaining a sample of the formation being drilled for geological information purposes.

Coriolis effect.   Apparent curvature of a straight trajectory over the rotating earth. A ball or rock thrown due south in the northern hemisphere will appear to curve to the west as viewed from the earth because of the earth's west to east rotation. The effect exerts a significant influence on the circulation of oceanic water.

coronas.   Produced by the reaction between magma and early formed crystals or between adjacent minerals in contact and characterized by rims and/or zones containing radiating mineral aggregates differing in composition from the original mineral. Coronas are sometimes called reaction rims. See *aureole*

corrasion.   Mechanical erosion (down cutting) performed by running water and using rock particles as tools.

correlation.   Establishment of the equivalence in geologic age and stratigraphic position of two or more sedimentary units in separate areas; criteria include fossil assemblages, geochemical constituents, mineralogic composition, rock structure and texture, and direct physical connection.

correlation (drill).   Charting graphically formations, noting depths and thickness from substance information gathered from one well and relating it to others; compares electrical well logs, radioactivity logs, and cores from different wells.

cosmic sediment.   Particles of extraterrestrial origin observed in deep sea sediments as black magnetic spherules.

countercurrent pipe exchanger (refin).   Heat exchanger; direction of cold oil is opposite to that of hot oil passing through pipes enclosed in a chamber in which cold medium is to be circulated.

country rock.   General term applies to rock surrounding igneous intrusions penetrated by mineral veins; in a wider sense, applies to rocks invaded by or surrounding an ore deposit.

covalent bond.   A type of linkage between atoms. Each atom contributes an electron(s) to a shared pair(s) that constitutes a chemical bond. (*See* ionic bond)

cracked distillate (refin).   Distillate obtained by cracking.

cracked fuels (refin).   Fuels consisting predominatly of cracked residue not necessarily blended with cracked distillate.

cracked gasoline (refin).   Product of cracking process composed predominantly of power fractions and pick-up; has a higher anti-knock quality than straight run gasoline; requires less anti-knock compound or blending agents to increase anti-knock value to a predetermined level.

cracking (refin).   Process whereby less volatile components of petroleum undergo complex chemical changes when heated to high temperatures and put under high pressures, either in the presence or in the absence of catalysts. Carbon-to-carbon bonds are broken under these conditions, producing several compounds with lower boiling points. In this way, molecules that have boiling points too high to fall into the gasoline fractions are broken down and are formed into new compounds that belong in the gasoline fractions. Thus complex molecules of high molecular weight are "cracked" or divided into simpler compounds.

cracking, catalytic (refin).   Oil refining process making use of a catalyst to effect conversion under less severe conditions of pressure than is otherwise possible. Temperatures used are essentially the same in both thermal and catalytic cracking. Catalyst is not consumed in the process which by its mere presence promotes conversion.

cracking coils (refin).   Refinery equipment used for cracking heavy petroleum products; consists of a coil of heavy pipe running through and over a furnace, so oil passing through is subjected to higher temperatures.

cracking, coking (refin).   A lengthy cracking process in which coke is produced as the bottom product.

cracking, liquid-phase (refin).   Decomposing petroleum fractions in the liquid condition.

cracking plant (refin).   Plant used for cracking petroleum oils in the manufacture of cracked gasoline.

cracking, reforming (refin).   High temperature cracking process utilizing straight run gasoline or naphtha as a charging stock. The cracked distillate has a much higher octane number than the charging stock.

cracking, selective (refin).   Process separating charge stock into several parts, each part being cracked separately under the most favorable conditions for that part.

cracking still (refin).   Still operating at high temperature and pressure to accomplish the cracking of petroleum.

cracking, thermal (refin).   Use of heat and pressure convert the largest molecules into smaller ones.

cracking, vapor phase (refin).   Decomposing petroleum fractions in the vapor condition.

cracking, viscosity breaking (refin).   Short time decomposition usually conducted at low cracking (860°F–900°F) for purpose of reducing temperature viscosity or pour point of a heavy straight run fuel.

crater (drill).   Term meaning "the hole is caving in"; refers to the result sometimes accompanying a violent blowout during which the surface surrounding the well bore falls into a large hole blown in the earth by the force of escaping gas, oil, or water.

craton.   Relatively immovable large segment of the earth's crust as continental nuclei. Central stable region common to nearly all continents and composed chiefly of highly metamorphosed Precambrian rocks. *See* continental shield.

creep.   Slow downslope movement of rocks and soil. A variety of mass wasting or mass movement of material caused by alternate expansion and contraction of soil either by alternate freezing and thawing of soil moisture or by alternate wetting and drying of expandable soil clay minerals. Below is a diagram of common evidences of creep. **(a)** moved joint block; **(b)** curved tree trunks; **(c)** downslope bending of strata; **(d)** displaced posts and fences; **(e)** broken walls; **(f)** displaced roads.

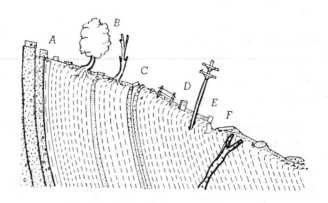

Cretaceous Period. Geological period, last in the Mesozoic era lasting over 70 million years; noted for thick chalk deposits especially in England and France. The name for the period is taken from the Latin word creta meaning chalk. During this period the present Rocky Mountain area was flooded with a shallow sea extending from the Gulf of Mexico to the Arctic Ocean. Many of the coal deposits in the western United States were formed at this time.

critical point. Thermodynamic state in which liquid and gas phases of a substance coexist in equilibrium at the highest possible temperature; no liquid phase can exist at a higher temperature.

critical pressure. Pressure necessary to condense a gas at the critical temperature, above which the gas cannot be liquefied no matter what pressure is applied.

critical reactor. Steady state condition of a reactor in which the neutron fission process is self-sustaining without the aid of external neutron sources.

critical size. Amount of fissionable material just large enough to sustain the chain reaction; smallest amount of atomic fuel to support an atomic "fire." With less fuel, losses of particles that cause atoms to split are too great, and the chain reaction will not start.

critical temperature. Temperature above which a fluid cannot exist solely as a liquid or as a vapor and therefore cannot be liquefied by pressure alone.

critical volume. Volume of unit mass at the critical temperature and pressure.

crooked hole (drill). A wellbore deviating from the vertical, inadvertently.

crop out (geol). Exposed at the surface; refers to bedrock.

cross-bedding. Structure of sedimentary beds characterized by parallel laminations lying at an angle to the planes of general stratification; may be of large scale

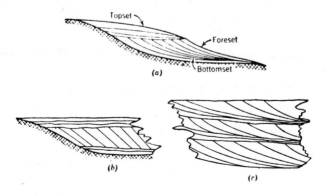

as in sand dunes or preserved eolian deposits, or of intermediate to small scale as with fluvial sediments (deltaic structure or ripple structure); determines paleocurrent direction. Diagrammatic representation of fluvial cross-bedding. (a) Common structure of water laid cross-bedding in which foreset beds are tangential with bottomset beds and are truncated by topset beds. (b)In "torrential" cross-bedding, considered to be developed by strong currents, the foreset strata are transected by both bottomset and topset beds. (c)Typical exposure of three sets of cross-beds showing the topset of one group becomes the bottomset of overlying unit.

crossover joint (drill).   A length of casing with one kind of thread in the coupling; used to make a change-over from one thread to a different one in a string of casing.

cross section (geol).   Geologic diagram of actual field exposure showing geologic formations and structures transected by a vertical plane. Cross section diagrams are commonly used in conjunction with geologic maps and contribute to an understanding of the subsurface geology. Formations, faults, veins, and so forth, are shown by conventional symbols or colors, and the scale is adapted to the size of the features present.

crown block (drill).   Block on the top of an oil derrick upon which rests the crown pulley and the other pulley assemblies. Carries the pulleys over which runs the line from the hoisting drum to the traveling block.

crude desalting (refin).   Procedure of washing crude oil with water to remove materials as dirt, silt, and water-soluble minerals. Water and oil mixture is so intimate that settling alone may not be sufficient to separate the two phases. Separation can be hastened by adding surface active chemicals, or by applying a strong electric field, or by both.

crude oil.   Consists almost entirely of compounds of carbon and hydrogen, with varying amounts of organic sulphur, nitrogen and oxygen compounds and ash. The hydrocarbons present in crude oil are of three types, paraffinic, naphthenic and aromatic. Oil varies considerably in appearance and color, from brownish yellow to black, some viscous, others limpid, while a few carry paraffin wax in suspension.

crude oil, pour point: Temperature below which oil will no longer flow unaided: an indication of viscosity or of ease of pumping with non-waxy crude oils; of little value with waxing oils because of the unpredictable wax crystallization dependent on the thermal history of the oil.

crude shale oil.   Applying to several kinds of organic and bituminous shales varying in mixtures of organic matter with shale and clay. The organic matter is mainly composed of a mineraloid called kerogen, of indefinite composition, insoluble in petroleum solvents, and of uncertain origin. These shales are called kerogen shales. The oil distilled from most of them does not occur as oil in the shale but is formed by heating, distillation of the vapors beginning at temperatures around 662°F (1,350 C), hence kerogen shales are classed as pyrobitumens.

crude still (refin).   Refinery equipment designed to permit physical separation of crude oil by the application of heat.

crude wax.   Paraffin wax with a crystalline structure recovered by dewaxing operation, excludes petroleum secured from residue. Oil remaining in crude wax may be recovered by sweating process.

crust.   Outer shell of the solid earth; its lower limit generally taken to be the Moho discontinuity; varies in thickness from approximately five to seven kilometers under the ocean basins to 35 kilometers under the continents.

cryohydrate.   Solid separating from a freezing saturated solution, contains the solvent and the solute in the same proportions as they were in the saturated solution.

crystalline.   Consists of atoms arranged in an orderly pattern or array; opposite of amorphous.

crystalline rock.   Rock, in the solid state, composed of crystals of various minerals, produced during the process of formation of the rock; orderly arrangement of atoms in a mineral; opposite of detrital or clastic rocks.

crystallization.   Formation of crystals from melts, solutions, or amorphous solids. Igneous rocks form from crystallization of magma, rock salt, from an aqueous solution, and many limestones from shell and skeletal remains.

cuesta.   Asymmetric ridge with one long, gentle, plainlike slope along the dip of the rock and one steep slope across it.

curie.   An amount of radioisotope emitting radiation at the same rate as one gram of radium. Unit indicating the strength of radioactive sources in terms of the number of disintegrations per second in the source. One curie is equal to 3.7×10^{10} disintegrations per second.

Curie point.   All ferromagnetic substances have a definite temperature of transition at which point the phenomenon of ferromagnetism disappears and the substance becomes merely paramagnetic. This temperature is called the "Curie Point" and is usually lower than the melting point.

curved path method (geoph).   Method of analysis and plotting of seismic data allowing for curvature of ray paths, results from increasing seismic velocities with depth in the earth.

cut back asphalt.   Asphalt rendered liquid by fluxing with a light volatile product as naphtha. Upon exposure to the atmosphere, the volatile product evaporates leaving the asphalt behind.

cutback products.   Heavier oils blended with lighter oils to bring the heavier oils to desired specifications.

cut oil.   Oil partially emulsified with water in the presence of air.

cuts.   Petroleum fractions obtained in the distillation of oil.

cycle of erosion.   Succession of events involved in the reduction of a region from its youthful stage to base level through normal processes of erosion.

cycle stock (refin).   Products taken from some later stage of a process and re-charged, to the process, at some earlier stage.

cycle time, drilling fluid (drill).   Time of a cycle, or down the hole and back, is the time required for the pump to move the drilling fluid in the hole. The cycle, measured in minutes, equals barrels of mud in the hole divided by barrels per minute.

cyclic compound.   Any compound containing a ring of atoms or a closed chain of atoms in its molecule.

cyclic sedimentation.   Deposition of different types of sediment in repeated regular sequence. (*See* cyclothem)

cyclization process (refin).   Method reshaping low-octane straight chain members of the gasoline family into high-octane ring-shaped aromatics. In combination with the dehydrogenation process, normal heptane, a low anti-knock member can be changed to toluene, also a high anti-knock blending compound but also used for many other purposes.

cyclone separator.   Conical vessel with a tangential inlet for a gas stream containing powdered solids or liquid droplets and normally provided with a centrally located over-head base withdrawal line. Powdered solids or coagulated liquids separated by centrifugal force and pass downwardly along the incline (conical) to a centrally located outlet. Usually a leg, known as a dip leg, connects to the solids outlet; in catalytic cracking, serves to convey the solids back to the dense-phase powdered catalyst.

cyclothem.   Repetitive succession of marine and non-marine beds deposits during a single stratigraphic cycle reflects regular changes in the environment of deposition, as storms and seasons.

cylinder stock (refin).   Residue remaining in a still after the lighter parts of a crude are vaporized; various steam cylinder oils are derived from it filtered and processed, to become bright stocks.

D

Dakota sandstone. Widespread, land-laid formation underlying the midcontinent Niobrara formation; one of the finest sources of artesian water in the world. This Cretaceous period sandstone is noted for its fossil plant remains, particularly in the ironstone concretions of Ellsworth County, Kansas.

Dalton's Law (gases). Pressure of a mixture of several gases, in a given space is equal to the sum of the partial pressure which each gas would exert if it were confined alone in space.

datum. Any numerical or geometrical quantity or set, serving as a reference or base for other quantities; frequently, a hypothetical reference line or plane used when surveying land, interpreting geophysical data, or drilling a well.

daughter isotope. End product of a radioactively decayed parent isotope.

day work (drill). Basis for payment on drilling contracts when payment by the foot drilled is suspended while drilling rig is used in taking extra cores, logging, or other activities delaying actual drilling. The compensation for use of the rig on a day's work is usually agreed upon and included in the drilling contract.

dead line (drill). End of the drilling line not reeled on the hoisting drum of the rotary rig; usually anchored to the derrick substructure and immobile as the travelling block is hoisted.

dead oil. (1) Products heavier than (1000 specific gravity) water, obtained from coal tar as carbolic acid. (2) oil in storage or elsewhere, free of any entrapped gas as found in live oil.

deasphalting (refin). Operation extracting asphalt components from a petroleum product. Propane deasphalting employs propane as a solvent to separate oil and asphalt. The liquid propane dissolves the oil and the asphalt settles out.

deblooming. Process by which the bloom of fluorescence is removed from certain petroleum oils by exposing oil to sun and atmospheric conditions in shallow tanks.

deblooming agents. Chemicals, as monotroasphalene and yellow coal-tar dyes, added to mineral oils to mask the fluorescence.

debris slide. Erosion from mass wasting by rapid downslope movement of a mass of unconsolidated material not moving as a single block, as in the case of a slump. Inertia may carry debris past the toe of a slope and onto a valley floor producing hummocky topography, superficially resembling glacial morainal topography.

debutinization (refin). Process of distillation in which the lighter components of a distillate are separated from pentane and heavier components.

decay constant. Rate of spontaneous disintegration of a radioactive element unaffected by heat or pressure.

declination. At any given location, angle between the geographical meridian and the magnetic meridian, as in the angle between true north and magnetic north.

decoking, hydraulic (refin). Removal of petroleum coke from vessels using high velocity water streams from special types of nozzles as cutting means.

decolorizing. (1) Process of removing suspended colloidal and dissolved impurities from liquid petroleum products by filtering, adsorption, acid treatment and other chemical treatment, redistillation, or bleaching. (2) Altering the color of oils by removing colloidal, suspended, or dissolved substances.

decomposition. Breaking down of minerals and rocks of the earth's crust by chemical activity; complex compounds are usually broken into simpler ones that are more stable under existing conditions. Decomposition is usually attended by mechanical disintegration to produce comprehensive changes called weathering.

deflation. Erosional process of the wind carrying off unconsolidated material as removal of material from a beach, river flood plane, desert floor, or other land surface by wind action; term derives from Latin "to blow away."

deflation basin. Shallow basin formed by wind erosion; also called "blow-out."

deflocculate. Separation of flocculated masses into individual particles; peptize. Has a profound effect on sedimentation of material consisting of clay minerals or clay-sized particles. Coarse-grained flocculated globules may deflocculate when encountering a change in water chemistry resulting in a change in grain size of the sediment load while being transported in suspension.

deformation. Distortion of the shape, structure, or orientation in space of rocks due to internal earth forces. Folds and faults are common products of deformation; production of gneiss, schist, and other foliated metamorphic rocks. A general term pertaining to physical changes in rocks, associated with mountain building.

degradation. General lowering of land surface by erosion processes, especially through removal of material by the combination of erosion of running water and mass wasting.

dehydrate. Process of removing water or moisture from a substance by heat, filtration, centrifugal action, with a desiccant, etc.

dehydration tank. Tank in which dry, warm air blown through oil removes moisture.

dehydrogenation (refin). Manufacture of ethylene and propylene by vapor-phase cracking and multireaction by catalytic reforming.

delay rental. Payment, commonly made annually on a per acre basis validating a lease in lieu of drilling.

deliquesence. Liquefaction of a solid substance by adsorption of moisture from the air; compounds as sodium chloride (NaCl) and calcium chloride ($CaCl_2$) exhibit this property.

delta. Alluvial deposits spreading off-shore of the body of water into which a stream empties. Precise shape and limit variable because its deposits are gradational with others generally not considered to be part of the delta. Not a single environment but a composite of several independent environments as beach, alluvial channel, swamp, and lagoons. As distributaries shift position, coarse sand deposition occurs in localities formerly receiving silts and clays. Lenses of sand become incorporated in the body of silts and clay. Marine delta actively extending outward from shore are constructive. Constructive deltas are further classified by shape as lobate if distributary lobes coalesce to yield the classical triangular shape or elongate if the deltaic growth perpendicular to shore is excessive over growth in other directions, thereby developing a pronounced extension off-shore. Deltas prohibited from regular offshore growth or extension by wave or tidal action are destructive. Tide dominated destructive deltas do not show a concise delta front but grade off-shore from a network of distributary channels to estuarine tidal channels to an ill-defined system of submarine channels and alternately subsequent to emergent sand bars. The submarine channels and associated sand bars are elongated parallel to the tidal current. Wave dominated destructive deltas may be either lobate or elongate deltas severely truncated by wave action.

delta/fan. *See* fan/delta.

delta front. Actively prograding edge of a delta; surface of steep slope immediately offshore of the nearby horizontal tidal flat and marsh deposits, area of most rapid sedimentation.

deltaic. Pertaining to deltas as in deltaic sediments, deltaic environment.

density. (1) Matter measured as mass per unit volume expressed in pounds per gallon (ppg), pounds per square inch per 1000 feet of depth (pis/1000 ft), pounds per cubic feet (lb/cu ft), and grams per cubic centimeter (g/cm3). (2) In drilling, the weight of a substance per unit of volume, as density of a drilling mud described by "10 lb per gallon" or "75 lb per cubic foot."

density current.   *See* turbidity current.

denudation.   Removal of surface materials by erosion.

deoiling petroleum (refin).   Operation of recovering oil remaining in petroleum or amorphous wax collected during dewaxing processes; accomplished by (1) centrifugal force, (2) solvent refining, (3) filtering, (4) sweating, or a combination of these.

depentanizer (refin).   Fractional distillation tower removing the pentane fraction from a debutanized or butane free fraction.

dephlegmation (refin).   A particular kind of fractionization separating vapor mixture into components by partial condensation; vapor is progressively cooled and successively lower and lower boiling condensates are collected.

dephlegmator (refin).   Type of fractioning tower as used in the manufacture of natural gasoline to separate oil vapors from gasoline.

depletion allowance.   Percentage of income from mining or petroleum operations not subject to taxation; basis for this allowance—resource being exploited is not renewable, hence when depleted cannot be regenerated in contrast to logging, farming, ranching, etc. Allowance is controversial and strong arguments are made both for and against its continuation.

deposits.   (1) Anything laid down. (2) Matter left behind when any agent of erosion loses its capacity to transport sediment. (3) Mineral matter precipitated by chemical or other agents as ore deposits.

deposition.   Laying down of sediment in suspension in wind, water, or ice, or out of solution in water. Deposition of suspended material may be called selective sorting, where coarse material is dropped first and fine material later, owing to gradual decrease in ability of wind or water to suspend sediment. Sorting does not occur when capacity to carry sediment is diminished rapidly or ended abruptly, as when a glacier melts, dumping or depositing sediment of all grain sizes at the same time. Deposition of sediment out of solution occurs a° a result of (1) evaporation, (2) change in water chemistry, e.g. river flowing into ocean, or (3) extraction of dissolved salts by organisms and subsequent sedimentation as fecal matter or skeletal material.

depression contour.   Contour lines connecting points of equal value as elevation, thickness, etc. By convention, a contour line closing on itself shown by a circle, has higher values inside the closure than outside as for a hill. If the value inside the closure of a contour is less than outside as in a pit or hole in the ground, the contour is called a depression contour and is identified by hatchure marks.

topographic depression contour   topographic contour

depropanization (refin).   Process of distillation with the lighter components of a distillate separated from the butane and heavier components.

depropanizer (refin).   Separation of a butane fraction by another distillation in a fractional distillation tower is called a depropanizer since its purpose is to separate propane and the lighter gases from the butane fraction.

depth in, depth out (drill).   Referring to measurements made in connection with running bits or other equipment in the hole and pulling it out again.

derrick.   Any one of a large number of types of load-bearing structures. In drilling work, the standard derrick has four legs standing at the corners of the substructure and reaching to the crown block. The substructure is an assembly of heavy beams used to elevate the derrick above the ground and provide space to install blowout preventers, casing heads, etc. The standard derrick has largely been replaced by the mast for drilling. The mast is lowered and raised without disassembly. For land transport, may be divided into two or more sections to avoid excessive length on the highway.

derrick man (drill).   Crew member with work station in the uppermost portion of the derrick while the drill pipe is being hoisted or lowered into the hole.

desalting.   Removal of salt from crude oil by washing with water or by an electrical method as an electrical dehydration.

desert pavement.   When loose material containing pebbles or large stones is exposed to wind action, the finer dust and sand are blown away and the pebbles gradually accumulate on the surface, forming a sort of mosaic protecting the finer material underneath from deflation.

desiccant.   Substance as salt having a hygroscopic nature used as a drying agent.

desiccation.   Removal of water. In sedimentary deposits, water originally filling pore spaces is forced out by compaction; evaporation of water from unconsolidated sediments, resulting in shrinking and compaction.

desiccation crack. Crack developing on floodplains and tidal flats as a result of the drying and shrinking of clay mineral rich muds; also called mud cracks. When preserved and incorporated into the statigraphic record, useful in establishing whether or not sedimentary beds have been overturned by faulting or folding.

desiccator, absorption test. A vessel with a tight-fitting cover and a well at the bottom to hold a hygroscopic material as silica gel, calcium chloride, used for drying the moisture absorbed by a salt or acid.

design factor (drill). Ratio of the ultimate load a member will sustain to the safe permissible load placed on it. Safety factors are incorporated in the design of casing, to allow for unusual burst, tension, or collapse stresses.

destructive distillation (refin). Process of distillation in the absence of air continuing until all liquids or fractions are vaporized and only a carbonaceous residue remains in the still, as in the cracking process when cracking is continued until only coke remains.

desulfurization (refin). Process removing sulfur or sulfur compounds from petroleum products usually by some chemical as sodium plumbite or sulfur dioxide.

detergent. Cleaning agent removing foreign matter by dispersing or washing action rather than by dissolving action; either oil-soluble or water-soluble.

detergent acids (soaps). Oil insoluble petroleum sulfonates found in certain acid sludges.

detrital. Referring to mineral and rock material remaining after the disintegration, weathering of igneous, metamorphic, or sedimentary rock as in detrital sediment. Differences between definitions of detrital and clastic or broken are subtle and rarely observed.

detrital rock. Sedimentary rock consisting of detritus-conglomerate, breccia, sandstone, siltstone, mudstone, and shale; essentially synonymous with clastic rock.

detritus. Mineral or rock material remaining after disintegration (weathering) of igneous, metamorphic, or sedimentary rock; usually implies movement of material after produced.

deuteric. Interaction between igneous minerals and the final fluid phases of the parent magma; alteration of early formed solid phases by remnants of the magma just before crystallization is completed; final reactions or alterations in the history of crystallization of an igneous mass as a batholith.

deuterium. Isotope of hydrogen with atoms containing one proton and one neutron in each nuclei in contrast to ordinary hydrogen with only one particle; slows downs neutrons passing through the nucleus and used as a moderator to adjust the speed of neutrons in atomic furnaces; also called heavy hydrogen or H_2.

development well (drill). A well drilled in proven territory for the purpose of completing the desired pattern of production; sometimes called an exploitation well.

deviation (drill). Inclination from the vertical of the well bore. The angle of deviation of angle of drift, in degrees, taken at one or at several points of variation from the vertical as revealed by a deviation survey.

deviation survey (drill). A survey to obtain the angle through which the bit has deviated intentionally or inadvertently from the vertical during drilling operations. Two basic kinds of deviation survey instruments exist, one reveals only the angle of deviation and the other reveals both the angle and the direction of deviation.

devil's pitchfork (drill). Fishing tool having flexible prongs for recovering a bit, under-reamer, cutters, etc. lost during drilling operations.

Devonian Period. Named for Devonshire, England, and proposed by Sedgwick and Murchison in 1839. Many types of sediments have been found in these rocks including tillites of glacial origin. There is evidence of considerable volcanism in some areas. Lake and stream deposits with typical fresh water fossil forms also occur. A wide variety of marine fossils including brachiopods (clam shell animals) and other clamlike fauna are found as well as large numbers of sponges and corals. Fossils of wingless insects are first seen in the rocks of this period.

Dewar flask. Container with double walls. The innerwalls are silvered and the area between them evacuated; generally used to store liquefied gases. (Pronounced dew-er)

dewaxing. Removal of wax from raw lube stock by either a mechanical or solvent process or using a solvent. Oil remaining in the wax is usually recovered by sweating process and from petroleum by centrifugation.

dewaxing propane. Process for utilizing propane as a refrigerant and wax anti-solvent for removing wax from wax-bearing distillates.

diabase. Gray, black, or deep green rocks of the basalt-gabbro clan with a characteristic texture exhibits rods or laths of plagioclase surrounded by coarser augite.

diabasic. Similar to diabase in its texture or mineralogy; refers to a true diabase or a rock resembling a diabase.

diabasic acid.   Acid containing two atoms of acidic hydrogen in a molecule, giving rise to two series of salts, normal and acid salts, as sulphuric acid (H_2SO_4) gives rise to normal sulphates and bisulphates.

diagenesis.   All post-depositional changes in a sediment occurring before completion of the transition of sediment to sedimentary rock. Changes may be physical or chemical to include compaction, cementation, recrystallization, replacement and authigenesis.

diameter (drill).   Distance across a circle measured through the center. In measurement of pipe diameters, the inside diameter, I.D., is that of the interior circle, whereas the outside diameter, O.D. is the measurement to the exterior surface of the pipe.

diamond core drill (drill).   Hollow bit used to obtain cores of rock being penetrated. Drilling surface resembles a donut. Rock is removed from around a central core. The core enters the hollow center of the bit and is protected until brought to the surface for analysis. Designed with diamonds embedded in the steel structure doing the actual cutting. These bits vary in diameter and may contain industrial diamonds weighing up to three carats each.

diamond drill (drill).   Form of rotary rock drill in which the work is done by abrasion instead of percussion; black diamonds are set in the head of the boring tool.

diapir.   Originally a fold or dome structure caused by uplift and piercing of a sedimentary salt dome. Term includes reference to the process of upward movement, intrusion of heavier country rock by lighter, natural earth material, salt or magma and the structures produced. Now term is used by some to designate the intruding mass as an irregular salt dome, an igneous intrusion of magma, or a mobile mixture of crystals and magma. Not all intrusions are diapirs, only those whose emplacement is powered by density contrast between the light intrusive mass and the heavy country rock. Used for a long time by sedimentary geologists and only recently by igneous geologists, sedimentary diapirs frequently lead to the formation of structures favorable to the accumulation of petroleum and natural gas; igneous diapirs rarely do.

diastem.   Minor or obscure break in sedimentary rocks involving only a short or insignificant time loss. A break in deposition of sediments accompanied by little or no erosion. Diastems are most common in the deposits of seas whose waters became so shallow that nothing could be deposited and little or nothing could be worn away. Similar to a disconformity but of lesser magnitude or importance.

diastrophism.   General term for processes deforming the earth's crust, as faulting and warping. The energy manifested by these movements is derived from within the earth and the effects contrast strongly with the purely superficial results of weathering and erosion. (*See* orogeny)

diatomaceous earth. An earthy very porous siliceous deposit consisting mainly of shells of diatoms.

diatom ooze. A modern deposit found chiefly under cold oceans of the Antarctic Circle at depths of 2,400 to 8,800 feet; contains considerable amounts of mineral matter and animal remains as well as diatom tests or shells. Countless billions of diatoms settle in shallow water but are hidden by coarser sediments.

diatreme. Carrot-shaped, pipeline igneous mass reaching into the crust through explosive activity of gaseous constituents within a magma.

die collar (drill). A collar or coupling of tool steel, threaded internally, used to retrieve pipe from the well on fishing jobs.

diesel index. An indication of the ignition quality of a diesel fuel calculated from the formula GA over 100 where G is API gravity and A is aniline point.

diesel squeeze (drill). Pumping dry cement mixed with diesel oil through casing perforations to recent water-bearing areas.

differential fill-up collar (drill). A device used in setting casing; runs near the bottom of the casing and automatically admits drilling fluid into the casing as required to cause the casing to sink rather than "float" in the well.

differential pressure (drill). Difference in pressure between the hydrostatic head of the drilling fluid column and the formation pressure at any given depth in the hole; can be positive, zero, or negative with respect to the hydrostatic head.

differential weathering. Weathering progressing at different rates in different places. Different rates controlled by mineral or chemical inhomogeneities in the weathering rock or by physical inhomogeneities (joints and faults) which weaken rock, locally. Differential weathering may be small scale producing pockmarked weathered surfaces with holes a few centimeters in diameter. It may be of moderate scale producing caves of several cubic meters volume as cavernous weathering or large scale resulting in great arches, buttes and pinnacles.

differentiates. Different kinds of igneous rocks formed as a result of magmatic differentiation.

differentiation. Process whereby two or more rocks of different composition are derived from a single body of magma.

dike. Sheet-like body of igneous rock filling a fissure in older rocks which it entered while in a molten condition. Dikes penetrate all types of country rock: igneous, metamorphic, and sedimentary. Dikes range from a few yards to 100 miles in length and several inches or hundreds of feet in width. Dikes range from coarse-grained to fine-grained and glassy, depending on the size of each dike, its composition, its depth of emplacement, and the rate at which it cooled. Dikes are discordant, that is, they cut across any pre-existing rock structure. In contrast, sills are concordant, that is, they are dike-like in intrusions which parallel pre-existing rock structures.

dilatancy (geoph).   Expansion of granular masses as sand when deformed due to rearrangement of the grains.

Dilchill dewaxing/deoiling (refin).   Removes waxes from various lube oil stocks to make high quality finished lubricants with very low pour points; can make specification wax of less than 0.5 percent oil content simultaneously. The charge is a wide range of virgin or treated waxy distillates and deasphalted residual oils covering virtually any viscosity cut from any crude source. The product is essentially wax-free lubricating oils with pour points as low as -20° F to -30° F. A three to 15 percent oil content slack wax is obtained by single stage filtration. Wax of controlled melting point and hardness containing less than 0.5 percent oil can be made simultaneously by adding a warm-up deoiling stage and recovery facilities for a third product.

diluted LP-gas.   Mixtures consisting principally of butane and propane mixed or diluted with air.

Di-Me solvent dewaxing and wax deoiling (refin).   Prepare low pour point lubricating oils by selective crystallization of soft and hard waxes from solutions of waxy oils in dichloro-ethane-methylene chloride mixture. Soft waxes can be separated from hard waxes in an optional second stage.

diolefins.   Open-chain hydrocarbons having two double bonds per molecule; particularly conducive to oxidation, gum formation, and loss of octane rating in storage.

dip.   (1) Angle at which rock structure is inclined down from a horizontal plane. Azimuth of direction of greatest dip is normally recorded with the dip angle. Dip direction is always perpendicular to strike direction. *See* also strike. (2) Angle formed by lines of total magnetic force and horizontal plane at the earth's surface; declared positive if downward.

dip and strike.   Technique for describing the orientation of a plane in space. In this way planes as joint and fault surface as well as sedimentary beds may be described, plotted on a map and their orientation in space dealt with quantitatively. Strike is the compass direction or azimuth of a horizontal line drawn on a plane. Dip is the direction and angle of maximum inclination of the plane with respect to a horizontal plane. By definition, strike and direction of dip are at right angles to each other. Symbols used to express dip and strike: **(1)** North strike with 30 degree dip to the east. **(2)** Horizontal. **(3)** Vertical. **(4)** East-west strike, originally horizontal surface has been rotated 160 degrees to the south until it is now nearly upside down and inclined 20 degrees to the north.

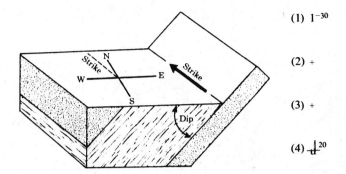

(1) 1^{-30}

(2) +

(3) +

(4) $\downarrow^{20}$

dip-slip fault. A class of fault in which the predominant movement has been parallel to the dip of the fault plane; includes both normal and thrust faults.

dip slope. Slope of an escarpment in the direction of the dip of the bed forming the escarpment.

dip throw. Component of the slip measured parallel with the dip of the strata.

direct lift. Lifting of particles from the stream bed into the low pressure, high-velocity portions of a strongly turbulent flow.

directional drilling. Although well bores are normally planned to be drilled vertically, many occasions arise when it is necessary or advantageous to drill at any angle from the vertical. Controlled directional drilling makes it possible to reach subsurface points laterally remote from the point where the bit enters the earth. Diagrams of typical applications of controlled directional drilling.

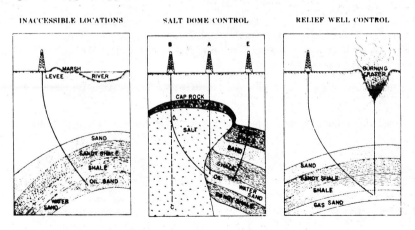

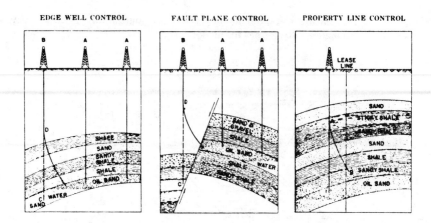

EDGE WELL CONTROL   FAULT PLANE CONTROL   PROPERTY LINE CONTROL

discharge.   Volume of stream flow, per unit of time, through a given cross-section of the stream.

disconformity.   A loss of stratigraphic record owing to erosion or non-deposition; orderly sequence of stratified rocks above and below the disconformity are parallel. Break is usually indicated by erosion channels with sand or conglomerate, indicating a lapse of time or absence of part of the rock sequence.

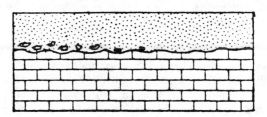

discontinuities.   Relatively abrupt changes in natural earth material, laterally or vertically. Zones in the earth where rapid changes in the velocity of earthquake waves occur because of changes in the elasticity and density of the rocks as in "M" discontinuity or the "MOHO."

discordance.   Lack of parallelism between contiguous strata.

discordant.   Igneous rock contact cutting across bedding or foliation of adjacent rocks.

discordant intrusions.   Intrusive bodies whose contacts cut across the bedding or other preexisting structure of the intruded rock as in dike, batholith.

discovery well.   First producing oil well drilled in a new structure or oil field.

disintegration.   Reduction of rock to small pieces mainly by mechanical means. Term can be restricted to the effects of purely physical agents as frost, rifting effect of roots, undermining and abrasion produced by wind, water, and ice; or include chemical actions, making it practically synonymous with weathering.

dislocation.   In a general sense refers to relative movements of rocks along a fault.

dismembered river system.   Drowned stream valley as a result of flooding, has its former tributaries entering the sea by separate mouths.

disperoid (drill).   A colloid or finely divided substance.

dispersed phase.   Scattered phase solid, liquid, or gas of a dispersion; particles are finely divided and completely surrounded by the continuous phase.

dispersion.   Fairly permanent suspension of finely divided undissolved particles in a fluid.

dispersion additive.   A material dispersing the solids contained in the tar-like residue removed by detergent addition; process called peptization.

displacement.   (1) Weight of fluid, estimated or actual, that is pushed aside by a body immersed or floating in the fluid. (2) Actual movement along a fault; same as slip.

disposal well.   A well through which water, usually salt water, is returned to subsurface formations.

dissection.   Erosion on flat upland areas shaping them into rugged hills, valleys, and ravines.

disseminated.   Scattered or diffused through the rock as a disseminated mineral is scattered through the rock rather than concentrated in masses or veins.

dissiminated deposits.   Natural occurrence or accumulation of ore minerals scattered throughout the rock, as copper minerals occurring in grains through prophyry; "impregnations" sometimes applies to deposits of this nature.

dissociation.   Splitting up of a compound or element into two or more simple molecules, atoms, or ions; applies usually to the effect of the action of heat or solvents on dissolved substances. The reaction may be reversible and not as permanent as decomposition; when the solvent is removed the ions may recombine.

dissolved gas drive (prod).   Gas in solution in the oil tending to come out of solution because of the pressure release at the point of penetration of a well. The

movement of oil produced by dissolved gas drive is analogous to the efferves-
cence that results from the uncapping of an agitated bottle of soda water. Dissolved gas drive is the least efficient type of natural drive.

Dissolved-gas drive

dissolved load.   Soluble products of weathering are carried from the site of disintegration, dissolved in ground water or surface runoff making their way to rivers
and streams. Dissolved load is carried by chemical means whereas particulate
load is carried by physical means.

distillate.   Applies to a liquid, cut, or fraction collected when condensing distilled
vapors as naphtha, kerosene, fuel oil, light lubricant oils, etc. Product of distillation is obtained by condensing the vapors from a still.

distillate hydrodesulfurization (refin).   Improves qualities of distillates, either
straight run or from coker, visbreaker, fluid catalytic cracker ranging from light
gasoline to heavy vacuum gas oil by removing sulfur, nitrogen, metallic contamination, etc; process hydrogenizes olefinic hydrocarbons and improves the color,
odor, and stability of these petroleum cuts. The range of catalysts also includes
those for aromatic hydrogenation to improve smoke point and cetane index.

distillate treating (refin).   Contacting a light hydrocarbon stream with a treating
agent as acid, caustic, etc., then separating the chemical phase for the treated hydrocarbon product. Process also removes wettable solids from hydrocarbons. The
charge is light hydrocarbons as propane, butanes, gasoline, kerosene, diesel fuel,

jet fuel, furnace oils, transformer oils, lube distillates and feeds or products for petrochemicals. The products are treated light hydrocarbons, free from carryover, stable, ready for blending, shipping and further processing.

distillation (refin).   Vaporizing a liquid and subsequently condensing it in a different chamber; separation of one group of petroleum constituents from another by means of volatilization in some form of close apparatus as a still, by the aid of heat. In refineries, it normally refers to a complete operation in which heating, vaporization, fractionization, and cooling are practical.

distillation flask.   A flask in which oil is distilled.

distillation range.   Difference between initial boiling point temperature and end point temperature secured by the distillation test.

distillation stripping (refin).   A fractional distillation operation carried out on each side-stream product immediately after it leaves the main distillation tower; removes the more volatile components and thus reduces the flash point of the side-stream product. Since strippers are short, they are arranged one above another in a single tower. Each stripper, however, operates as a separate unit.

distributary.   An outflowing branch of a river that does not rejoin it as one of several branches into which a stream divides as it flows across a delta.

distribution law.   A substance distributes itself between two immiscible solvents so that the ratio of its concentration in the two solvents is approximately a constant and equal to the ratio of the solubility of the substance in each solvent; requires modification if more than one molecular species is formed.

disulfied oils.   Oils obtained by oxidizing the mercaptans extracted from the light petroleum distillates to disulfides separating from the extract as an oily layer.

disulfides.   Colorless liquids completely miscible with hydrocarbons and insoluble in water. The lower members, when pure, possess a nauseating sweet odor, particularly clinging and penetrating.

divergence loss (seis).   Part of the transmission loss, due to the spreading of seismic or sound rays in accordance with the geometry of the situation.

divergent plate boundary.   Two crustal plates moving away from each other share a divergent plate boundary. Plate movement is at right angles or nearly so, to the plate boundary. New oceanic crust rises and solidifies to fill the space left as the plates diverge. Divergent plate boundaries are sometimes referred to as zones, areas or centers of spreading and are frequently marked by volcanic eruptions and earthquakes. The Mid-Atlantic Ridge is an example of a divergent plate boundary.

divide.   Higher land separating two adjacent drainage basins; crest of uplands that divides flowing water into separate channels.

doctor test.   Method of detecting certain sulfur compounds as hydrogen sulfide and mercaptans in gasoline, kerosene, stoddard solvent, etc. A quantity of doctor solution and a pinch of pure sulfur are mixed with the test sample under test conditions. If the sample or sulfur discolor, the test is reported sour, otherwise it is reported sweet.

doctor treatment.   Refining process using sodium plumbite to remove sulfur compounds, particularly in light distillate; before removal the distillate is termed sour, and after removal, sweet.

dolomite.   (1) A mineral, $CaMg(CO_3)_2$ resembling calcite. (2) A crystalline sedimentary rock consisting of the mineral dolomite, of mixtures of the minerals dolomite and calcite, predominantly dolomite or of impure calcite in which a significant but irregular amount of calcium has been replaced by magnesium. Some geologists favor a primary origin of dolomite, while others favor a replacement of calcite by dolomite, as the mode of formation.

dolostone.   A sedimentary carbonate rock rich in magnesium, includes dolomite.

dome.   (1) A circular or elliptical structure in which the strata dip from a central point rather than from an axis and which is favorable for the accumulation of oil. (2) A roughly symmetrical uplift or upwarp, the beds dipping in all directions from the crest of the structure; a more or less circular local or regional upwarp resembling an inverted bowl. Diagram illustrates how oil accumulated in a dome-shaped structure. The dome is circular in outline.

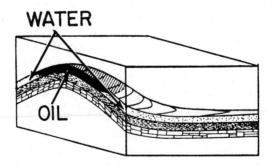

double (drill).   Stand of drill pipe consisting of two lengths joined together.

double board (drill).   Working platform of the derrick man, located at a height on the derrick equal to two lengths of pipe joined together.

double decomposition.   Simple exchange of parts of two substances to form two new substances.

double solvent extraction (refin).   Method of solvent refining employing two solvents not miscible in each other.

doughnut (drill).   Ring of wedges supporting a string of pipe; a threaded, tapered ring used for the same purposes.

downcomer (prod).   Pipe receiving downward flow.

downflow (refin).   Soaking drum in which hot oil enters at the top and is removed from the bottom.

downwarp.   Segment of the earth's crust broadly bent downward; a basin, if circular or elliptical.

dragfold.   Minor folding of a weak bed lying between folded strong beds.

dragon's blood.   Bright yellow powder used for titrating benzol to determine chemical purity.

drainage basin.   Entire area from which a stream and its tributaries receive their water.

drainage net.   Pattern of tributaries and master streams in a drainage basin.

draw works (drill).   Nerve center of the drilling rig; principal parts include hoisting drum, brakes, clutches, catheads, and control panel. The mud pump may also be part of the completed draw works unit. In drilling operations, lowers and raises pipe and casing inside the drilled hole. (See Fig. p. 92)

dresser sleeve (drill).   Slip type collar joining plain pipe together.

dressing (drill).   Sharpening, repairing and replacing parts, especially drilling bits and tool joints to make various items of equipment ready for reuse.

drill collar (drill).   Adds weight above the bit to increase its cutting power; by placing heavy and rigid drill collars immediately about the bit, they reduce the tendency of the drill pipe to bend and divert the bit off a vertical course. Drill collars are heavy shafts of steel, each about 30 ft. in length and weigh anywhere from 1/2 to 2 tons according to diameter, depending upon the weights required, two to twenty or more drill collars may be used at one time.

drill column (drill).   Connected stands of drill pipe in the well during rotary drilling operations.

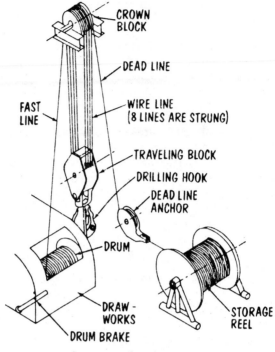

Rotary rig hoisting system.
DRAW WORKS

drill pipe (drill).   In rotary drilling, the heavy seamless tubing rotating the bit and circulating the drilling fluid. A string of steel pipe, screwed together and extending from the rig floor to the drill collar and bit at the bottom of the hole. The drill pipe transmits the rotating motion from the derrick to the bit and conducts the drilling mud from the surface to the bottom of the hole. Usually made in lengths, "stands" or "joints" of 20, 30, or 40 feet. Its diameter varies from two and three-eighths inches to seven inches, and may weigh as much as 25 pounds per foot.

drill pipe protector (drill).   Rubber or ball bearing steel ring attached to each joint of drill pipe to reduce friction and subsequent wear between drill pipe and casing.

drill stem test (drill).   Test taken by means of special equipment to determine whether or not oil or gas in commercial quantities has been encountered in the well bore.

drill string (drill).   A "string" or column of drill pipe. One to thirty collars may be used at the bottom of the drill string. The drill string is also referred to as the drill stem.

driller (drill). Employee directly in charge of a drilling rig and crew. Operation of the drilling and hoisting equipment are his main duties.

drilling block (drill). Usually a lease or a number of leases of adjoining tracts of land constituting a unit of acreage sufficiently large enough to justify the expense of drilling a wildcat well.

drilling cable (drill). In cable-tool drilling, a cable to which is attached the string of cable tools; coils on the drum of the bull wheel and passes over the crown pulley to attach to the stationary or nut end of the temper screw.

drilling fluid (drill). Circulating fluid bringing cuttings out of the well bore.

drilling in (drill). Act of penetrating the productive formations. Term correctly used only when casing is set and cemented on the top of the pay section and the hole, then deepened into the oil formation.

drilling mud (drill). Main function of the drilling fluid are (1) to cool and lubricate the drilling bit and by reason of the pressure at which it leaves the water course in the bit, to assist in the actual drilling, (2) to remove the cuttings from the bit and carry them to the surface, (3) to have thixotropic or gelling properties so that in the event of a pump failure, the mud will gel and hold all the cuttings in suspension, (4) to plaster up the open walls of the well in order to counteract any tendency for a loose formation to cave into the well, (5) by its own weight, to control the pressure in the well and to prevent a well running wild in the event of oil or gas being met under pressure pending completion of control equipment at the surface, and (6) to reduce friction in the drilling string. The clay mineral, bentonite, is frequently a major constituent. Barite or tiff may be added to increase the specific gravity of the mud thereby improving its capacity to lift drill cuttings.

drilling out. Drilling out of the residual cement normally remaining in the lower section of casing and well bore after casing has been cemented.

drilling tool string (drill). Consists of the drilling bit, drill collars and drill pipe, a string of steel pipes screwed together and extending from the rig floor to the drill collar and bit at the bottom of the hole.

drips (refin). Oil coming through the cloths of the paraffin-wax presses during pressing and "drips" into the trough below; applies also to filter drainings too dark in color to be included in filter stock.

drive pipe (drill). Pipe threaded so that ends of the pipe meet in the coupling; drives through the surface soil to the bedrock and in soft formations not permitting under-reaming.

dropping point test.   Test recording the temperature at which a grease passes from a semi-solid to a liquid state, under test conditions.

drowned river valleys.   Result of downwarps along the coastline or eustatic sea-level rise following deglaciation causing marine invasion and valleys to become deeply indented bays and estuaries; recognized by shallow waters in the estuary, dendritic form of the tributary valleys, accordant bottom levels, valley cross sections and extension into an existing river system landward.

drum, flash (refin).   A drum or tower into which the heated outlet products of a preheater or exchanger system and conductor, often go to release pressure; purpose is to allow vaporization and separation of the volatile portions for fractionation elsewhere.

drumlins.   Smooth, elongated hills, composed largely of till. Ideal drumlin shape has an asymmetric profile with a blunt nose pointing in the direction from which the vanished glacier advanced and with a gentler, longer slope pointing in the opposite direction; range from 25 to 200 feet in height, average somewhat less than 100 feet. Most are between a quarter-mile and a half-mile in length, usually several times longer than they are wide. In most areas, drumlins occur in clusters of drumlin fields. Diagram illustrates a portion of the Coniston valley immediately south of Coniston Water where the valley widens to more than twice its previous breadth, and drumlins are formed in profusion, in contrast to the narrower upper valley where there are none.

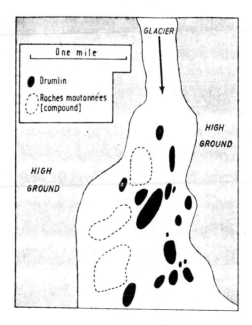

druse. A crust of small crystals lining a cavity; an irregular opening in ore or rock containing mineral crystals.

dry gas. Natural gas produced without liquids, also a gas treated to remove all liquids.

dry hole. An exploratory or development well found to be incapable of producing either oil or gas in sufficient quantities to justify completion as an oil or gas well.

dry hole contribution. Cash contribution from those interested in seeing a well drilled in a particular area, supporting a test well, payable to the operator if the venture is a dry hole.

dry point (refin). Temperature, in a still or flask, when the last liquid fraction has been vaporized and the remaining residue is dry. During distillation test, the end point corresponding to the dry point.

dry processes (grease). A process dehydrating the grease during the initial manufacturing stage. The exact amount of water necessary is added later.

dry sand. Formation, below the production sand into which oil has escaped due to careless well drilling; a non-productive oil sand.

dual completion (prod). A well in which two separate formations are producing at the same time. Production from each zone is segregated by running two tubing strings with packers or running one tubing string with a packer and producing the other zone through the annulus.

ductility. Capable of being pliable and drawn into thin sheets or threads. In asphalt, expressed as the maximum number of centimeters a test specimen will stretch without breaking; does not apply to emulsified asphalt.

dump (refin). In acid treating, each separate portion of acid put on the oil is called a "dump." With small amounts of acid, the action is not overly vigorous, the total acid may be put in one "dump." When large amounts of acid are used, particularly fuming acid, as in white oil treating, it must be divided up and applied in small dumps.

dump bailer (drill). Cylindrical vessel used to lower cement or water into a well likely to cave in if a fluid is poured from the top.

dune. A hill or ridge formed by wind, from sand or other unconsolidated granular material.

dune, barchan. Crescent shaped dunes with points or wings directed downwind;

common where the wind direction is constant and moderate and little irregularity in the surface over which the sand is moving. The steady wind sweeps the ends of the dune ahead, forming the crescent points.

barchans

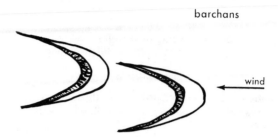

dunes, longitudinal.   Formed where sand is in short supply and the direction of the wind is constant; characteristically linear, nearly symmetrical cross sections, and extend parallel to the dominating winds.

longitudinal

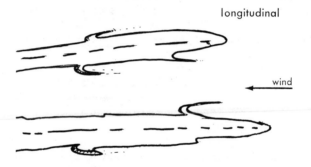

dune, parabolic.   Long, scooped-shaped dunes looking rather like barchans in reverse, their horns point upward, rather than downward.

parabolic

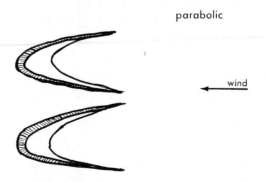

dune, seifs.   A variety of longitudinal dune showing some semblance to a barchan with one wing missing.

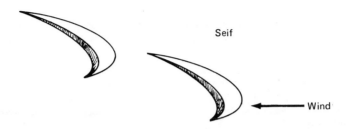

Seif

Wind

dune, transverse.   Dunes aligned perpendicular to wind direction. Found along coasts and lake shores. These may be very long, but rarely attain heights of more than 15 feet.

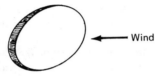

Wind

duo-sol process (refin).   A double solvent refining process employing propane and selecting a mixture of phenol and creosol to accomplish solvent extraction.

Duotreat oils.   Produces food grade white oils, meeting rigid FDA specifications by reducing aromatic content in lubricating oils. The charge is raw distilates from naphthenic crudes and dewaxed raffinates from Mid-Continent type crudes.

duplex pump (drill).   A reciprocating type pump having two pistons or plungers; most commonly used mud or slush pump.

dynamic metamorphism.   Metamorphic textural and mineral changes produced by rock deformation principally by faulting and folding.

E

earth flow (geol).   Variety of erosion by mass wasting. Slow plastic downhill flow of clay-mineral rich rocks or soil when sufficient water is added causing the slope to fail.

earthquake.   A group of elastic waves in the solid earth, generated when internal earth stress exceeds crustal strength causing rocks to break, fracture, or slip along fault planes or zones. Deep focus earthquakes, lower crust or upper mantle are generated by rapid changes in rock volume related to phase changes. The sudden fracture or volume change causes shock waves to travel through the adjacent portion of the earth and an earthquake results. The point of origin of shock waves within the earth is the focus. Body waves, p and s, move away from the focus in all directions. Those traveling fastest and arriving at seismographs first are primary or p, while those traveling more slowly and arriving second are secondary or s. S and p waves reach the surface first at the epicenter where they generate new waves called surface or L. S and p waves are rarely felt by those who experience an earthquake. L waves are felt and cause damage if large enough. Illustration of a type of simplified instrument recording earthquakes. A. General construction of a seismograph. B. A seismogram showing the p and s waves.

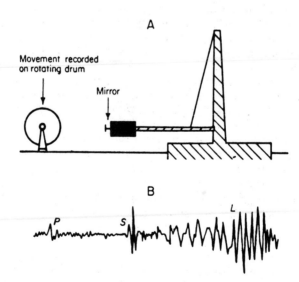

earth tide.   Periodic movement of the earth's crust caused by tide producing forces of the moon and sun.

eccentric bit (drill).   Modified form of chisel used in drilling with one end of the cutting edge extended further from the center of the bit than the other.

echelon faults.   A series of short, parallel faults at an angle to the direction of stress producing the faults.

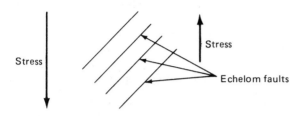

echo (seis).   An acoustic signal reflected or otherwise returned with sufficient magnitude and time delay to be detected as a signal distinct from that directly transmitted.

echo sounder (geol/oceanography).   An instrument emitting pulses of sound and recording them after they are reflected off the bottom of a body of water as ocean, bay, lake, or river, thereby measuring the depth of water. Sound pulse is in the audible or slightly less than audible sound frequency. As frequency of sound is lowered, the pulses penetrate the bottom and are reflected off the sedimentary beds. Instruments of these lower frequencies are "sub-bottom profilers." The distinction between sub-bottom profilers and echo sounders is arbitrary and depends on the degree of penetration into the bottom sediments obtained or desired. Echo sounder is also a fathometer.

ecology (geol/biology).   Study of all parameters (biological and physio-chemical) of an environment and their interrelationships.

eddy current (oceanography/marine geology).   A circular movement of water adjacent to and related to a major linear movement of a fluid; usually applied to large scale oceanic gyres, but may be applied to the atmosphere and to small scale gyres in streams, lakes, and estuaries.

edge water.   Water holding oil and gas in the higher structural positions in wells; usually encroaching on a field after much of the oil is recovered and the pressure greatly reduced.

edgewise conglomerate.   A conglomerate whose individual clasts are highly flattened and are stacked parallel to each other but at an angle to the bedding. Clasts are usually calcareous but need not necessarily be so. Indicative of a high energy environment as a swift stream or a beach. (*See* shingle beach).

effective grain size (geol/hydrology).   (1) Diameter of a sphere having the same hydrologic characteristic of an irregularly shaped sediment particle with all other parameters as density being equal. (2) Grain size or diameter of equidimensional particles in an assemblage which hypothetically would exhibit the same permeability and porosity of a natural earth material.

effective porosity (geol).   Actual volume of fluid a saturated material, rock or sediment, will yield under any specified hydraulic condition relative to the total volume of the material; opposed to actual void volume relative to the total volume of the material.

effervesce.   Bubble, foam, or hiss through application of an acid to any carbonate as limestone, dolomite, etc.

efficiency, thermal.   Ratio of heat in generated steam to heat supplied to a boiler in the form of fuel is the thermal efficiency of a boiler. Ratio of work done to heat energy received by an engine is the thermal efficiency of an engine.

efficiency, volumetric.   Ratio of volume delivered from the discharge of a pump or compressor to volume displaced by the plungers or pistons of the pump.

efflorescence.   Powdery or fluffy growth of crystals caused by evaporation of salt-rich solutions, commonly seen in desert areas where high saline ground water or soil moisture evaporates leaving a fragile white encrustation. In man-made structures, efflorescence indicates leaks in dams, pipes, or tanks.

effluent.   Flowing out of. Streams flowing out of a lake are effluent streams. Sediments brought to a lake, bay or ocean by a stream are effluent sediments or stream effluent sediments in the immediate vicinity of the stream's mouth. Sewage leaving a pipe or conduit is effluent sewage.

effluent refrigeration alkylation (refin).   Combination of propylene, butylenes and amylenes with isobutane in the presence of strong sulfuric acid produces high octane branched chain hydrocarbons, alkylate, for use in aviation gasoline and motor fuel. Effluent refrigeration is essentially a means of obtaining economically a very high concentration of isobutane in the reaction mixture with minimum isobutane fractionation. Designed to process a mixture of propylene and butylenes.

effusive.   Designating igneous rock formed from lava poured out or ejected at the earth's surface.

Eh (geol).   Total oxidation-reduction potential of an aqueous environment measured by electrodes and expressed on an arbitrarily established scale.

ejecta.   Pyroclastic material thrown out or ejected by a volcano, includes ash, volcanic bombs and lapilli.

elasticity.   Property of solids; when deforming stresses are removed, solids return to their former volumes and shapes.

elastic limit.   Maximum stress a material bears without exhibiting plastic strain or rupture.

elastic rebound.   Relief of elastic strain occurring when stressed material ruptures. Springing back of rocks, after rupture, to a position of no stress.

electric log.   A graphic recording of various electric properties of a sediment or rock; obtained by lowering electrodes into hole drilled in sediment or rock; also called *"E" log*.

electrical distillate treating (refin).   Treatment of most distillates, from naphtha to lube oil boiling ranges, with agents as caustic, acid, or water. Process also extremely effective for removing suspended solids from distillate streams. The charge is petroleum distillates.

electro drilling (drill).   Method of drilling; bit is rotated by an electric motor located at the bottom of a drilling string, or lowered into the well on a cable. In the latter case, the motor also drives a small fluid pump for circulation of the fluid clearing the bit of cuttings.

electrolyte.   A substance dissociating into positive and negative ions when in solution or in a fused state and conducting an electric current. Acids, bases, and salts are common electrolytes.

electron.   Smallest unit of negative electricity; has a mass much smaller than that of atomic nuclei and an electric charge equal in magnitude to that of a proton but of opposite sign. An elementary negative particle has a mass of 0.107×10^{-28} gram and a charge of 4.80×10^{-10} statcoulomb.

electron shell.   A group of electrons in an atom, all having approximately the same average distance from the nucleus and the same energy.

electron theory.   Theory that all matter consists of atoms comprising a positive nucleus and a number of negative electrons detachable from the atom under certain conditions, leaving it positively charged.

electrovalence.   Number of electrons an atom must gain or lose to acquire stability.

electron volt.   Unit measuring energy in nuclear physics, abbreviated to ev. Energy gained by one electron when acted upon by a potential difference of one volt. The energy of nuclear reactions is often expressed in millions of electron volts, abbreviated mev.

electrostatic desalting (refin).   Removal of salts, solids and formation water from unrefined crude before subsequent processing; charge is crude oil and product is crude oil with most water-soluble and solid contaminants removed as in chlorides, sulfates, bicarbonates, sand, silt, rust, and tar.

element.   A material consisting of a single kind of atom; 101 elements are known, hydrogen, the lightest and mendelevium, the heaviest. Each element has a unique atomic number, i.e., all atoms of a given element contain the same number of protons and no other atom of any other element contains the same number of protons.

elementary particle.   One of the 34 particles from which all matter is made up. The list includes photon, two types of neutrino, electrons, mu meson, two pi mesons, two K particles, proton, neutron, lambda particles, three sigma particles, and two xi particles, each with its antiparticle.

elevation.   (1) Height of a position above mean sea level usually measured in feet or meters. (2) In general, the height of a position above any convenient reference plane or altitude. (3) In drafting, a verticle cross-section or a verticle profile or a side view of an object or structure.

elevator (drill).   A clamp gripping a stand or column or casing tubing, drill pipe, or sucker rods so that it can be raised or lowered into the hole.

Elliott tester.   Closed type of tester for determining the flash and fire points of an oil.

elution.   Process of moving a substance through a packed porous bed or chromatographic column by means of a slow moving stream of liquid or gas.

elutriation.   Separation of fine material from coarser material by pouring the fines while suspended in water. Laboratory technique for washing or cleaning sand grains, or well cuttings before examination or analysis. For some geologists, term refers to the natural sorting or washing process occurring in streams, resulting in clean sand and gravel remaining in the beds of streams while fine sediments are transported downstream.

elutriation test.   Test to determine the fineness of sand or other fine asphalt filler. Filler is brought into suspension, weighed and passed through a 200 mesh sieve, the quantity retained on the sieve carefully weighed.

eluvial placers.   Placer minerals concentrated near the decomposed outcrop of the source. Deposit of rain wash.

emanations.   Vapors and gases given off by magma, may or may not reach the surface of the earth.

emergence.   Raising of land relative to sea level or lowering of sea level relative to land; exposure to the atmosphere of sediments and surfaces once covered by the ocean. Whether land rose, sea level fell or a combination of both occurred is irrelevant to the definition, but may be of critical interest to petroleum geologists. Term is applied to local changes as well as those of a global scope.

emulsification.   Process of emulsifying or finely dispersing a liquid into a second liquid in which it is completely or partly immiscible.

emulsified asphalt.   Finely dispersed asphalt in water, percentage may be as high as 45.

emulsion.   An intimate mixture of two liquids not miscible with each other, as oil and water. Water-in-oil emulsions has water as the internal phase and oil as the external, whereas oil-in-water emulsions reverses the order. A true emulsion may include a third component, usually a solid insoluble in both liquids.

emulsion polymerization.   Process of polymerization performed by first emulsifying the raw materials, called the monomers, and then executing the reaction of polymerization in the emulsified state.

end cuts (refin).   Same as heavy fractions.

end moraine.   Terminal moraine; an accumulation of glacial drift at the end or margin of a glacier.

endosmosis.   Transmission of a fluid inward through a porous septum or partition separating it from another fluid of different density.

endothermic.   Heat absorbing; an endothermic reaction is one in which heat is absorbed as in the melting of ice or the vaporization of a gasoline.

end point.   (1) Ending of some operation where a definite change is observed. In titration, this change is frequently a change in color of an indicator added to the solution or the disappearance of a colored reactant. (2) The highest temperature indicated by a thermometer inserted in a flask during a distillation test. Generally, the temperature at which no more vapor can be driven over into the condensing apparatus.

enechelon.   A series of relatively short structures parallel but staggered with re-

spect to each other. Usually refers to faults or folds but may also relate to topographic features controlled by faulting or folding. (*See* echelon faults) Diagram illustrates a series of enechelon faults.

Engler degree.   A measure of viscosity. Ratio of the time of flow of 200 ml of the liquid tested through the viscosimeter, devised by Engler, to the time required for the flow of the same volume of water, given in number of degrees Engler.

enthalpy.   Sum of the thermal internal energy plus flow work, displacement energy, or the heat effect in a constant pressure process when only compression or expansion occurs. Change or enthalpy is the change in the internal energy plus the work of compression or expansion.

entrainment (refin).   A relatively non-volatile contaminant material carried over by the "overhead" effluent from a fractioning column or a reactor vessel; that portion of a finely divided catalyst escaping the cyclones of a reactor or regenerator of cracking units is called "carryover."

eolian.   In geological processes, material and structures pertaining to wind.

eolian placers.   Placers concentrated by wind action.

eolian sands.   Sediments of sand size or smaller transported by winds.

eon.   The longest division of geologic time recognized in geochronology.

epeiric seas.   Shallow inland seas with restricted communication to the open ocean and having depths of less than 250 meters (137 fathoms).

epeirogenic movement.   Broad uplift or depression of areas of land or of sea bottom in which the strata is not folded or crumpled, but may be tilted slightly or may retain their original horizontal attitude; refers to relatively gentle deformations of large segments of the earth's crust involving upwarp or downwarp or both by tilting.

eperiogeny.   Broad uplift and/or subsidence of the whole or large portions of continental areas of ocean basins.

epicenter.   Position on the earth's surface directly above the focus of an earthquake; point of origin of surface or "L" seismic waves.

epiclastic (or detrital sediment).   Consists of the physically weathered (disintegrated) and insoluble products of chemically weathered (decomposed) material from older rocks (igneous, metamorphic, or sedimentary) which has been transported and deposited by running water, streams, ocean currents, wind, or ice. In the main, it consists of minerals not readily soluble, as quartz, feldspar and clay minerals transported as particles.

epicontinental. Located on a continent, as an epicontinental sea.

epicontinental sea. Shallow sea lying far in upon a continental mass.

epigenetic deposits. Deposits formed later than the rocks that enclose them.

epithermal. Applying to hydrothermal deposits formed at low temperatures and pressures.

epoch. Unit of geologic time corresponding to the time required for a series of rocks to be deposited; division of a period.

epm (equivalents per million). Chemical weight unit of solute per million unit weights of solution. The epm of a solute in solution is equal to the ppm (parts per million) divided by the equivalent weight.

equilibrium constant. Product of the concentrations or activities of the substances produced at equilibrium in a chemical reaction divided by the product of concentrations or activities of the reactive substances.

equipotential surface (geoph). An imaginary topographic surface connecting points of equal potential of any given force field. These surfaces may be drawn for studying variations in potential of the earth's gravity and magnetic fields as well as electrical fields both natural and induced. These plots are useful to the geophysicist when interpreting gravity, magnetic or electrical data in an effort to locate ore deposits or potentially oil rich structures.

equivalent circulating density (drill). For a circulating fluid the equivalent circulating density in pounds per gallons equals the hydrostatic head (psi) plus the total annular pressure drop (psi) divided by the depth (ft) and by 0.052.

equivalent weight or combining weight. Atomic or formula weight of an element, compound, or ion divided by its valence. Elements entering into combination always do so in quantities proportional to their equivalent weights.

era. One of the major subdivisions of geologic time, comprising one or more periods; most universally recognized eras are Paleozoic, Mesozoic, and Cenozoic.

erosion. Principally the mechanical transportation of sediment by wind, water, ice, or mass wasting. Includes the cutting and carving away of land by mechanical wear effected by the pounding and grinding of sediment or bedrock by particles in transport. When water is involved, mechanical transport and wear may be supplemented by chemical decomposition and solution transport. Some authors include weathering as a part of erosion. Others separate the two and consider weathering to take place in situ, exclude transportation and preceed erosion (*See* weathering). Illustration shows the relation of various erosional landforms to the structure and dip of the strata from which they have been carved or eroded. (See **Fig. p. 106**).

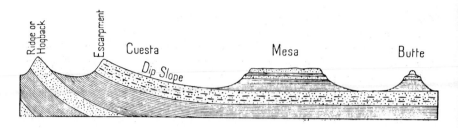

erratic (glac).   Large rock fragment, different from the bedrock on which it lies, transported to that location by glacial ice.

erratic block (glac).   Large rocks detached from their original formations and carried varying distances by glacial ice.

eruptive rocks.   Rocks formed from molten magma and either intruded into older rocks or extruded at the earth's surface through volcanic activity or eruption.

escarpment.   Steep slope or cliff separating gentle sloping areas; extended line of cliffs or bluffs, high steep face of a mountain or ridge. Example of a development and recession of an escarpment by stream erosion.

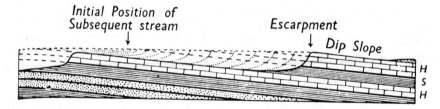

esker.   Serpentine ridge of gravel and sand which marks melt water channels under or within decaying ice sheets, in which streams deposited glacial sediment released by melting of the glacier.

ester.   A group of acid constituents which may be present in new or used lubricating oils. A product of the reaction between alcohol and a fatty acid, as in ethyl alcohol and acetic acid produces ethyl acetate, an ester, and water.

estuarine.   Processes, materials or structures associated with estuaries.

estuary.   Portion of a stream valley influenced by the tide of the body of water into which it flows; characterized by water whose salt content is between that of fresh and marine environments, and by a distinct population of animals and plants.

etched.   Rough frosted surface as of minerals or sand grains, resulting from chemical interaction with water.

ethane.   Second member of the gas family having two carbons and six hydrogen atoms, and a boiling point of-120°F; one of the chief constituents found in most commercial grades of natural gas.

ether solubles.   Substances soluble in a specific ether. In a chemical analysis, various liquids as ether, chloroform, benzol, etc. are used as solvents to determine percentages of certain materials.

ethylene.   Homologous series of unsaturated open-chain hydrocarbons, olefins, of which ethylene is the lowest member.

ethylene glycol.   (1) Colorless fluid with characteristic sweet odor and taste, derived from the cracking of petroleum. (2) A thick, clear liquid obtained from petroleum lowering the freezing point of water; used as a refrigerant and as a permanent type anti-freeze fluid.

eugeosyncline.   A term of wide usage before plate tectonics gained wide acceptance. Attempts have been made to modify the definition to suit plate tectonic theory but many geologists advocate abandoning the term. In general terms, the fine grained, siliceous deposit in deep water of the outer continental shelf and continental slope as it extends along an entire continental margin.

eustatic.   (1) Pertaining to a simultaneous worldwide change in sea level, as from the melting of continental glaciers, but not to relative change in level resulting from local coastal subsidence or elevation. (2) Pertaining to a land mass that has not undergone elevation or depression.

eutectic.   Referring to a mixture of solids in a specific ratio by weight percent (eutectic ratio) which exhibits a minimum melting temperature (eutectic temperature). The melt has the same composition as the bulk composition of the mixed solids.

euxine deposition.   Deposition in a nearly isolated sea basin where for lack of circulation and mixing, the deep waters are deficient in oxygen and toxic to all life but anaerobic bacteria, and where hydrogen sulfide is produced from muds rich in organic matter.

evaporate.   Process changing a liquid into a vapor by means of heat. Boiling represents a high rate of evaporation.

evaporator (refin).   Usually a vessel receiving the hot discharge from a heating coil and by a reduction in pressure, flashes off overhead the light products and allows the heavy residue to collect in the bottom.

evaporites.   Sedimentary rock composed of minerals that were precipitated from solution during the evaporation of the liquid in which they were dissolved. Rock salt and gypsum are the most abundant evaporites. Anhydrite, simply gypsum without the water, is an evaporite composed of the mineral of the same name.

evaporite basin.   A shallow bay in a region where evaporation is high and rainfall is low. Formation of such a basin requires periodic inundation with sea water followed by a restriction of the flow.

event (seis).   Any recognizable change in phase or amplitude on a seismogram. Change may be related to reflection, refraction, diffraction, or interference of seismic waves; used for seismograms of earthquakes or in exploration geophysics using man made energy sources.

evolution.   (1) Steps in formation, or historical development, as evolution of a continent or of a batholith. (2) A theory of development of life in which changes take place gradually from generation to generation. Evolutionary change takes place through mutation and accommodation of the organism to its environment.

excess reactivity.   Amount by which the neutrons of one generation exceed those of the previous generation in a reactor. If the excess reactivity is large, the intensity of the chain reaction will increase rapidly with time.

exchanger (refin).   Closed-coil heat exchangers, header and tube type are used throughout a refinery with a possibility of economically conserving heat. Stills are equipped with exchangers through which the hot vapors from the still are run countercurrent to the feed stock going to the battery, thus partly condensing the vapors and getting the fraction obtained by partial condensation, and at the same time heating up the feed stock and thereby conserving fuel.

exfoliation.   Large scale process of physical weathering by successive, commonly curved layers of rock flake off from larger rock masses. This process commonly occurs where plutonic granite is exposed at the surface.

exothermic.   Chemical reactions in which one of the products is heat as in combustion of gasoline.

exotic.   A material originating elsewhere than its current location.

expansion drum (refin).   Large cylinder or drum of varying size and construction to which the vapor line of a still is connected and in which vapors first expand on being driven over from the still.

exploration.   Work involved in seeking new deposits of economic value, including metallic and nonmetallic ores, natural gas, petroleum, building materials, and other natural earth resources. Exploration includes geophysical, geochemical, and field reconnaissance techniques.

export white (kerosene).   Darkest shade of the standard colors of kerosene corresponding to standard white.

exposure. An outcrop of rock exposed at the surface of the earth. Term embraces all earth materials appearing at the surface not hidden by vegetation, water, or the works of man.

exsolution. Separation of individual solid minerals from a solid solution when temperature is lowered.

extension fracture. Natural earth fracture, joints or faults occurring as a release of compressive stress. Fractures are oriented parallel to the compressive forces; produced by tension at right angles to the compression.

extinction. (1) Position of an anisotropic mineral, or man-made substance on the stage of a petrographic microscope at which the mineral appears dark (black) when polarizing plates (nicols) are crossed. (2) Disappearance of a species or larger unit of life as a result of its inability to cope with its environment or adapt to environmental change.

extraction (refin). Process of separating a material by means of a solvent into a fraction soluble in a solvent extract and an insoluble residue.

extractive distillation (refin). Separation of different components of mixtures, having similar vapor pressures, by flowing a selective and a relatively high boiling solvent down a distillation column as the distillation proceeds.

extreme pressure additive. An additive imparting extreme pressure characteristics to a lubricant, commonly referred as an EP additive.

extreme pressure grease. Grease withstanding extreme high pressures. Lead base grease is called extreme pressure grease because of its excellent load carrying characteristics.

extrusion. The emission of magmatic material, generally of lavas at the earth's surface; the structure or form produced by the process of lava flow or certain pyroclastic rocks.

extrusive rock. Igneous material forced through other rocks and flowed out upon the surface of the earth or the sea bottom before solidifying; same as volcanic rocks. Occasionally solidified molten lava is referred to as lava in which case it is also equivalent to extrusive rock.

F

Faber viscosimeter. Special viscosimeter designed to test used oils containing solids and diluents without disturbing these contaminants.

fabric. Relative size, shape and spatial arrangement of grains or crystals in a sediment deposit or a rock.

facet. Plane surface abraded on a rock fragment.

faceted spur. End of a ridge truncated by erosion or faulting.

facies. A set or composite of features in a rock, (mineral, chemical, biological or structural) which in aggregate define or describe the conditions of formation of the rock.

Fahrenheit temperature scale. Temperature scale with the freezing point of water at 32° and the boiling point at 212° at standard atmosphere pressure.

fan, alluvial. Terrigenous counterpart of a delta, typical of arid and semiarid climates, but may form in almost any climate if conditions are right. A fan marks a sudden decrease in the carrying power of a stream as it descends from a steep gradient to a flatter one.

fan, deep sea. Fan-shaped hemipelagic deposit, consisting principally of turbidites. Deep sea fans resemble alluvial fans and result from the loss of energy to carry sediment by ocean bottom currents; found along the continental slope and continental rise environment as well as along the slopes surrounding offshore, continental borderland basins. Ultimate source of most of the sediment is terrigenous, either directly from influx of riverine sediments or indirectly when continental shelf and slope deposits fall in massive gravity flows when triggered by slope overloading, storm wave activity or earthquakes. When terrigenous sedimentation processes are quiet or absent, pelagic sediment beds are deposited. Pelagic units, rich in organic constituents may serve as petroleum source beds from which gas and oil may migrate to collect or be trapped in turbidite reservoir beds. Most geologists divide deep-sea fans into 3 parts: **(1)** Inner-fan, closest to shore with fan surfaces showing the steepest gradients of the entire fan and incised with a deep, simple system of relatively straight valleys. **(2)** Middle-fan starts where valleys of the inner fan begin to meander and split into distributary channels. The middle fan surface is characterized by a concave upward bulge. **(3)** Outer-fan, surface returns to a concave upward gradient and is the distal channeled portion of the fan. Channels are numerous, broad, shallow and frequently braided.

fan-delta complex. Combined structure of a delta grading into a submarine fan;

occurs most frequently when a major river discharges into the ocean where the continental shelf is narrow or where the head of a submarine canyon is close to shore.

fan, submarine. *See* fan, deep sea.

fan, washover. A small sediment fan on the lagoon side of an emerged longshore bar, or barrier beach or barrier island, formed when storm waves break over a portion of the bar; occur intermittently in time and space but owing to the low energy environment of the lagoon, they may be preserved for a long time and may play a significant role in the filling in of a lagoon.

fan shooting. A technique in seismic exploration geophysics. Seismometers are placed in a fan-shaped array to detect anomalies in refracted wave arrival times related to small, more or less, circular rock structures as salt domes.

fanglomerate. Consolidated gravel and associated sediment deposited as an alluvial fan.

fast line (drill). End of the drilling line affixed to the drum or reel; so called because it travels with greater velocity than any other portion of the drilling line.

fathometer. *See* echo sounder.

fatty acids. A large group of aliphatic monocarboxylic acids, many of which occur as glycerides in natural fats and oils.

fault. A fracture or fracture zone in rock along which one side has been displaced relative to the other side, some are clean sharp breaks, others are composed of subparallel faults along which displacements have been distributed. The terms shear zone or fault zone are often applied to closely spaced subparallel structures along which there has been displacement or movement. Some faults are knife-like breaks, others, because of the frictional effects of rock masses sliding over one another, break or crack, brecciate the rocks on either side of the rupture. Still others pulverize the rock in the fault zone to clay-like powder called gouge. Displacements along faults may vary from a few inches to many miles.

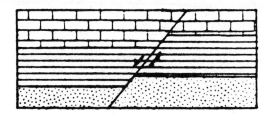

Fault–separation at sediment layers (beds).

fault block.   Rock body bounded on at least two opposite sides by faults, usually longer than it is wide; when depressed relative to the adjacent regions, called a graben, down thrown block, fault embayment, fault basin, fault trough, or rift valley; when elevated, called a horst, or upthrown block.

fault breccia.   An assemblage of broken country rock fragments associated with a fault. Grain size may vary from clay size to boulder, poorly sorted. The material of which a fault zone may consist unless the grain size is too small, in which case the material is called gouge.

fault creep.   Slow and continuous displacement along a fault.

fault dip.   Maximum inclination of the fault plane or shear zone, as measured from a horizontal plane.

fault, dip slip.   A fault in which the displacement is parallel to the dip of fault plane or surface. When established on basis of separation, in which case either a component of slip is parallel to the dip of the fault plane or it may *appear* possible that there is dip slip displacement, the fault should be called dip separation fault or apparent dip slip fault. *See* fault displacement and fault separation. Normal and reverse faults are examples of dip slip faults.

fault displacement.   Actual direction of movement of the earth along a fault plane or surface. Usually defined in terms of direction of movement relative to the dip and strike of the fault plane.

fault escarpment.   Escarpment or cliff resulting from a fault; may be caused directly by fault displacement or by erosion along a fault.

fault fissure.   Fissure produced by a fault even though afterwards, filled by a deposit of minerals.

fault footwall.   If a fault surface is inclined, the upper side is the hanging wall and the lower side is the footwall. The mass of rock over the hanging wall is the hanging wall block and the mass of rock below the footwall is the foot wall block.

fault hanging wall.   *See* fault footwall.

faulting.   Movement of bedrock along a fault.

fault-line scarp.   A fault scarp modified by differential erosion so it no longer marks the exact location nor necessarily the sense of faulting. *See* fault scarp.

fault, normal.   Hanging wall moved down relative to the foot wall.

fault, oblique slip.   A fault, in which the displacement is oblique to the dip or

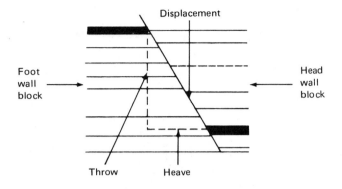

Normal fault

strike of the fault plane or surface. When named on basis of separation the fault should be called an oblique separation fault. *See* fault displacement and fault separation.

fault plane.   Surface along which dislocation or faulting has taken place.

fault, reverse.   If a hanging wall has gone up relative to the foot wall, the fault is called a reverse fault. Below is a diagram of cross section of a reverse fault and a cross section of a normal fault.

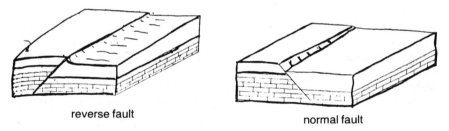

reverse fault   normal fault

fault scarp.   Cliff associated with dip slip or oblique slip. Below are diagrams of fault scarps and fault line scarps. Faulting originally produces fault scarps (Fig. 1 and 4). Erosion destroys these (Fig. 2) and eventually produces fault-line scarps reversing the topography (Fig. 3) or causing the scarp to migrate (Fig. 5).

1

Strong   Weak   Strong

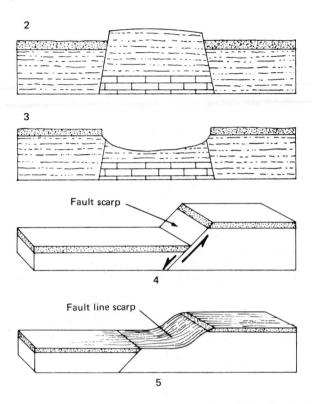

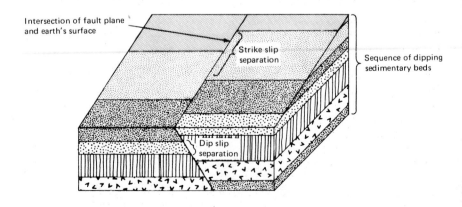

fault separation. Apparent displacement of a fault. Displacement can be measured only on the basis of disrupted linear features. When planar features are disrupted, it is frequently impossible to determine the actual movement along the fault. Illustration of an indeterminate displacement. Viewed from above, the fault shows strike slip separation (strike separation fault) and viewed in cross-section, the fault shows dip slip separation (dip separation) fault. *See* fault displacement.

fault slip.   Used interchangeably with fault displacement.

fault, strike.   Compass direction or azimuth of the intersection of the fault surface, or the shear zone, with a horizontal surface.

fault, strike slip.   A fault with horizontal displacement or displacement parallel to the strike of the fault plane or surface. When named on the basis of separation in which case, either a component of slip is parallel to the strike of the fault plane or it may *appear* possible that there is strike slip displacement the fault should be called a strike separation fault or an apparent strike slip fault. *See* fault displacement and fault separation. A transform fault is an example of a strike slip fault.

fault, transform.   A strike slip fault related to the relative movement of crustal plates; may occur to compensate for varying rates of spreading along a divergent plate margin or when crustal plates are moving laterally with respect to each other rather than converging or diverging. The San Andreas Fault is classified as a transform fault by some authors, on this basis. In general, transform faults occur to release horizontal stress developing when plate divergence and/or convergence are locally non-compensating or irregular.

fault traps.   Reservoirs formed by faulting, e.g, breaking or shearing and off-setting of strata. The escape of oil from a trap is prevented by non-permeable rocks that have moved into a position opposite the porous petroleum bearing formation or by impervious fault gouge. The oil is confined in traps of this type due to the tilt of the rock layers and the faulting.

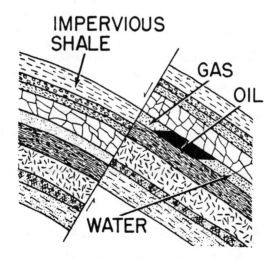

fault zone.   A system of closely spaced faults pulverizing or brecciating the bedrock. Zones may vary greatly in thickness as with the San Andreas Fault zone, which exceeds one mile thick at some localities.

fecal pellets (*also spelled faecal*).   Preserved and/or fossilized excrement, primarily from marine invertebrates. Ovoid pellets averaging one millimeter in diameter found widely in calcareous and argillaceous sediments and sedimentary rocks. In modern sediments, they consist of clay minerals and organic wastes bound together by mucus. Pellets experience alteration during fossilization and frequently appear as glauconite pellets in shale or replaced by calcite in limestones.

fee land.   Land operated for oil, gas, etc., under right of ownership as contrasted with land operated under lease.

feed-off (drill).   To lower the bit continuously or intermittently by allowing the brake to disengage and the drum to turn; its rate to the unwinding of the cable from the drum is the feed-off rate.

feed water (refin).   Water supplied to a boiler to make up for evaporation loss.

felsic.   A term constructed from "fel" denoting feldspars and feldspathoids and "si" for silica; used as a descriptive term for light colored rocks (almost exclusively igneous) made principally of silica rich minerals; may be applied to silica rich minerals apart from their associations in rocks as in quartz, feldspar (orthoclase in particular), feldspathoids and muscovite.

felsite (petrology).   A group of fine grained (microcrystalline in whole or part) igneous rocks consisting of felsic minerals and usually light in color.

femic.   A term constructed from "fe" for iron and "m" for magnesium; describes minerals containing these two elements as essential components as in olivine, most pyroxenes and amphiboles, as well as the common igneous accessory minerals as magnetite, ilmenite and hematite; synonymous with ferromagnesian. A rock, generally igneous, consisting principally of femic minerals is called mafic. Some authors do not draw this distinction between femic and mafic but rather use the terms interchangeably.

fence diagram (geol).   A diagram portraying at least three geologic sections in three dimensional perspective; constructed to show relationships between wells and subsurface formations. Sections frequently close on each other to give an appearance of a fence with the wells as fence posts, hence the name fence diagram.

ferric.   Designation for iron salts in which the iron is trivalent; iron in a state of higher oxidation.

ferro.   A combining prefix used to indicate presence of, or connected with iron.

ferrofining (refin).   Mild hydrotreatment to improve the color and stability of lubricating oils. The charge is distillate, or solvent refined lubricating oils. The product is a complete range of finished lubricants with low colors and good oxidation stabilities.

ferrous.   Designation for iron salts in which the iron is divalent; iron in a lower oxidation state.

ferruginous.   Consisting of, or containing a significant amount of iron; may denote an iron-rust color.

field.   (1) A geographical area in which a number of oil or gas wells produced from a continuous reservoir; generally referring to the surface area, although at times it may refer to both the surface and the underground productive formations. In a single field, there may be several separate reservoirs at varying depths. (2) In geology, a region of interest to be mapped and/or from which samples are brought for laboratory analysis. (3) Region in space affected by a force, such as a magnetic field.

fill.   (1) Deposition of sediment in the low energy portion of the bend of a river. (2) Any material deposited by any erosional agent to fill, completely or partially a valley, sink hole, or other depression. (3) For engineers and engineering geologists, material used to raise the surface of the land artificially, as in highway construction or a natural deposit of material with potential use as an artificial fill by reason of its composition or location.

filler (lubricating oil).   Graphite and other dry lubricants added to lubricating oil act as a filler to impart body or to smooth out rough bearing surfaces.

filling density (LPG).   Ratio in percentage of the weight of LPG to the weight of water held in a tank at 60°F.

filling the hole (drill).   Pumping fluid into the wellbore as the pipe is withdrawn in order to maintain the fluid level in the hole near the surface; avoids danger of blowout and caving of the wall of the wellbore.

fill terrace.   A flat topographic surface developed by the aggradation or deposition of sediment rather than by erosional degradation or cutting; usually found in valleys or along shore lines, may be formed by fluvial, glacial or other processes.

film strength.   Property of an oil enabling it to maintain an unbroken film over lubricated surfaces under operating conditions, thus avoiding the scuffing or scoring of the bearing surfaces.

filter.   Porous material in which solid particles are largely caught and retained when a mixture of liquids and solids is passed through it.

filter cake (drill).   A compact solid or semisolid material remaining on a filter after pressure filtration of mud with the standard filter press. Thickness of the cake is reported in thirty-two seconds of an inch. Filter cake may also refer to the layer of concentrated solids from the drilling mud forming on the walls of the borehole opposite the permeable formations.

filter pressing.   Squeezing out of a residual magma from the interstices of a mesh of crystals, a process similar to squeezing water out of a sponge.

filter press (refin).   Apparatus employed to separate wax and oil in paraffin wax distillates; consists of a series of canvas covered plates separated by narrow rings. The distillate is run into a narrow bore extending the entire length of the press, and is forced up into the space between the plates formed by the rings. The canvas covering the plates is easily penetrated by oil but impenetrable to wax, so that the oil drips down from the canvas into a trough beneath the press. A mechanical separation of the oil and wax is thus effected on the principle of filtration.

filtered cylinder oil.   Oil filtered to give a lighter and more transparent color. When filtered to certain requirements, it is marketed as bright stock.

filtered stock.   Lubricating oil filtered or refiltered to improve certain characteristics.

filtering clay.   Clay of various kinds used to filter petroleum products.

filtration.   Process separating substances in petroleum products by passing them through a filtering medium. The substances include solids, liquids as water, not soluble in oils. The removal of colored substances from oil by percolation through clay.

filtration percolation.   Treatment of oil with clay by gravity flow of oil through a fixed column of clay.

finger board (drill).   A rack located in the derrick to support the stands of pipe while stacked in the derrick.

fines.   Rock or mineral in too fine or pulverulent condition to be treated in the same way as ordinary coarse material.

fingerlake.   Lake occupying long, narrow rock basin or a dammed river valley; frequently related to continental glaciation.

firedamp.   Combustible gas, damp formed by the decomposition of coal or other carbonaceous matter, and consisting chiefly of methane (CH_4).

fire point.   Lowest temperature, under specified conditions, in a standardized apparatus, at which a petroleum product vaporizes sufficiently rapidly to form an air-vapor mixture above its surface burning continuously when ignited by a small flame.

firn.   Old snow that has become granular and compacted as the result of various surface metamorphosis, mainly melting and refreezing but also includes sublimation.

first arrival. First seismic event recorded on a seismogram. Term used in conjunction with exploration refraction seismology since only first arrivals are considered. Refraction seismograms are interpreted assuming that first arrivals represent shortest time paths not necessarily coinciding with shortest distance path.

fish (drill). Any lost article in a well which must be retrieved.

Fischer-Tropsch process. Synthetic process making many petroleum products from coke oven, natural gas, or coal. These products include gases, liquid fuels, lubricating oils, and wax. The coal, coke-oven gas or natural gas is partly steam oxidized to yield hydrogen and carbon monoxide which are then reacted catalytically into petroleum products.

fishing (drill). Operation on a drilling rig for retrieving from the well bore sections of pipe, casing or items which may have been inadvertently dropped in the hole during drilling operations. Principal tools and methods used in fishing are: **(1)** Mill—cuts away irregularities and permits useful contact. **(2)** Spear—goes inside and holds a fish by friction. **(3)** Overshot—goes over and grabs the outside of a fish. **(4)** Washover pipe—goes over a fish to permit circulation in the annulus between the fish and wall of the hole. **(5)** String shot—permits back off of the drill string above the fish. **(6)** Accessory devices as jars to loosen and retrieve a fish and safety joint to permit release of fishing string.

fishing jars (drill). Jars used with fishing tools to recover lost tools or casing from a hole; have a longer stroke than those used in drilling.

fishtail bit (drill). Flat fishtail shaped steel bit used in rotary drilling; varying in width from a few inches to several feet; used for drilling sand, gravel and clay, not hard formations.

fissility. Property of splitting easily along closely spaced parallel lines, usually exhibited by shale.

fission. Splitting of a nucleus into two more-or-less equal fragments. Fission may occur spontaneously or may be induced by capture of bombarding particles. In addition to the fission fragments, neutrons and gamma rays may be produced during fission.

fission fragments. Two parts, approximately equal in weight, in which nuclides such as a uranium or plutonium atom split in the process of fission; usually highly radioactive, disintegrating several times before reaching stability. In slow neutron fission, the fragments are seldom equal in mass but generally fall into a heavier group with masses around 140 and a lighter group with masses around 95.

fission products.   Nuclides produced by the fission of a heavy element nuclides as uranium-235 or plutonium-239. Thirty-five fission product elements from zinc through gadolinium have been identified from slow neutron fission.

fissure.   Extensive break, crack, or fracture in rocks. A cleft; a narrow chasm made by parting substances; frequently used in conjunction with volcanic eruptions or faulting.

fixation.   Process by which a fluid or gas becomes or is rendered firm or stable in consistency and evaporation or vaporization prevented.

fixed carbon.   Carbonaceous product of oxidation of lubricating oils characterized by its inability to be dissolved by any solvent. In coal, coke and bituminous materials, the solid residue other than ash obtained by destructive distillation, determined by definite prescribed methods.

fixed oil.   Oil which cannot be distilled without decomposition or changing its chemical composition as vegetable and animal oils.

fjord.   Narrow, deep, steep-walled inlet of the sea formed either by the submergence of a mountainous coast or by entrance of the sea into a deeply excavated glacial trough after the melting away of the glacier.

flag (drill).   Tying a piece of cloth or other marker on a bailing or swabbing line to enable the operator to know the depth at which the swab or bailer is operating in the hole.

flame arrestor.   Assembly of perforated plates or screens enclosed in a case and attached to the breather vent on petroleum storage tanks.

flame structure.   A load cast. A sedimentary structure caused by deformation penecontemporaneously with deposition; usually a minor disruption of bedding caused by dewatering, compaction and flow under the weight of overlying sediment. Frequently part of an underlying layer squeezed upward and into an overlying layer in an irregular fashion resembling tongues of flame. Some horizontal slip or flow may accompany the upward movement.

flare.   An arrangement of piping and a burner to dispose of surplus combustible vapors; situated around a gasoline plant, refinery, or producing well.

flash.   Sudden release in pressure resulting in partial or complete vaporization; sudden burst of light; momentary blaze.

flash and burning points.   Flash point is the temperature at which the vapors rising off the surface of the heated oil will ignite with a flash of very short duration when a flame is passed over the surface. When the oil is heated to a higher temperature, it ignites and burns with a steady flame at the surface. The lowest temperature at which this will occur is known as the burning point. A measurement of the hazard involved in handling and storing petroleum and petroleum products.

flashing (refin).   Separation of products by releasing the pressure in a hot oil as it enters a vessel; the lighter fractions vaporize or "flash off" and the heavy oils drop to the bottom.

flash test.   Test determining the flashing point of oil.

flash tower (refin).   Vessel used for the separation of oil fractions in a flash distillation process.

flash vaporization (LPG).   A system withdrawing LPG from storage and consisting of (1) first stage regular, (2) intermediate pressure system, and (3) a second stage regulator, producing gas for use in appliances. Liquid from the storage tank is flashed into a vapor in the intermediate pressure system and the second stage regulator provides the low pressure necessary for using the gas in the appliances.

fleet angle (drill). Side angle at which a rope approaches the sheave; angle between the center line of the sheave.

Flexicoking (refin).   Integrates conventional Fluid Coking with coke gasification. Alternatively, a non-integrated gasifier (outboard gasifier) can be added to existing Fluid Cokers. Any pumpable hydrocarbon feedstock, including tar, sands, bitumens can be fed to the unit. Feedstocks, as vacuum residue having high metals and/or carbon content are particularly attractive. The products are light ends and liquid yields, the same as those in conventional Fluid Coking and lend themselves to upgrading via hydrotreating technology. The gross coke products from the reactor is gasified to produce a coke gas containing hydrogen sulfide.

Flexicracking.   Applied to the catalytic conversion of a wide variety of virgin and cracked gas oils, deasphalted oils to lower molecular weight products as olefins, high octane gasoline, middle distillates and other desirable products.

flint.   A dark-gray to black variety of chert, essentially silica with some water; chert and flint are used interchangeably.

float.   Bits or pieces of valuable mineral or fossil found on the surface of the ground some distance, usually downhill, from their place of origin as a result of weathering and erosion.

float collar (drill).   A collar inserted one or two joints above the bottom of the casing string containing a check valve permitting fluid to pass downward through the casing but preventing it from passing upward. The float collar prevents the drilling mud from entering the casing while it is being lowered, thus allowing the casing to "float" during its descent, and decreasing the load upon the derrick. The float collar also prevents the back flow of cement during the cementing operation.

float shoe (drill). A short-heavy, cylindrical steel section with rounded bottom attached to the bottom end of the casing string; contains a check valve and functions in the same manner as the float collar but in addition serves as a guide shoe for the casing.

float test. Test applied to viscous fluids and semi-solid bituminous materials; offers a general indication of the nature and consistency of the substance.

float viscosity. A test used largely for determining the viscosity or consistency of semi-solid bituminous materials.

floater. Storage tank constructed so the roof floats by means of pontoons on the surface of a liquid as crude oil.

floating in casing (drill). Necessity on very deep wells to float the casing as it is lowered into the well by means of a float valve made of cast iron drilled out after the casing has been cemented.

floating/floating grains (sedimentology). Terms used more or less interchangeably; floating defines the general condition, whereas floating grains is used by sedimentary petrographers in reference to sand sized grains, usually as seen in thin sections. Floating is the condition of a sedimentary particle being completely surrounded by much finer material, matrix or cement, to the end that the floating particle is not in contact with other sedimentary particles of a like size.

floc. Any small, tufted or flake-like mass of matter floating in a solution, as produced by flocculation or by precipitation.

flocculant. A substance, added to a suspension, causing or enhancing flocculation.

flocculate. To aggregate into lumps, as when fine or colloidal clay particles in suspension in freshwater clump together upon contact with salt water and settle out of suspension; a common depositional process in estuaries. The force of attraction promoting flocculation is electrostatic.

flocculation Loose association of particles in lightly bonded groups, non-parallel association of clay platelets. In concentrated suspension as drilling fluids, flocculation may be followed by irreversible precipitation of colloids from the fluid.

flood basalt. Highly fluid basaltic lavas extruded from fissures onto the earth's surface in very large volumes, horizontal layer, upon horizontal layer with occasional small departure from the horizontal. Most authors state that flood basalts are restricted to tholeiitic, olivine poor, basalts of subcontinental origin but some include olivine basalts, as well. The Columbia River and Snake River plateaus of the United States and the Deccan Plateau of India are classic examples of tholeiitic flood basalt occurrences. They are thought to occur when a continent overrides a hot spot.

flood plain.   A valley cut by a river, covered by riverine sediments deposited during the present river regimen and subject to inundation when the river overflows its banks during flood stage; often the site of meander scars and/or oxbow lakes.

flooding (drill).   (1) Process of drowning out a well with water. (2) Process by which oil is sometimes driven from the sand into the well by water introduced under pressure into a key well.

floorman (drill).   A member of the drilling crew whose work station is on the derrick floor. On rotary drilling rigs, there are normally two floormen on each crew, with three used on heavy duty rigs.

floor pulley (drill).   A sheave pulley or snatch block secured in place at or near the floor on which a hoisting rope runs for the purpose of changing the direction of the pull on the rope.

flotation (mineral dressing).   A technique for concentrating economically valuable minerals. Reagents are added to water creating a froth which will float some finely crushed mineral varieties while other mineral varieties sink.

flow banding.   Structure of igneous rocks, especially common to silicic lava flows caused by the movement or flow of magmas or lavas and evidenced by the separation of mineralogically unlike layers.

flow bean.   A plug in the flow line at the well head with a small hole drilled through it where oil flows and keeps a well from flowing at too high a rate.

flow breccia.   Fragments of solidified lava welded or cemented by lava, as at the top of a lava flow.

flow by head.   A well flowing oil at irregular intervals.

flow chart.   Chart made by a recording meter showing the rate of oil or gas production.

flow line (prod).   Pipeline leading away from a flowing well conducting the oil or gas; surface pipes through which oil travels from the well to storage.

flow line pressure (prod).   Pressure on the downstream side of the flow nipple.

flow point.　Measure of the fluidity of semi-solid petroleum products as asphalt; temperature at which the flow of these products becomes visible.

flow structure (igneous rocks).　Any sort of banding produced as molten rocks move. In some cases, minerals carried by the magma or lava are sorted into different layers, show crumpling produced during flow. In other deposits, mineral grains merely are aligned in bands, or the rock may consist of layers and stringers of glassy and grainy material.

flow tank (prod).　Tank to which oil is piped from the well and where gas and water may be separated from the oil before the latter goes into a stock tank. A lease storage tank to which produced oil is run.

flow treater (prod).　A single unit acting as an oil and gas separator, an oil heater, and an oil and water treater.

flue gas.　Gases from the combustion of fuel, the heating effect of which has been substantially spent and discarded to the flue stack. Its constituents are principally carbon dioxide, carbon monoxide, oxygen and nitrogen.

flue gas analysis.　Analyzing flue gas to determine the degree or extent of fuel combustion. Results indicate whether more or less oxygen is required for complete combustion.

fluid cracking process (refin).　A catalytic process employing a finely divided, fluidized catalyst in the form of slurry.

fluid inclusion.　Liquid included in a cavity in a mineral. Volume of inclusion is usually very small.

fluid injection (prod).　Injection of gases or liquids into a reservoir to force oil toward and into producing wells.

fluidized.　A finely divided solid material said to be fluidized when suspended in either dense or dilute turbulent phase by an aerating gas or vapor, so that the suspension can flow like a liquid from a point of high pressure.

fluorescence.　(1) Instantaneous reemission of light of a greater wave length than that of light originally absorbed. (2) All oils exhibit more or less fluorescence, also called "bloom"; aromatic oils being the most fluorescent. Fluorescent colors of crude oils range continuously from yellow through green to blue; rapidly reduced by aging. Fluorescence is observed under ultraviolet radiation, in petroleum normally having a wave length of 2,537 and 3,650 angstrom units.

fluorograph.　Method for quantitative determination of significant hydrocarbons contained in samples of soil and a technique for interpreting the structural meaning of various degrees of fluorescence observed.

fluorologging. A petroleum well logging technique where well cuttings are examined for fluorescence, usually under ultraviolet radiation related to trace occurrences of oil. Logs are prepared by plotting intensity of fluorescence against depth or origin or cuttings and used as an indication that the well is approaching a potential reservoir bed or producing horizon.

flushing. Force migration of oil or gas by moving water. Water may be from natural sources as ground water or may be pumped into the ground from a well.

flush production. First yield from a flowing oil well during its period of greatest production.

flute. An irregularity on a scalloped rock surface as on a drape or flowstone wall of a cavern or on the rippled or welted surface of a sedimentary bed.

flute cast. A variety of sedimentary sole mark, conical to subconical and sharply defined; rounded to bulbous on one end, broadening and diminishing until it merges gradually with the normal bottom of a sandstone bed at the other end; used in determining the paleocurrent direction responsible for the transportation and deposition of the sandstone bed.

fluting. (1) Large grooves or gutterlike channels carved into bedrock by glacial action. (2) Weathering process producing a welted or corrugated surface on the exposed surfaces of some granites and gneisses.

fluvial deposit. Mud, sand, gravel or rock deposited by stream or river action.

fluviatile. Belonging to a river or stream.

fluvioglacial. Pertaining to streams of glacial melt water or deposits made by them.

fluviomarine. Deposits carried into the sea by rivers or streams and resorted, redistributed by waves and currents, and mixed with the remains of marine animals.

flux oil. Oil of low volatility suitable for blending with bitumen or with asphalt to yield a product of softened consistency or greater fluidity. Selected residual fuel oils may be used for this purpose.

flux. (1) Bituminous materials, in which predominating constituent is bitumen, used in combination with asphalts for the purpose of softening the latter. (2) Substance as sodium tetraborate added to a silicate melt to lower its viscosity. (3) Passage of a chemical element or compound across a physical or chemical boundary.

flysch. A term defined before the impact of plate tectonics on geologic concepts and theories. Originally applied to extensive sedimentary deposits on the northern and southern borders of the Alps. Deposits contain a wide variety of rock types but sandy and calcareous shale predominate. Various attempts have been made to redefine and/or reinterpret the term while some geologists advocate that it be discarded.

foam (drill).   Two-phase system similar to an emulsion where the dispersed phase is a gas or air.

foaming agent (drill).   A chemical used in gas wells, oil wells producing gas, and drilling wells in which air or gas is used as the drilling fluid to lighten the water column so as to assist in the unloading of water.

focus (seis).   In seismology, the source of a given set of elastic waves. The true center of an earthquake, within which the strain energy is first converted to elastic wave energy.

folds.   Rock strata subjected to pressures beyond their elastic limit may yield to slow plastic deformation by bending or folding into more or less symmetrical series of folds with alternating crests and troughs. The principal types of folded structures are monoclines, anticlines, synclines, domes, and basins. Diagrams illustrate various types of folds; (a) monocline, (b) symmetrical anticline and syn-

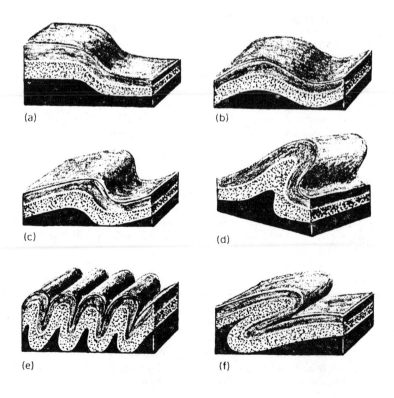

(a)

(b)

(c)

(d)

(e)

(f)

cline, (c) asymmetrical anticline and syncline, (d) overturned anticline, (e) isoclinal folds, (f) recumbent folds.

folds, drag.   Weak layers thrown into minor undulations. Their asymmetry indicates the relative displacement of adjacent layers. Because the same type of relative movement or shear takes place between beds in the folding of sediments, it is frequently possible to use drag folds to determine in which limb of a syncline or anticline a considered exposure lies.

foliation.   A planar, layered to laminated structure occurring almost exclusively in metamorphic rocks. When caused by parallel orientation of mineral plates and rods, it is referred to as schistosity. When caused by segregation of minerals into light and dark colored bands, it is called gneissic foliation by some and schistosity by others who detect parallel orientation of crystallographic axes of equidimensional grains using the petrographic microscope.

fool's gold.   Mineral pyrite, composed of iron sulfide.

foots oil (refin).   Trapped oil remaining in the wax after the dewaxing process and removable by sweating process, so named because the oil goes to the foot or bottom of the sweating pan.

foot shoe (drill).   Special shoe designed to help float a string of casing when it is being lowered into a well.

footwall.   Side of a fault lying beneath the inclined fault plane. The mass of rock beneath a fault plane or vein of ore.

forble board (drill).   Board or platform located in the upper part of the derrick at a suitable elevation so that a man can attach and detach the elevators from the drill pipe when it is being handled in stands of four joints each.

fore arc basin.   A basin in the fore arc region lying between the volcanic arc and the trench of a convergent crustal plate boundary.

foredeep.   Long, narrow, crustal depression or furrow, bordering the convex or ocean side of a folded orogenic belt or island arc.

foreland.   Promontory or cape; a point of land extending into the water from the shoreline.

fore reef.   Steeply inclined distal or seaward surface of a reef extensively littered with broken reef material, talus.

fore set beds.   Series of inclined layers accumulated as sediment rolls down the steep frontal slope of a delta.

foreshock. A lesser earthquake preceding a larger earthquake by a relatively short time, no more than a few weeks and originating at or near the focus of the larger earthquake. Magnitude is arbitrary in that the foreshock of a major earthquake may be of greater magnitude than that of a minor principal earthquake elsewhere. The full sequence of seismic events, foreshocks, principal shock and after shocks must be recorded until a single event can be identified as a foreshock.

foreshore. In general, portion of the shore environment lying between high and low tide levels, grading seaward into the surf and near shore zones. Contains gently dipping, seaward and horizontally laminated deposits of medium to coarse grained sand with occasional lenses of conglomerate or gravel.

fork (drill). (1) Appliance used in free fall systems of drilling serving to hold up the string of tools during connection and disconnection of the rods. (2) A branch or tributary stream to a main river.

form (crystallography). All the faces of a crystal related in the same way to the crystallographic axes by orientation and intercept. All faces of a crystal having identical symmetry notation.

formation. A lithologically distinctive product of essentially continuous sedimentation selected from a local succession of strata as a convenient unit for mapping, description and reference. Each formation is a given name based on the person who studied the formation, the geographic locality of formation outcrop or on the fossils found in the formation.

formation factor (geoph). In electrical resistivity techniques of well logging, ratio of the electrical resistance of electrolyte saturated rock and electrical resistance of the electrolyte alone; used to estimate subsurface porosity because of observed inverse relationship between porosity and formation factor.

formation fracturing. Method of stimulating production by increasing the permeability of the producting formation. Under extremely high hydraulic pressure, a fluid as distillate, diesel fuel, crude oil, dilute hydrochloric acid, water, or kerosene is pumped downward through production tubing or drill pipe and forced out below a packer or between two packers. The pressure causes cracks to open in the formation and the fluid penetrates the formation through the cracks. Diagram of a fracturing treatment. A packer is set to separate the producing formation from that above it. (See Fig. p. 129)

formation map. Map showing rock formations by use of color or by various lines, generally used where strata have steep dips; essentially synonymous with geologic map.

formation pressure. Pressure exerted by formation fluids, recorded in the hole at the level of the formation with the well shut in.

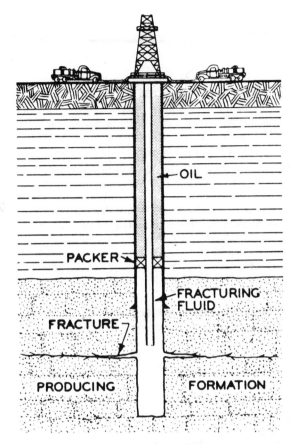

formation fracturing

formation testing.   Testing of a formation to determine its potential productivity before installing casing in a well. The conventional method is the drill stem test. Incorporated in the drill stem testing tools are a packer, valves or ports which may be opened and closed from the surface and a pressure-recording device. The tool is lowered to the bottom on a string of drill pipe and the packer set, isolating the formation to be tested from the formations above and supporting the fluid column above the packer. A port on the tool is opened to allow the trapped pressure below the packer to bleed off into the drill pipe, gradually exposing the formation to atmospheric pressure and allowing the well to produce, sampling and inspecting the well fluids.

Formex aromatic extraction (refin).   Extraction of aromatics from catalytic reformate or from steam-cracked liquids. Benzene toluene xylene (BTX) can be recovered in high quality at low processing cost.

founder. To sink under water either by depression of the land or elevation of sea level; usually applies to large crustal masses, islands or significant portions of continents.

fossils. Structural record, partial or complete, of life which is pleistocene in age or older. Fossils include (1) original material as preserved in amber or tar. (2) casts and molds. (3) petrified or permineralized materials. (4) tracks, trails, imprints and coprolites. Fossils are usually found in sedimentary rocks, although they sometimes turn up in igneous and metamorphoric rocks. Fossils account for almost the entire volume of certain rocks, as coquina and limestones formed from ancient reefs. Fossils are the single most important source of information about depositional environment and sedimentary rocks.

fossil fuels. Remains of once living organisms that can be burned to produce energy; such remains include petroleum, natural gas and coal.

fourable (drill). Section of drill pipe casing or tubing consisting of four joints screwed together. A stand of drill pipe made up of four twenty-foot lengths screwed together and having a tool joint at each end of the stand.

fourable board (drill). A platform installed in the derrick at an elevation of 80 to 120 feet above the derrick floor. The derrick man works on this board while the pipe is being hoisted from or lowered into the well bore.

fraction (refin). A common term for the products obtained by distillation as the gasoline fraction, kerosene fraction, gas oil fraction, and lube oil fraction. A portion of distillate having a particular boiling range separated from other portions in the fractional distillation of petroleum products.

fractional crystallization. (1) Reverse of partial melting; occurs as magma cools, the mafic minerals crystallize when the temperature is high, while the felsic minerals crystallize later at lower temperatures. (2) Controlled crystallization of saline waters; different salts are crystallized out at different temperatures.

fractional column (refin). An apparatus arranged to separate the various fractions of petroleum by single distillation, the condenser or "column" being tapped off at different points along its length so as to separate the various fractions in order of their condensing temperature.

fractional condensation (refin). Separation of vaporizing oil coming off during distillation by condensing vapors in stages, partial condensation. The oil highest in boiling point will condense first and may be removed in the liquid state allowing the portion still in the vapor state to pass one.

fractional distillation (refin). Dividing a liquid into fractions by the process of distillation. By changing the temperature in the still, each product in the crude oil is vaporized. The vapors can then be changed to a liquid by the process of condensation.

fractional extraction (solvent).   Solvent action extracting a fraction or part from the body of a liquid.

fractional fusion (refin).   Consists of separation of liquid and amorphous wax from solid crystalline by the gradual application of heat; sweating.

fractionate (refin).   Separate a mixture into fractions having more or less fixed properties, but which are not necessarily definite components.

fractionation (refin).   Separation of a mixture by some process, as distillation into a series of fractions, the constituents of which unavoidably overlap to an appreciable extent. Each fraction may then be resolved by the same process into a new series of fractions, the overlapping of the constituents of each fraction reduced. If this procedure is resorted to a sufficient number of times, a nearly complete separation of the components of the original mixture is obtained.

fractioning tower (refin).   A unit dividing or separating petroleum into fractions or parts. The hot charge from the furnace or pipe still enters the flash chamber where the vapors flow upward and condense on the fractioning plates. The condensate from a number of these plates may be collected and withdrawn as a cut or fraction. This cut or fraction may be run through a side stripper to closer fractionate the product.

fracture.   (1) Mineral grain breaks or splits in a direction not parallel to a cleavage direction. Fracture is best observed in large pieces of a mineral and generally is not easily determined in thin sections. Types of fractures that may be identified microscopically are conchoidal, uneven, splintery or ragged. (2) An irregular break in solid rock.

fracture cleavage.   Joints so closely spaced that they resemble rock cleavage as may develop parallel to metamorphic foliation or schistosite. If only one set of joints develops the fracture cleavage produces flat plates, but if more than one set occurs, the result is pencil-like rods or small polygonal blocks. Fracture cleavage is developed best in dense homogenous rocks and is not related to orientation of constituent mineral grains.

fragmental.   Usually applied to sedimentary rocks consisting of broken fragments of older rock; may imply transportation of fragments from their place of origin before being incorporated in the fragmental sedimentary rock. Some authors use this term interchangeably with detrital, others do not. *See* detrital.

Frasch Process.   Process for exploiting sulfur deposits by pumping superheated water into the deposit and pumping the melted sulphur out of the ground.

Frass breaking point.   Cracks first appear at this temperature in a test sample when stretched under test conditions at low temperature in a Frass tester.

free acid.   Unreacted acid occurring in new and used lubricating oils and greases in varying quantities, generally being proportional to the degree of oxidation of the material.

free carbon.   In tars, organic matter insoluble in carbon disulphide.

free air anomaly (geophysics).   Difference between the observed force of gravity at a given station and the theoretical force of gravity at sea level corrected only for the elevation of the station above or below sea level. This type of reduction of gravity data is popular because it requires a minimum of calculations. Major changes in density of earth materials will be reflected even though anomalies as those caused by variations in topography have not been removed. Gravity geophysical techniques are of importance to the oil industry because they can delineate areas of density change, as salt domes, serving as traps for petroleum.

free energy.   A thermodynamic measure of the ability of a system to do work or a measure of the maximum work obtained from a given process or reaction.

free fall (drill).   An arrangement by which, in deep drilling, the bit is allowed to fall freely to the bottom at each drop or down stroke. The process of operating the drill, often called Russian, Canadian and Galician free fall.

free point indicator (drill).   A tool designed to measure the amount of stretch in a string of stuck pipe and in so doing to indicate the deepest point at which the pipe is free. The free point indicator is lowered into the well on a conducting cable. Each end of a strain gauge element is anchored to the pipe wall by friction springs or magnets and as strain is put on the pipe, a very accurate measurement of its stretch is transmitted to the surface.

friable.   Easily crumbled or pulverized; easily reduced to powder.

frost heave.   Upward expansion of a soil surface caused by the freezing of internal water droplets.

frost wedging.   Physical weathering caused when water in intergranular pores, intercrystalline pores, in joints or other openings freezes, expands and breaks the rock.

frozen pipe (drill).   Applied to pipe rendered immovable in a hole owing to cavings settling around the outside of the pipe.

fuel gas.   Composed principally of hydrogen and carbon compounds in a gaseous state. Essentially, gas differs from liquid and solid fuels in having molecules at a lower ratio of carbon to hydrogen atoms.

fuel oil.   Liquid or liquefiable petroleum product burned for the generation of heat in a furnace or fire box, or for the generation of power in an engine.

fuel oil, diesel.   High grade fuel oil free of water, tarry deposits, paraffin, and other impurities; used as a fuel for diesel engines.

fuel ratio.   (1) Heating capacity of a fuel as compared with another fuel taken as a standard. (2) Coal fuel ratio is the fixed carbon content divided by the total volatile content present in a test sample of coal.

fuel soot.   Pulverized, dust-like form of carbon forming during incomplete combustion of fuel.

fugacity.   Tendency of a substance to escape or disappear chemically from the phase in which it is present.

fuller's earth.   Fine earth resembling clay; chemically similar to clay, but having a decidedly higher percentage of water; by filtering process, removes basic colors from oils.

fully refined wax.   Crystalline wax refined to meet certain requirements for color, oil content, odor and taste; used in ointments, foods, etc.

fumarole.   A small volcanic vent for the escape of interior gases to the atmosphere. Hole or vent in a volcanic region emitting steam or hot gases.

furfural.   A solvent obtained from oat hulls and used in connection with selective solvent extraction by the furfural solvent refining process of lubricating oils; also used to recover butadiene during the hydrogenation of butenes. An extractive solvent extensively employed for refining a wide range of lubricating oils and diesel fuels.

furnace oil.   A cut from the crude between the kerosene fraction and the light lubricating oil.

furrow cast.   Reverse reproduction, cast of a furrow on the underside of the sedimentary bed deposited as the furrow was buried; used in determining paleoslope or paleocurrent directions at the time of deposition of the sediments.

fusion.   (1) A process in which nuclei of light atoms fuse together to form a heavier atom with release of energy. The energy of the sun is released by a fusion process. (2) Melting of a solid, usually as a result of the application of heat.

G

gage (prod). Measuring or ascertaining the capacity of something as a barrel or tank.

gage pressure. Refers to pressure per square inch shown by a pressure gage to distinguish from absolute pressure.

gage tables (prod). Graphs showing the contents of a tank for each one-eighth inch or one-sixteenth inch of an oil contained in the tank. Tables of temperature corrections are often also prepared to serve as gage tables for reducing the contents of the tank to a standard volume.

gager (prod). An oil field worker charged with gathering samples of oil, testing them to determine the quality as to gravity and freedom from water and measuring the quantity of oil run from the producer's tank to the pipeline.

gal. Unit of measure for the acceleration due to gravity; equal to one centimeter per second per second. In geophysics, gravity field anomalies are expressed in gals or milligals (0.001 gal).

galvanometer (seis). An instrument for measuring a small electric current, for detecting its presence or direction by the movement of a magnetic needle, wire, or coils in a magnetic field.

galvanometric camera (obsolete seismic instrumentation). An instrument especially adapted to the peculiar needs of reflection shooting. The electrical leads from the vacuum tube amplifiers connected to the seismopickups and amplifying the feeble electric currents generated by the seismic waves are connected to the galvanometers positioned in the camera so as to reflect beams of light on a strip of moving photographing paper on film.

gamma. Small unit of magnetic field intensity describing the earth's magnetic field; defined as being equal to 10^{-5} oersted.

gamma eye (drill). An electrical instrument enclosed within a short length of tubing; designed to be lowered into a well for the purpose of detecting if there are oils or gas saturated formations through which the drill has passed.

gangue. Pronounced "gang"; worthless rock or vein matter in which valuable minerals occur.

gap (geol). A ravine or notch cut by stream erosion completely or partially through an elongated hill, mountain ridge, or mountain chain. If a stream is still flowing through the ravine, it is a water gap. If the hill or ridge was uplifted faster than the stream could cut down, the resulting notch with a floor elevated above the base of the ridge is a windgap. Term originally coined in the Valley and Ridge province of the Appalachian Mountains in the Commonwealth of Pennsylvania.

garbet rod (prod).   Short rod attached to lower end of a traveling valve of a well pump and to the standing valve to pull it out of its seat when repairs are necessary.

gas.   A state of matter in which the molecules are practically unrestricted by cohesive forces. A gas has neither definite shape nor volume.

gas, acetylene.   An unsaturated hydrocarbon, colorless gas of garlic odor, slightly soluble in water, more highly soluble in alcohol; used as a general anesthetic, for acetylene welding and cutting materials, in organic synthesis and as an illuminant.

gas black.   Form of lamp or carbon black made by introducing a cold iron surface into a luminous gas flame.

gas cap drive reservoir.   Normally found in discontinued, limited, or essentially closed reservoirs, where a quantity of free gas overlies oil in a reservoir. Recovery is aided by the gas cap expanding into the oil zone as oil is produced, thus a tendency to maintain pressure.

Gas-cap drive

gas carbon.   A compact variety of carbon obtained as an incrustation on the interior of gas retorts; used for the manufacture of cabon rods, pencils, for electric arc, and for plates of voltaic batteries.

gas, carbureted.   Gas treated with oil vapor to increase its luminosity.

gas coal.   Any coal yielding a large quantity of illuminating gas in distillation; usually low in sulphur and other impurities.

gas, coke oven.   Rich coal gas obtained as a byproduct of coking ovens where the volatile matter forming the coal is driven or distilled off leaving the coke. The gas is usually processed for the removal of benzene, benzol, ammonia, etc.

gas column.   The difference in elevation between the highest and lowest portions of various producing zones.

gas cut.   Fluffy mixture of gas-bearing drilling mud recovered in testing.

gas energy.   Work done by expanding gas while overcoming the resistance of the flow of oil through the sand of a natural oil reservoir. The energy stored within compressed gas.

gas, illuminating.   Gas containing hydrocarbon vapor to produce a luminous flame; may be coal gas obtained by the distillation of bituminous coal in retorts, or produce gas carbureted with oil vapor.

gas, liquefied.   Gas liquefied by high pressure; when the pressure is released, reverts to a gaseous state. Propane and butane, two gases produced in petroleum refining are used for this purpose.

gas, naphtha.   Naphtha separated from gas, either by simple compression or by absorption system.

gas, natural.   Gaseous hydrocarbon compounds originating within the earth, probably of organic origin, usually associated with petroleum. Gas coming from oil wells is generally rich in combustibles as methane and ethane.

gas oil.   A petroleum of medium gravity following kerosene and precedes the light lubricating oils; used for enriching carbureting water gas to increase its luminosity or as a high grade fuel oil.

gas-oil ratio.   Number of cubic feet of gas produced with a barrel of oil. As produced from an oil well, the number of cubic feet of gas at atmospheric pressure divided by the number of barrels (42 gallons) of oil will give the oil ratio. In case water is being produced with the oil, the total barrels of liquid may be used as the division and gas-liquid ratio is obtained.

gas reversion (refin).   Combination of thermal cracking or reforming of naphtha with thermal polymerization or alkylation of hydrocarbon gases carried out in the same reaction zone.

gas spurts.   Little heaps observed in the surface of certain geological strata containing organic matter; believed to be due to the escape of gas.

gas trap.   Device for separating and saving gas from the flow and lead lines of producing oil wells. The mixture of oil and gas is allowed to flow through a chamber large enough to reduce the velocity of the mixture to the point at which the oil and gas tend to separate. The gas seeking the top of the chamber is drawn off free of oil while the oil is discharged at the bottom. Also called gas separator.

gas water.   Water passing through coal and absorbing the impurities of the gas.

gas zone.   A formation containing capillary or supercapillary voids, or both, full of natural gas under considerably more than the atmospheric pressure.

gaseous.   In form or nature of gas; pertaining to gases.

gaseous fuel, dew point.   Temperature at which the gas becomes saturated with water vapor at the existing atmospheric pressure.

gaseous fuel, specific gravity.   Comparison of the density of the fuel to the density of the dried air of normal carbon dioxide content at the same temperature and pressure.

gasoline.   An indefinite name for a light distillate of petroleum covering a wide range of gravities and boiling points. A refined petroleum naphtha, by its composition suitable for use as a carburant in internal combustion engines.

gasoline acidity test.   A test to determine the presence of sulfuric acid and other free acids, organic compounds or mineral acids.

gasoline, casing head.   Extremely light and volatile gasoline obtained from wet casing head natural gas.

gasoline compression plant.   An installation of compressors in a plant used to compress wet gas into natural gasoline.

gasoline, cracked.   Gasoline obtained by high temperature and high pressure distillation of petroleum. The heavier components of the crude are "cracked" or decomposed to furnish a greater yield of gasoline naturally contained in crude.

gasoline gum.   A product of oxidation, the first visible evidence is the discoloration of the gasoline. As oxidation continues, the member causing discoloration is further changed by oxidation to gasoline gum. Therefore when gum forms in gasoline, it is decayed gasoline caused by the same action of oxidation accounting for decayed wood or rotten fruits and vegetables.

gasoline, polymer.   Polymerization of normally gaseous hydrocarbons to hydrocarbons boiling in the gasoline range.

gasoline, straight run.   Gasoline obtained by the fraction distillation of crude petroleum; gasoline naturally contained in the crude and consisting of a series of consecutive distillates with uniform boiling points.

gasoline trap.   A trap especially designed and built for removing gasoline light oils, natural gasoline condensate and water from natural gas under pressure.

Gay-Lussac's law of combining volumes.   If gases interact and form a gaseous product, the volumes of the reacting gases and the volumes of the gaseous products are in simple proportions expressed by small whole numbers.

geanticline.   A broad, linear, crustal upwarp, with dimensions measured in tens or hundreds of miles.

gel.   (1) A state of a colloidal suspension in which shearing stresses below a certain finite value fail to produce permanent deformation. The minimum shearing stress producing permanent deformation, known as the shear or gel strength of the gel. Gels commonly occur when the dispersed colloidal particles have a great affinity for the dispersing medium, called lyophilic. (2) In drilling, a term used to designate highly colloidal, high yielding, viscosity-building commercial clays as bentonite and attapulgite clays.

gel cement (drill).   Cement having a small to moderate percentage of bentonite added as a filter and/or reducing the slurry rate.

gel strength.   Ability or measure of the ability of a colloid to form gels. Gel strength is a pressure unit usually reported in pounds per hundred square feet. A measure of the same interparticle forces of a fluid as determined by the yield point except that gel strength is measured under static conditions, yield point under dynamic conditions. The common gel-strength measurements are initial and the 10-minute gels.

genus (pl. genera).   One of the taxonomic or scientific classifications of plants and animals. Groups of similar species belonging to the same genus or related genera form families. For animals, in descending order, the system is kingdom, phylum, class, order, family, genus, species.

geo-.   A prefix meaning earth, as in geology, geophysics.

geocentric.   Relative to the earth as a center; measured from the center of the earth.

geochemistry.   Application of chemical procedures and principals toward a better understanding of geology. Includes the study of (1) abundance and distribution of the elemental and isotopic constituents of the earth, (2) migration (fluxes) of elements and compounds within and/or between the hydrosphere, lithosphere, and atmosphere, (3) chemical reactions involved in the formation and deterioration of rocks and minerals, (4) thermodynamics of changes of state within the earth and of natural, earth chemical reactions, and (5) organic chemistry of fossil fuel formation.

geochronology.   Time as applied to earth history, including classification of geologic time in eons, eras, periods, epochs, and ages. Determination of the age of rocks, geologic processes or events in geologic history in years before Present (B.P.) using radioactive, paleontologic, and correlative age dating techniques.

geode.   A nodule or irregular ball of rock containing a hole partly filled with crystals, minerals in layers or both. Quartz is the most common mineral to fill a geode, calcite is second in abundance, while dolomite, selenite and sphalerite may also occur.

geodesic line.   A line of shortest distance between any two points on any mathematically defined surface; also termed a geodesic.

geodesy.   Scientific measurement and determination of various dimensions of the earth, including density, weight, shape, gravitational, and magnetic variations. It especially refers to surveying of areas that are so large, an allowance must be made for the earth's curvature.

geoid.   Figure of the earth as defined by the geopotential surface nearly coinciding with mean sea level over the entire surface of the earth. The figure of the earth considered as a mean/sea level surface is extended through the continents.

geological age.   Age of a given stratum in years before Present (B.P.); subdivision of geologic time in which it was formed.

geologic column.   Worldwide standard diagramatic representation of the sequence of strata in chronological order with the oldest on the bottom. The column may be portrayed in the most general way showing only the largest stratigraphic unit. Portions of the column may be shown in great detail, with formations subdivided, lithologies represented by standard symbols and relative resistance to weathering and erosion reflected. A hypothetical example of a detailed geologic column is given below.

Lithology	Period	Formation
	Silurian	Petroleo
	Ordovician	Dinero

geologic cross section. Downward projection of surficial geology along a vertical plane. Vertical walls of streams or vertical road cuts are naturally occurring geologic cross sections. When a geologic cross section cannot be seen it must be constructed using geometric or trigonometric techniques.

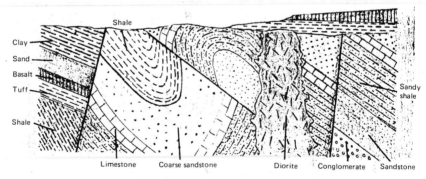

GEOLOGIC CROSS SECTION

geologic map. A map plotting geologic information. The distribution of the formations is shown by means of symbols, patterns, or colors. The surficial deposits may or may not be mapped separately. Folds, faults, rock types, mineral deposits, and other geologic occurrences are indicated by appropriate symbols. Example of a geologic map with appropriate symbols.

GEOLOGIC MAP

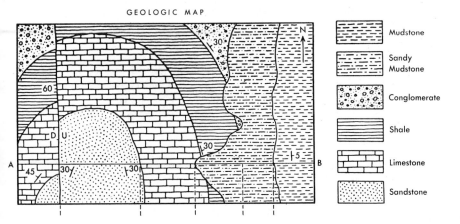

geologic map symbols.

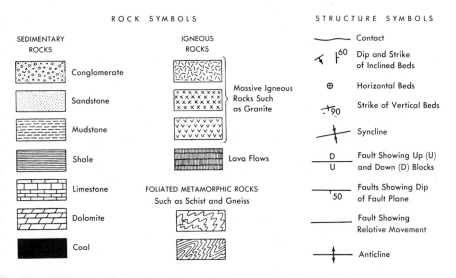

geologic rock unit.   Stratigraphic unit defined solely on the basis of lithology and/ or structural features. Geologic rock units may cross time boundaries, varying in age from one portion of the unit to another; also called rock statigraphic unit or simply rock unit.

geologic time-rock unit.   Stratigraphic unit defined on the basis of lithology and/or structural features and representing a specific interval of geologic time; also called time-stratigraphic unit or simply time-rock unit.

geologic time unit.   Subdivision of geologic time corresponding to a time-rock unit. *See* geological time table for names of major geologic time units; also called time units.

geological oceanography.   Study of floors and margins of the oceans, including description of submarine relief features, chemical and physical composition of bottom materials, interaction of sediments and rocks with air and sea water, and action of various forms of wave energy in the submarine crust of the earth.

geological time table.   After Van Eysinga (1975).

ERA	PERIOD	EPOCH	MILLIONS OF YEARS
CENOZOIC	QUATERNARY	RECENT	1.8
		PLEISTOCENE	
	TERTIARY	PLIOCENE	3.2
		MIOCENE	20
		OLIGOCENE	15
		EOCENE	15
		PALEOCENE	10
MESOZOIC	CRETACEOUS	UPPER	76
		LOWER	
	JURASSIC	UPPER	54
		MIDDLE	
		LOWER	
	TRIASSIC	UPPER	35
		MIDDLE	
		LOWER	
PALEOZOIC	PERMIAN	PROVINCIAL SERIES	50
	PENNSYLVANIAN MISSISSIPPIAN	NO FORMAL SUBDIVISION	65
	DEVONIAN	UPPER	50
		MIDDLE	
		LOWER	
	SILURIAN	UPPER	40
		MIDDLE	
		LOWER	
	ORDOVICIAN	UPPER	65
		MIDDLE	
		LOWER	
	CAMBRIAN	UPPER	70
		MIDDLE	
		LOWER	
PRE-CAMBRIAN			4,000

(GEOLOGIC TIME)

geology.   Science treating the origin, composition, structure and history of the earth, especially as revealed by rocks and of processes which form, alter and destroy rocks; included is the study of the origin and evolution of living organisms, especially in prehistoric times.

geomagnetic equator.   Great circle on the earth's surface that is everywhere equidistant from the geomagnetic poles; that is, the equator in the system of geomagnetic coordinates.

geomagnetic pole.   Point where the axis of a centered dipole most nearly duplicating the earth's magnetic field would intersect the surface of the earth.

geomorphology.   Branch of both geography and geology dealing with forms of the earth, general configuration of its surface, and changes that take place in the evolution of land forms. Geomorphic cycles of an arid region.

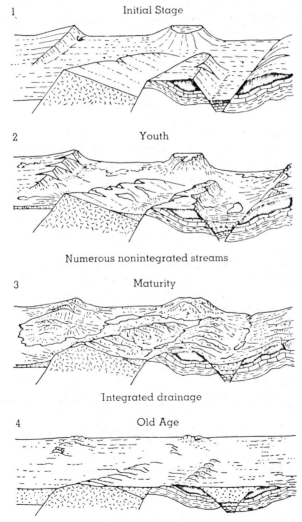

1   Initial Stage

2   Youth

Numerous nonintegrated streams

3   Maturity

Integrated drainage

4   Old Age

Disintegrated drainage

geophone (seis).   An instrument for detecting seismic waves, transforming sound signals into electrical signals. An important component of a seismograph.

geophysics.   Application of principles and procedures of physics toward a better understanding of the composition, structure and genesis of the earth. Magnetic, gravity, and electrical potential fields of the earth and seismic waves are studied as well as the physical properties of rocks and minerals as magnetic susceptibility, electrical conductivity, density and elasticity. In its broadest sense, geophysics includes the fields of seismology, geodesy, meteorology, and physical oceanography. Geophysical techniques may be applied to the search for economic mineral deposits and fossil fuel resources. *See* geophysics, exploration.

geophysics, exploration.   Application of geophysical techniques to the discovery of ore and energy resources. Technique can be subdivided into four groups: **(1)** Magnetic, in which studies are made of anomalies in the earth's magnetic field caused by variations in rock and mineral magnetic properties. **(2)** Gravity, in which studies are made of anomalies in the earth's gravitational field caused by variations in rock and mineral densities. **(3)** Electrical, in which earth potentials are studied as they are generated naturally in the earth or in which the resistivity of rocks are measured by inducing a potential into the earth and, **(4)** Seismic, in which shock waves are generated by explosion or other means. Earth structure is determined by the ways in which the shock waves are reflected or refracted in their passage through the earth. Geophysical studies of thermal anomalies are gaining in importance as geothermal energy resources become popular.

geosphere.   Solid and liquid portions of the earth, lithosphere and hydrosphere.

geosyncline.   A large, elongated and thick deposit of continental shelf and slope sediments. Before the concept of plate tectonics was accepted, these deposits were thought to be laid down in great troughs, geosynclines which subsided as fast as sediments were deposited. The concept of plate tectonics shows that geosynclines are not related to a single trough but rather are elongated wedge of continental shelf and slope sediments which occasionally may occupy continental shelf basins (fore arc, and back arc basins) and fore arc trenches. Geosynclinal sediments are deformed when deposited along convergent plate boundaries. The term geosyncline will mean therefore something altogether different in the scientific literature published prior to the mid 1960's than it does in most publications of the 1970's.

geothermal.   Of, or pertaining to, the heat of the earth's interior.

geothermal gradient.   Change in temperature of the earth with depth, expressed either in degrees per unit depth, or in units of depth per degree. The steady increase of temperature with depth within the earth at a rate approximating 30°C per kilometer in the upper crust of the earth.

geothermometry.   Techniques or procedures for measuring earth temperatures, including **(1)** direct measurements by thermometer, thermocouple or optical pyrometer, and **(2)** indirect techniques using mineral melting temperatures, thermoluminescence, and phase changes.

geyser.   Hot spring erupting hot water and steam periodically. Diagram of the anatomy of a geyser. The superheating of ground water in "traps" causes periodic eruptions.

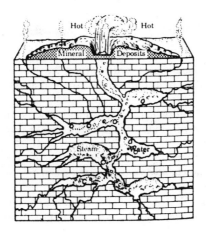

glacial action.   Grinding, scouring, and polishing effect by glacial ice upon rock fragments, and the accumulation of rock debris from such action. Diagram of the characteristic assemblage of features seen on a recently glaciated area of low relief.

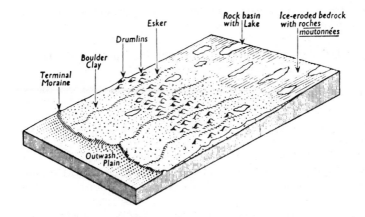

glacial cirque.   All mountain systems which are or have been centers of local glaciations exhibit numerous examples of a type of alpine valley terminating their heads in rocky amphitheaters known as "glacial cirques."

glacial drift.   Sediment and rock fragments transported by glaciers or icebergs and deposited on land or in the sea.

glacier.   A mass of ice formed by the recrystallization of snow; flows or has flowed at some time in the past under the influence of gravity.

glacier, alpine.   Valley glaciers nourished on the flanks of high mountains and flowing down the mountain sides.

glacier, continental.   Usually reserved for great ice sheets obscuring mountains and plains of large sections of a continent, as Greenland and Antartica.

glacier, Piedmont   Formed when alpine glaciers emerge from their valleys and spread out to form an apron of moving ice on the plains below.

glaciofluvial.   Related to melt water flowing out of a glacier. Formerly glacial sediments are reworked and deposited by the melt water.

gland (prod).   A device designed to close a stuffing box; has tabular projections embracing the rod, extending into the bore of the box, bearing against the packing, to prevent leakage.

glauconite.   A green phyllosilicate mineral, a common constituent in some marine sedimentary rocks; considered by some to be a variety of the clay mineral illite, while others describe it as closely related to micas; normally found associated with fecal pellets.

global tectonics.   General term for the process of crustal movement and for the deformation of plates of lithosphere.

globule.   Small glob or spherical particle often applied to particles of liquids found in rock cavities or suspended in other liquids.

glyceride.   An ester of glycerol (glycerin) as the searin glyceryl tristearate.

glycol.   A colorless, odorless compound $(C_2H_6O_2)$ having a sweet taste.

gneiss.   A banded metamorphic rock generally resembling granite; it contains quartz, feldspar, and mica in grains large enough to be seen with the unaided eye. Its bands may be straight, wavy or crumpled and of uniform or variable thickness.

go-devil.   A scraper with self-adjusting spring blades, inserted in a pipe-line car ried forward by the fluid pressure, clearing away accumulations from the walls of pipe.

Go-Fining & Residfining (refin).   Serves as an initial desulfurization step in the production of low sulfur fuel oils to reduce the sulfur content of the total fuel oil pool by about 50 percent. High boiling distillate products as virgin vacuum gas oils, visbroken gas oils, thermal and cat cycle oils, along with coker gas oils can be more than 98 percent desulfurized via Go-Fining, which is also used to pretreat cat cracker feeds to improve crackability and reduce sulfonyl pollution. Reductions in the sulfur content of the fuel oil pool of 90 percent and higher can be achieved via Residfining of atmospheric tower bottoms.

gooseneck (drill).   A return bend having one leg shorter than the other attached to the top of the rotary swivel for making slush hose connections. The vapor pipe line on a still.

GOR (Gas Oil Ratio).   Gas-oil ratio expressed in cubic feet of gas per barrel of oil (cu ft/bbl).

gossan.   (1) Ferruginous deposit filling the upper parts of mineral veins or forming a superficial cover on masses of pyrite; consists principally of hydrated oxide of iron and has resulted from the oxidation and removal of the sulfur as well as the copper, etc. (2) Rust-colored oxidized capping of an ore deposit; oxidized equivalent of aggregated sulfide material. Gossan usually consists of ferric oxide and quartz or jasper, sometimes with manganese dioxide, clay minerals, etc.

gouge.   Crushed and abraded material occurring between the walls of a fault.

graben.   A fault trough or a fault block downthrown relative to both margins; geological structure, usually long section of the earth's crust depressed between two faults.

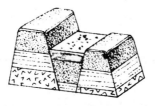

grab sample.   (1) A sample of rock or sediment taken without regard to any predetermined sampling scheme; a single sample taken at random. (2) An ocean or lake bottom sediment sample taken with a device equipped with spring loaded jaws snapping closed upon contact with the sediment surface; called a grab sampler.

grade.   A condition of erosional and depositional equilibrium in a river. A stream at grade has just the slope and velocity to carry its load of sediment without depositing or eroding more.

gradation.   Process by which stream erosion reduces a stream bed or topographic surface to grade through corrasion, transportation and aggradation.

graded bed.   A depositional feature produced as particles settle from turbidity currents. When consolidated, a deposit displays a gradation in grain size from fine in the upper portion, to coarse in the lower. Sample of graded bedding.

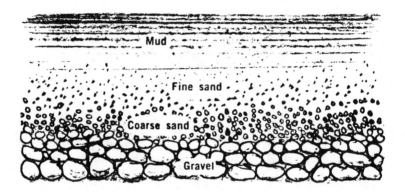

graded sediment.   Sediment consisting principally of grains of the same size range.

grade scale.   A means employed by sedimentologists to treat quantitatively the distribution of sediment particles according to size. Several grade scales have been constructed including the following: (1) Atterberg, (2) Phi, (3) Tyler, (4) Udden, (5) Wentworth. Each scale uses a convenient reference size and is subdivided into size classes.

gradient.   Change in pressure exerted by a fluid for each foot of fluid height. Fresh water exerts a gradient pressure of .433 psi per foot calculated in this manner, 4330 psi gage at 10,000 feet less surface pressure of 0 psi is 4,330-0 over 10,000 equals, 433 psi per foot, (2) slope of a stream bed, usually expressed in feet per mile.

Graham's law.   Relative rates of diffusion of gases under the same conditions are inversely proportional to the square roots of the densities of those gases.

grains (or particles).   The individual particles which form a sediment, ranging in size from smaller than 0.0039 millimeter for clay particles to larger than 256.0 millimeters for boulders.

gram atomic weight. Mass of an element in grams, numerically equal to its atomic weight.

granite (geol). Coarse-grained, intrusive igneous rock composed of orthoclase and quartz. Accessory ferro magnesian minerals are common. A common type of rock found in the ancient basement rocks beneath sedimentary sections.

granitoid. Designation of the texture of a coarsely crystalline, igneous rock. Suffix "-oid" means like; a granitoid rock is one resembling granite.

granization. Process converting various solid rocks to rock of granitic composition and structure without the intervention of a magmatic stage.

granule. Rock fragments of overall gravel size, larger than very coarse sand, 2.0 millimeters but smaller than pebbles, 4.0 millimeters.

grapestone. A cluster of small calcareous pellets, resembling grapes, stuck together in incipient cementation shortly after deposition.

grasshopper (prod). A tool used to align and join pipes prior to welding. One part of the grasshopper slips under the joints, forming a cradle while the other part is braced over the top.

gravel. A deposit of worn fragments ranging from small pebbles to cobblestones and boulders (2 to 256 millimeters). Many of the pebbles are quartz, but a great variety of hard rocks and minerals may be represented. Soft rocks like shale are rare constituents of gravel.

gravel packing. Placing of gravel, coarse sand, mesh 8 to 15, opposite a producing sand to prevent or retard the movement of loose sand grains into the well along with the oil. Usually forced through perforations under pressure.

gravimeter (geoph). Instrument for measuring the acceleration due to gravity or force of gravity by recording the force of attraction between the mass of the earth and a reference mass. Most gravimeters do not have a range of scale to measure the full gravity force field hence the difference in field strength is measured between a standard station and a location where the force of gravity is unknown and gravity at the new station is evaluated by simple addition or subtraction once the appropriate corrections have been made.

gravitational constant (geoph). G or γ in the law of universal gravitation by formula $F = G \frac{m_1 m_2}{r_2}$ where F is the force of attraction between two masses, m_1 and m_2 separated by a distance r. G may be evaluated by measuring the force between two unit masses a unit distance apart and solving the gravity formula for G.

gravity. (1) Ratio of the weight of a volume of any liquid to the weight of an equal volume of distilled water at 60°F. (2) The attractive force which the earth exerts upon all objects.

gravity (API).   Specific gravity or density of oil expressed in terms of a scale devised by the API (American Petroleum Institute). The lighter the oil, the greater the gravity, other factors being equal, the higher the API gravity, the better price the oil will bring. The formula is

$$\text{Degrees API gravity} = \frac{141.5}{\text{specific gravity}} = 131.5$$

gravity, anomaly.   Difference between the calculated theoretical value of gravity and the observed value of gravity at a given station or set of stations. Anomalies in excess of theoretical gravity are called positive, those less than theoretical gravity are called negative.

gravity flow.   (1) Movement of glacial ice resting on a sloping surface under the influence of the downslope components of gravity. (2) Mass wasting involving flow rather than slipping or rapid failure as earth flows or rock glaciers. (3) Underwater current flowing down slope because of the high density it exhibits as a result of a concentrated sediment load.

gravity, specific.   Density expressed as the ratio of the weight of a volume of substance to the weight of an equal volume of another standard substance. In the case of liquids and solids, standard is water. In the case of natural gas or other gaseous material, standard is air.

Gray's tester.   Instrument used for determining the flashing point of heavy oils.

grazing angle (seis-ocean).   Angle formed by sound ray path with the reflecting surface; usually applies to sound rays reflected from the bottom. Conventionally, the angle is measured from the horizontal.

grease.   A lubricant consisting of lubricating oil and soap of varying proportions; like a sponge filled with water, grease is a matrix of soap filled with oil.

grease channeling.   A property of grease; which prevents the grease from being used under certain conditions due to lack of cohesion, as in lime base grease channels when used to lubricate certain gear actions.

grease classification.   As the base of the grease provides the predominant characteristics, the most frequently used means to classify greases is by the alkali in the soap used in the manufacture of grease, as calcium base greases, sodium greases, aluminum base greases, lead base greases, etc., and mixed based grease when two or more alkali are used.

grease (lime soap).   Soap for lime (calcium) soap grease is made by mixing fats and an equal amount of lubricating oil with the proper amounts of lime (calcium hydroxide) at temperature of 300°F

grease (lithium soap). Prepared by heating lithium stearate and mineral oil until the mixture is melted; mixture is then poured into pans, allowed to cool, and homogenized into a smooth grease.

grease plant. An auxiliary department of certain refineries in which lime and soda soaps of various fatty products, as inedible lard and tallow oils, horse oils, etc., are compounded with non-viscous parrafin oils or other petroleum products into semi-fluid or pasty lubricants technically called greases.

grease (soda soap). Made directly in a grease kettle. Fat and lye solution, sodium hydroxide, are mixed in the kettle for two-hours at about 300°F to form the soap. Oil is then stirred into the soap.

grease (sodium soap). Greases made in much the same way as lime soap greases except that caustic soda, sodium hydroxide is used in place of lime, and the kettle is heated to a higher temperature, usually about 350°F, but as high as 500°F if used for certain brick greases.

green acid (soaps). Water soluble petroleum sulfonic acids.

green crude. Crude refined from a green crude or refined from a bluish or black crude given a green color or cast.

green oil. Color of certain petroleums. High grade crudes, the Pennsylvania for example, have a greenish cast. Thus the designation "green oil" carries with it an implication of high quality.

greyhounds (drill). Stands of drill pipe consisting of one or more lengths with a tool joint on each end and used to make up lengths less than one regular stand while drilling.

griefstem (drill). A heavy square pipe working through the square hole in the rotary table and rotating the drill stem in rotary drilling; generally called kelly or kelly joint.

grit. A hard, very densely cemented sandstone, whole grains are coarser than sand size but finer than gravel.

groin. A long narrow structure constructed with its large dimension at right angle to a shore which it has been designed to protect. Groins interrupt longshore currents and trap long shore drifting sediment in an attempt to enlarge beaches and reduce shore line erosion.

groundmass.   Fine-grained or glassy matrix of a porphyritic igneous rock. The fine grained base of a porphyry in which the larger distinct crystals are embedded; glassy or crystalline.

ground moraine.   Moraine with low relief, consisting mainly of unsorted till, deposited as a widespread veneer over a bedrock surface.

ground water.   (1)Underground water including that which is held in the unsaturated zone of aeration and accumulated below the water table in the zone of saturation. (2) Portion of underground water below the water table in the zone of saturation.

grout (drill).   Liquid cement used in connection with drilling operations.

guar gum (drill).   A naturally occurring hydrophilic polysaccharide derived from the seed of the guar plant. The gum is chemically classified as a galactomannan; slurries made up in clean, fresh or brine water possess pseudoplastic flow properties.

guide fossil.   Any fossil particularly useful in characterizing and identifying a stratigraphic unit. Time rock units require useful guide fossils to be abundant, of short stratigraphic range, wide geographic distribution, and common to a wide variety of environments. Some authors equate guide fossils and index fossils.

Gulfining (refin).   Hydrodesulfurization of heavy distillate gas oils. The charge is gas oils including vacuum distillate. Products are low sulfur fuel oil or catalytic cracking unit, charge stock.

Gulfinishing (refin).   Color improvement, neutralization and impurity reduction in lube oils, waxes and specialty oils. The charge is solvent extracted or raw paraffinic neutrals, bright stocks, naphthenic distillates, etc. The products are a full-range of finished lube oils and specialty products.

gum (drill).   Any hydrophilic plant polysacchrides or their derivatives when dispersed in water, swell to produce a viscous dispersion or solution. Unlike resins, they are soluble in water and insoluble in alcohol.

gumbo.   When wet, a sticky, puttylike clay material; associated with soils in western and southern states and with clay horizons encountered when drilling for oil or sulfur in Texas. When dry, may be hard and brittle; also associated with glacial deposits. *See* gumbotil.

gumbotil.   Clay rich product of decomposition by weathering of a glacial till. Plastic and sticky when wet, hard and brittle when dry. Some authors extend the definition to the weathering products of drift rather than restrict definition to till.

gummy. Viscous but without lubricating power, resinous, sticky but not oily.

gunbarrel (prod). Basically a vertical separator, usually a tall large diameter vessel permitting extended settling time for oil and water separation due to difference in specific gravity.

gun perforating (drill). Method of completing producing wells by shooting steel bullets through the casing into the producing formation. The mechanism for firing the bullets is operated electrically, and the effect is to open up the reservoir rock to the well bore at the exact depth from the greatest production, as indicated by various logging and sample data.

gusher. A flowing oil well, especially is the discharge if with force and does not require pumping. A gusher may flow by heads intermittently or continuously.

guyot. Deeply submerged volcanic cone with a broad flat top, common in the Pacific Ocean. A flat-topped submarine mountain or seamount whose summit is supposed to have been exposed to wave action and planed away to the surface of the ocean.

H

half-life. Length of time required for half of the atoms of a radioactive isotope to disintegrate. This type of disintegration, with a characteristic half-life, gives rise to an exponential form of decay of the radioisotope. Half-lives range from fractions of a second to billions of years.

halides. A compound formed by one of the halogens with any other element. A chemical compound containing a halogen.

halite. Mineral name for sodium chloride, or common salt, NaCl. Colorless to gray, occurring as cubic crystals with perfect cubic cleavage, in three planes at right angles to each other. Its hardness is 2.5. Beds of common salt occur interstratified with sedimentary rocks. In many places, the salt is associated with gypsum.

halmyrolysis. (1) Chemical interaction between suspended sediment particles and sea water including sediment temporarily at rest on the bottom. Many of the products, reactants and chemical reactions of halmyrolysis and diagenesis are the same, the principal difference being that halmyrolysis is predepositional and diagenesis is post-depositional. (2) Some authors equate halmyrolysis with submarine weathering.

halo. A ring shaped region, partial or complete, surrounding a central area. A geochemical halo has a chemical composition in marked contrast to the central area. A geophysical halo has a field polarity of opposite sign to the central area or a markedly different intensity.

halogen. Elements in Group VII of the periodic table including fluorine, chlorine, bromine and iodine.

halogenation. Process of chemically combining with a halogen as bromine, chlorine, fluorine or iodine.

hand specimen. A piece of rock trimmed to a size that can be held conveniently in the hand, usually one by three by four inches. Hand specimens are placed in study or reference collections, for analysis with hand lens or the unaided eye.

hanging wall. Mass of rock above a fault plane, vein, or bed of ore and opposite of a footwall. Side of a fault enclosed over a fault plane.

hang rods (prod). To hang sucker rods in the derrick or rod hangers rather than to lay them on a rack.

hard asphalt (pitch).   Solid asphalt with a normal pentration of less than ten.

hardness.   Ability of a mineral to resist being scratched. Hard minerals scratch soft minerals. Hardness is controlled by the strength of bonding between a mineral's constituent atoms. The hardness of a mineral is expressed in terms of the Mohs scale; 1, talc; 2, gypsum; 3, calcite; 4, fluorite; 5, apatite; 6, orthoclase; 7, quartz; 8, topaz; 9, sapphire; 10, diamond.

hardpan.   Hard impervious layer, composed chiefly of clay, cemented by a combination of silica, alumina and ferric oxide; does not become plastic when mixed with water, and definitely limits the downward movement of water and roots. Results from precipitation of minerals out of ground water in the "B" soil zone.

hard rock . (1) A general term referring to igneous and metamorphic but not sedimentary rock. (2) In engineering and engineering geology, refers to rock which must be drilled and blasted to be removed; too hard to be ripped.

hard rock geology.   Geology of igneous and metamorphic but not sedimentary rocks.

hatch.   An opening into a tank, usually through the top deck.

hay tank.   A tank or enclosure containing hay-like material used to filter oil out of water.

HDC Unibon (refin).   HDC Unibon process gives refiners a hydrocracker flow scheme with flexibility of matching output to market conditions. When processing vacuum gas oils, this permits users to maximize naphtha production with no production of distillate fuels during that part of the year when gasoline demand is high; and to maximize distillate fuels with minimum naphtha production at other times. Furthermore, products changeover is readily achieved by modifying operating conditions, without altering hardware.

HD oil.   Heavy duty oil.

HD supplement 1 oil.   An HD oil having an additional level of approximately seven per cent, being between the lowest of approximately four per cent and highest approximately eleven per cent.

head.   (1) Pressure of a fluid upon a unit area due to the height at which the surface of the fluid stands above the point where the pressure is taken. Often expressed in pounds per square inch and is also stated in feet. A head of pure water at 60° F and 100 feet high equals a pressure of 43.31 pounds per square inch. Heads due to other liquids are proportional to the specific gravity of same. (2) Circular end plate of an oil still. (3) The action of a well when flowing intermittently.

head, hydraulic.   Height of a fluid column, usually considered water, maintaining a pressure on the surface. This pressure head may be given in terms of the pressure per square area or simply as the height of a water column in feet or inches.

headache post.   A post set under the walking beam at one side of the derrick floor. The beam rests upon the headache post, preventing it from falling upon the workmen when the pitman is disconnected from the crank.

header.   A common manifold in which a number of pipe lines are united; usually refers to the U-bend connection between two consecutive tubes in the coil.

headward erosion.   Lengthening of a river or stream valley away from its mouth. Deepening and extending a stream valley head by water as it concentrates from sheet flow to rills.

heat balance.   Law of heat energy states that the actual amount of heat generated by a fuel equals the amount utilized plus the amount lost; that the energy can neither be created or destroyed.

heat content (combustible).   Available quantity of heat as a result of the combustion or burning of a combustible and recorded in terms of British thermal units (Btu). Gasoline has approximately 19,000 Btu's per pound, considerably higher when compared to TNT and other explosives.

heat convection.   Transfer of heat by movement of material because of density difference created by thermal gradients. Fluids which have been heated to become lighter than their surroundings will rise or float up through their surroundings, carrying the heat with the rising fluid.

heat exchanger (refin).   A unit heating a liquid by transferring the heat from another liquid or a vapor. At a refinery, crude oil is preheated by flowing through pipes surrounded with hot residue coming from the still.

heat flow.   Movement of heat from the hotter interior of the earth to the cooler crust where the heat is dissipated into space. Movement is principally by conduction or radiation. Heat flow approximates 1.2 times 10^{-6} calories per centimeter squared per second within the earth's crust.

heat of condensation.   *See* heat of vaporization.

heat of crystallization.   Amount of heat evolved when crystals form from a melt without change in temperature. The same amount of heat is required to melt crystals at constant temperature called heat of melting or heat of fusion. This is not the same as the heat change occuring when crystals precipitate out of solution, (heat of solution) but the two are frequently confused.

heat of dissociation (geol).   (1) Heat loss or gain with the breakdown or dissociation of a phase into two or more simpler phases. (2)(geochem)Change in heat energy associated with the separation or dissociation of molecules into their ionic constituents. Usually occurring in aqueous solutions and involve gain, loss or exchange of electrons.

heat of melting/fusion.   *See* heat of crystallization.

heat of solution.   Loss or gain of heat when a solid dissolves in a liquid; not a standard value and subject to great variability according to the concentration or activity of the solute already in solution and temperature of the system. An identical quantity of heat is involved when crystals precipitate out of solution but opposite in sign to heat of solution, provided of course solution concentration and ambient temperature are identical in both cases.

heat of vaporization.   Heat needed to convert the unit mass of a liquid into vapor at its boiling point without a change in temperature. The same amount of heat is released when the vapor condenses to a liquid at constant temperature called the heat of condensation.

heat, specific.   Quantity of heat required to raise the temperature of a unit weight of material through a temperature difference of one degree. Oil industry uses the English system in which the specific heat is defined as the number of British thermal units (Btu).

heat transfer coefficient.   Rate at which heat is transferred per hour, per unit surface, per degree of temperature difference; may be expressed as the Btu's of heat transmitted per hour through an area of one square foot of one inch thickness of a substance.

heat unit.   Unit quantity of heat, as calorie or British thermal unit. The heat required to raise a unit mass of water through one degree of temperature, within a specified temperature range.

heating value.   Calorific content, or heating value of a fuel is the total amount of heat developed by the complete combustion of a unit weight or unit volume of a fuel.

heave (geol).   (1) Component of displacement along a fault plane which is horizontal and at right angles to the strike of the fault plane. The horizontal component of dip separation of a fault. (2) Uplift of the surface of the earth because of expansion of one or more soil constituents, usually temporary. When uplift is caused by freezing soil moisture it is called frost heave.

heaving (drill).   Partial or complete collapse of the walls of a hole resulting from internal pressures due primarily to swelling from hydration or formation gas pressures.

heaving plug (prod).   A plug set at the bottom of a well to prevent unconsolidated sand from heaving.

heavy fraction (refin).   Last products recovered from the crude during distillation; also known as heavy cut or last cut.

heavy minerals.   Accessory minerals of high specific gravity found in rocks. Most authors restrict usage of this term to detrital sedimentary minerals but others use the term without discrimination to rock type.

heavy oils (refin).   All products distilled or those processed from the crude beginning with the first lube oil distillate. Those preceding the first lube oil distillate are classed as light oils.

heavy oil cracking (refin).   To convert residual crude fraction into light olefins, gasoline and furnace oil. Products are motor gasoline, furnace or diesel fuel blending stocks, light olefins, LPG, hydrogen, fuel gas and carbon black feedstocks. The heavy oil cracking process can be designed to handle wide variety of residual fractions from crude oil. The operating conditions may be varied to maximize either high octane gasoline and olefins for alkylation or furnace and diesel fuel oil.

heavy petroleum spirit.   A water white petroleum distillate used as a thinner in slow setting paints and varnishes and related products; has a minimum initial boiling point of 340° F and a maximum end point of 485°F and has other properties as described in ASTM.

heavy water.   Water in which the hydrogen of the water molecule consists entirely of the heavy hydrogen isotope of mass two (deuterium); written D_2O; density, 1.1076 at 20°C.

height of wave.   Vertical distance separating a wave crest and wave trough immediately following or preceeding. Twice the wave amplitude if the wave approaches sinusoidal form. Usually applied to waves on lakes and oceans but some authors use the term to describe sedimentary bed forms, particularly large sand waves.

helrazee (drill).   An electro-magnet fishing tool for recovery of pieces of broken bits and cutters, etc., lost during drilling operation.

hemipelagic.   Pertaining to sedimentary deposits and rocks consisting of a mixture of pelagic and terriguous components and reflect continental slope and rise environments.

Hempel flask. Hempel, a well-know German chemist, modified the English distillation flask by extending the neck so that it would serve as a tower. It is the use of this flask which has suggested the method for making fractional distillations. The common method for the analysis of petroleum makes use of this flask, having a long neck from which there is a side tube for the escape of hot vapors into a condenser.

Henry's Law. Amount of gas absorbed by a given amount of liquid at a given temperature, is directly proportional to the pressure of the gas.

heptane (Heptyl hydride; Menthyl hexane). One of the liquid paraffin series of petroleum hydrocarbons of the saturated order; forming a part of gasoline and is a volatile and flammable liquid. Specific gravity equals 0.694, boiling point equals 173°F. Any of nine isomeric hydrocarbon C_7H_{16}, normal heptane occurring in petroleum.

herringbone cross-lamination. Thin layers of sand crossed-laminated in opposite directions in alternating layers by frequently shifting currents in shallow water.

Herschel demulsibility number. A number indicating the ability of an oil to separate from water under specified conditions.

heterocyclic. Pertaining to, containing or designating a ring composed of atoms of different kinds.

heteromorphism. Two magmas of identical compositions may crystalize into two different mineral aggregates as a result of different cooling histories.

hexadecane. A paraffin obtained from petroleum. Formula $C_{34}H_{21}$; specific gravity 0.792.

hexahydro-benzine. Hydrocarbons of the naphthalene group. Formula C_9H_{18}; specific gravity 0.7812.

hexane. A light and volatile paraffin hydrocarbon fluid forming a part of gasoline. Specific gravity equals 0.660; boiling point equals 60°C; highly inflammable.

hexylene. Hydrocarbon obtained from petroleum; formula C_6H_{12}.

hiatus. Literally a "gap" in rock sequence. Geological formations normally present, are missing owing either to the fact that they were never deposited or were eroded prior to deposition of the immediate overlying beds. The hiatus of an unconformity refers to the time interval not represented by rock or to rocks known to be missing by comparison with other areas.

HF alkylation process (refin). A catalytic process used to produce alkylate. HF indicates that hydrogen floride is used as the catalyst instead of sulfuric acid.

high angle fault.   Fault plane dip exceeding 45°; usually applied to dip slip or oblique slip faults.

high flash solvent.   A volatile thinner of high-solvency power, obtained by the destructive distillation of coal tar.

high grade.   (1) *n*, rich ore. (2) *vb.*, selectively remove high grade ore to the detriment of the mining procedure. Most frequently applied to gold mines and usually implies unethical if not illegal activity.

high grade metamorphism.   Metamorphism taking place at high temperature and pressure; sometimes called high rank metamorphism.

high line (drill).   A specially rigged rope used to convey pipe, drill tools or other equipment from the derrick to the derrick wall or other locations outside the derrick. The line when pulled tight provides a suspended track on which a carriage travels for conveying the pipe, drilling tools or other equipment.

high pH mud (drill).   A drilling fluid with a pH range above 10.5; a high alkalinity mud.

high resolution profiling (geoph).   A seismic technique for measuring the thickness of unconsolidated sediment deposits on the ocean floor. If the underlying bedrock is soft sedimentary rock, seismic technique will penetrate several tens of meters (up to a few hundred meters in exceptional cases). The technique is called high resolution because it uses a signal frequency of 3.5 kHz (sometimes 7 kHz) in order to detect deposits and beds as thin as one meter. High resolution is obtained at the expense of depth of penetration; in general the longer the wave length of the signal, the deeper the penetration of energy and the deeper below sea floor, the geophysicist can gain information of geologic significance but at the same time the thicker a bed or deposit must be to be detected.

high solvency naphthas.   Special naphthas characterized by their high solvent power or low precipitating tendency for various resins, oils, and plastics used in paint and varnish manufacture. High aromatic content is often conducive to solvency.

high, structural.   A high related to an anticline, dome or horst, or other tectonic uplift of scale similar to the preceding examples. The high may be (1) Topographic high of the earth's surface, (2) Structure contour high on a subsurface structure contour map, or (3) Reflected by an isopach contour low showing thinning of sediment over a topographic high.

high yield drilling clay (drill).   A classification given to a group of commerical drilling-clay preparations having a yield of 35 to 50 bbl/ton and intermediate between betonite and lowyield clays. High yield drilling clays are usually prepared by peptizing low yield calcium montmorillonite clays or, in a few cases, by blending some bentonite with the peptized low yield clay.

Hillman test.   Test made on kerosene to predetermine the stability of the color during storage.

hill creep/hillside creep.   *See* creep.

hitch on (drill).   Connection of the drilling cable to the walking beam by means of the clamps on the temper screw; used during cable drilling operations.

hogback.   Created by differential erosion of a sequence of homoclinally dipping beds where irregular ridges (hogbacks) develop along the strike of resistant beds and valleys erode along the strike of soft beds. Dip of beds may result from faulting or folding.

hog still (refin).   A refinery colloquialism for a simple form of tower still in which only very light products as benzene are separated by use of steam.

hold down (prod).   A clamp used on rodline posts to keep the rod from moving in any direction but back and forth.

hole (drill).   Wellbore. In general terms, an opening, made purposely or accidentally in any solid substance. Mouse hole and rat hole are shallow bores under the derrick in which the kelly joint and joints of pipe are temporarily suspended while making connections. Rat hole also refers to hole of reduced size in the bottom of the regular well bore.

hollow reamer (drill).   A tool for straightening a crooked drilled hole.

Holocene (Recent).   Refers to present geological epoch, dating back to end of Pleistocene about 11,000 years ago when the last great glaciers began to retreat.

holocrystalline.   Applies to igneous rocks consisting entirely of crystals with no glassy constituents; no limitations on the grain size of the rock nor on the presence or absence of crystal faces on the crystalline grains.

homocline.   A group of beds all dipping in the same direction, cause is either indeterminate or of no consequence within the context of use of the term. Application is limited to local occurrences and small regions, rarely exceeding a few square miles.

homogeneous.   Of uniform or similar nature throughout; a substance or fluid having at all points, the same property or composition.

homogeneous reactor.   Uranium fuel is mixed with a liquid moderator, can be light or heavy water or a molten metal, instead of being formed into metallic rods.

hook.   A spit or narrow cape of sand or gravel whose outer end bends sharply landward.

Hook's law.   A description of the behavior of many elastic objects; states the distance an object stretches is proportional to the force exerted on it, provided it is not stretched beyond the elastic limit.

hopper (drill).   Shaped like large funnel through which solid materials may be passed and mixed with a liquid injected through a connection at the bottom of the hopper. It is used for purposes of mixing cement slurry, mixing clay and oil or water to form a drilling fluid, etc.

horizon.   (1) A surface between two beds. (2) A zone of material as a soil horizon or a producing horizon.

horn.   A spire of rock formed by the headward erosion of a ring of cirques around a single high mountain. When the glaciers originating in these cirques finally disappear, they leave a steep, pyramidal mountain outlined by the headwalls of the cirques. The classic example of a horn is the famous Matterhorn of Switzerland.

horn socket (drill).   In well boring, an implement ot recover lost tools, especially broken drill poles, etc; consists of a conical socket, the larger end downward sliding over the broken part.

horst.   A block of earth's crust, generally long compared to its width, that has been uplifted along faults. An area uplifted between parallel faults. Diagram of a

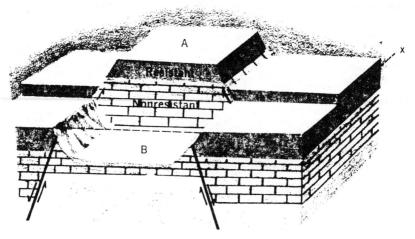

horst. (A)Extensive erosion may remove the more resistant cap rock allowing a valley to develop in the non-resistant strata. (B) Block B is still a horst despite the inversion of relief.

hot mix asphalt.   A mixture of stone, coarse sand and asphalt heated in a hot mix asphalt plant to a temperature of approximately 350°F, from where it is transported to make asphalt pavement. When the asphalt is mixed with somewhat larger aggregate and at a temperature under 250°F, termed a warm mix.

hot spot.   A locality of anomalously high heat flow from the earth's mantle; usually marked by basaltic volcanic activity. Hot spots appear to remain relatively constant in position as the earth's crustal plates migrate over them resulting in chains of volcanoes, only the end most of which is still active. Volcanism on the Hawaiian Islands, hot spring activity in Yellowstone park and some of the Icelandic volcanism are attributed to hot spots. The origin of hot spots is not well understood and subject to much debate.

Houdry process (refin).   Charging stock is heated to cracking temperature in a furnace or pipe still. The vapors pass to a separator where they are separated from any remaining residue and then to a series of three reactor chambers, one is operated as a cracking chamber and the other two contain a solid catalyst. Here catalytic action takes place and the gases are separated from the oil vapors which pass to the fractionator where the gasoline and heavy oil fractions are separated.

humic.   Material is carbonaceous and derived from plants; found in soils or sediments of all kinds.

humus.   Dark colored well-decomposed organic soil material consisting of the residues of plants and animals in soil horizon-a.

Hycracking.   To produce high-quality gasoline chemical naphtha including steam cracking feed and BTX precursors, jet fuel and midbarrel products by catalytic hydrocracking any combination of virgin and cracked gas oils. The development of new catalysts allows heavy feeds up to vacuum gas oils to be completely converted into either naphthas or high yields of midbarrel products in the same unit by merely changing fractionation cutpoint and reactor temperature. The Hycracking process is fixed-bed catalytic hydrocracking using catalyst having high activity maintenance in the presence of nitrogen and sulfur compounds.

Hydrofrac.   A copyrighted name of an operation where producing formations are fractured by hydraulic pressure to increase production.

hydrate.   (1) A substance containing water combined in the molecular form as $CaSO_4 \cdot H_2O$. (2) Action of adding water to a substance.

hydrated grease.   Grease made with a soap containing an alkali having water in chemical combination as hydrated lime, made by adding water to quick lime. Calcium base grease is an example.

hydrated lime. Same as slacked lime, made by adding water to quick lime and used in the manufacture of soap for making hydrated grease.

hydration. (1) Substance taking up water by means of absorption and/or adsorption. (2) Chemical addition of water to the minerals of a rock to form new minerals, chiefly hydrous silicates and hydrous oxides. Carbonation frequently occurs together with hydration, as when feldspar decomposes and converts largely to kaolin.

hydraulic gradient. Rate of charge of head (pressure) in a given direction causing underground water to flow. When the aquifer is enclosed, the hydraulic gradient may be enough to bring water to the surface spontaneously in artesian flow. When artesian conditions do not obtain, the gradient approximately parallels topography and is reflected by the configuration of the water table. Water flows underground away from hill and toward valleys where the water may flow on the surface from springs or stream beds.

hydraulic head. Pressure caused by the weight of a column of liquid considered in terms of its height. Although head refers to distance (height) it is convenient to consider it in terms of the pressure exerted by a body of liquid at rest. Fresh water has a head of 0.433 psi per foot of height (one kilogram/square centimeter/ten meters). The hydraulic heads of other liquids may be determined by comparing their specific gravities with that of water.

hydride. A combination of hydrogen with a radical group of an element.

hydrocarbon. Organic compounds of hydrogen and carbon, whose densities, boiling points, and freezing points increase as their molecular weights increase. Although composed only of carbon and hydrogen, hydrocarbons exist in a great variety of compounds, owing to a strong affinity of the carbon atom for other atoms and for itself. The smallest molecules are gaseous; the largest are solids. Different hydrocarbons mixed together make up crude petroleum.

hydrocarbons, aromatic. Unsaturated ring compounds. Naphthenes found in petroleum have five to six carbon atoms in their rings but the basic structure of the aromatics is always a six member ring. To this ring, various side chains are attached to form a variety of aromatic hydrocarbons. The various aromatic hydrocarbons have different names as benzene, toluene, xylene, naphthalene and anthracene.

hydrocarbon chain. Hydrocarbon series as the paraffin series in which the hydrocarbon atoms are bound to the carbon atoms in a chain form or with the succeeding carbon atoms aligned along a straight line with hydrogen atoms on either side.

hydrocarbons, naphthene. Saturated hydrocarbons similar to paraffins in their properties but the carbon atoms in each molecule are arranged in a closed ring. Therefore naphthenes are also known as cycloparaffins. They occur in naphthas and higher boiling fractions. Naphthenes can be changed into higher octane aromatics by catalytic dehydrogenation.

hydrocarbons, olefinic. Unsaturated hydrocarbons. If two hydrogen atoms can be added to an olefin, it is a mono-olefin; if four can be added, it is a diolefin. Olefins are almost entirely formed by cracking. Diolefins are also present but are so chemically unstable that they react with one another to form a gum therefore must be removed from gasoline or prevented from overacting. The mono-olefins have names corresponding to those of the paraffins, having the endings-ene, or -ylene (ethylene) and propene (propylene). Diolefins also have names corresponding to paraffins but end in -diene as in butadiene.

hydrocarbons, paraffinic. Saturated hydrocarbons containing the maximum number of hydrogen atoms. Chemically inactive under most conditions, but react chemically when broken apart. Their carbon atoms form chains. Those without branches, straight-chain structure are normal paraffins. Those with branches are isoparaffins. Paraffins with four or less carbon atoms are gaseous at ordinary temperatures; those with five to fifteen are liquids; those with over fifteen, are waxes. The lighter gaseous paraffins, methane, ethane are components of natural gas. The heavier gaseous paraffins, propane, butane and isobutone are components of liquified petroleum gas (LPG).

hydrocarbons, ring compound. A hydrocarbon molecule in which the carbon atoms are arranged in ring form to which one hydrogen atom is connected to each carbon atom in radial position; also known as a "cyclic compound."

hydrocarbons, saturated. Hydrocarbon molecules unable to absorb more hydrogen atoms, as the paraffin series.

hydrocarbon series. Hydrocarbons have been divided into various series, differing in chemical properties and relationships. The four that comprise most of the naturally occurring petroleums are the normal paraffin (or alkane) series, the isoparaffin series (or branched-chain paraffins), the naphthene (or cycloparaffin) series, and the aromatic (or benzene) series. Crude oils are referred, according to their relative richness in hydrocarbons of these groups, as paraffinic-base, naphthenic-base, or mixed-base (naphthenic-paraffinic) oils. The aromatics are rarely the dominant group. The naphthenes include the complex residues of the high-boiling range 750°F) petroleum.

hydrocarbon, unsaturated. Hydrocarbon compounds having the ability to absorb additional hydrogen than naturally existing in this compound. The olefins or ethylenes are unsaturated hydrocarbons and form new compounds when additional hydrogen is supplied.

hydrochloric acid.   Acid made from hydrogen and chlorine; also called muriatic acid.

hydrocodimer.   Saturated blending agent of high octane number produced by the hydrogenation of codimer.

hydrocol process (refin).   A catalytic process for making gasoline and other products from natural gas, crude oil, fuel oil and coal.

hydrodesulfurizing (refin).   Process in which the principal purpose of heating the oil with hydrogen is to remove sulfur.

hydrofining (refin).   Treatment of hydrocarbons; those boiling in the gasoline range at a temperature below that at which substantial pyrolysis or decomposition occurs in the presence of catalysts and of substantial quantities of hydrogen, to affect a partial saturation and refining of the hydrocarbon oil and to accomplish essentially the same results as would be obtained by the conventional methods of finishing as treatment with sulfuric acid and caustic and water washing.

hydroformate (refin).   Product obtained in the hydroforming process.

hydroformat, heavy (refin).   Cut from tolulene extraction plant boiling above toluene but still in the gasoline range and containing mainly xylenes and trimethyl-benzenes.

hydroformate, light (refin).   Cut from toluene extraction plant boiling below toluene or toluene concentrate and containing naphthenes and benzene.

hydroformer bottoms (refin).   Residue fractions from hydroforming reaction boiling over 400°F and containing alkylated naphthalenes and heavy aromatics.

hydroforming (refin).   Process of passing naphthas over a solid catalyst at elevated temperatures and moderate pressures in the presence of added hydrogen or hydrogen containing gases to form high-octane motor fuel, high grade aviation gasoline or aromatics.

hydrogen ion concentration.   A measure of the acidity or alkalinity of a solution, normally expressed as pH.

hydrogen sulphide (H_2S). A colorless gas with a characteristic foul odor; very soluble in water, one volume of water dissolving four point three tenths volumes of gas at 0°C and one atmosphere pressure; generally even more soluble in hydrocarbons than in water. Hydrogen sulphide has a decided corrosive effect upon metals, whether as a free gas or in solution in petroleum or reservoir waters.

hydrogenation.   Chemical addition of hydrogen to a material, which may be either nondestructive or destructive. In the former, hydrogen is added to the molecules only if and where unsaturation with respect to hydrogen exists. In the latter, operations are carried out under conditions resulting in ruptures of some of the hydrocarbon chains or cracking, the hydrogen adding on where the chain breaks have occurred.

hydrogenation process (refin). A catalytic process supplying hydrogen atoms to combine with carbon atoms in petroleum products. When providing hydrogen by this process to unite with these carbons, additional members of the gasoline family are obtained. The hydrogenation process, equivalent to cracking is termed hydrocracking.

hydrogenic sediments. Sediments precipitated from solution in water.

hydrogenize. Introduction into a compound or producing a chemical reaction in a compound by the introduction of hydrogen in the presence of a catalyst.

hydrogenolysis. Cleavage of the molecule of hydrogen.

hydrologic cycle. Cycle that water follows in the earth's atmosphere and hydrosphere. Water evaporates from the ocean, is carried by atmospheric circulation to the continents where it falls to the ground as rain and snow eventually to run off and return to the ocean.

hydrology. Science related to the earth's water. In geology, often restricted to the study of rivers, streams, and ground water.

hydrometer. An instrument commonly used to determine the specific baume or API gravity depending upon the scale on the stem of the instrument.

hydrometer, Baume. A hydrometer graduated in "Baume degrees" with division 10 signifying floating in pure water.

hydrometer, specific gravity. A hydrometer indicating the specific gravity or the relation of the weight of the given liquid per unit volume to the weight of a unit volume of water; used for all liquids, oils included.

hydrometer, thermometric. Hydrometer with any scale, provided with a thermometer to take the temperature of the liquid at the same time that the density is measured. This is of importance, as the density of a liquid varies with the temperature.

hydrometer, Twaddel. A hydrometer used for liquids heavier than water and marked with the Twaddel scale, which when multiplied by 0.005 gives the specific gravity.

hydrophile.   A substance, usually an emulsion or in the colloidal state, which attracts water or to which water adheres.

Hydrophilic-Lipophilic Balance (HLB).   One of the most important properties of emulsifiers; an expression of the relative attraction of an emulsifier for water and oil, determined largely by the chemical composition and ionization characteristics of a given emulsifier. The HLB of an emulsifier is not directly related to solubility, but it determines the type of an emulsion that tends to be formed. It is an indication of the behavior characteristics and not an indication of emulsifier efficiency.

hydrophobe (drill).   A substance, usually in the colloidal state, not wetted by water.

hydrophone (seis).   An electroacoustic transducer responding to waterborne sound waves and delivering essentially equivalent electric waves.

hydrosphere.   Irregular spherical zone of the earth lying between the lithosphere and atmosphere and containing the vast majority of the earth's liquid water.

hydrostatic head.   Pressure exerted by a column of fluid, usually expressed in pounds per square inch. Fresh water has a head of 0.433 psi per foot of height (one kilogram, one square centimeter/ten meters). The hydrostatic heads of other liquids may be determined by comparing their specific gravities with that of water.

hydrostatic pressure.   Pressure developed within a porous rock, directly proportional to the depth of the superimposed column of fluid. In many oil fields containing several oil and gas sands it is found that closed-in initial pressures increase with depth and are proportional to the vertical depths below the outcrop of the strata in which the accumulation occur.

hydrostatic reservoir pressure.   Causes water confined in an aquifer, when penetrated by a well, to rise above the top of the water-bearing formation until the static water level is reached. Such water is said to be artesian, and if it flows out over the top, the well is a flowing artesian well.

hydrothermal.   Pertaining to or resulting from the activity of hot aqueous solutions originating from magma or other sources deep in the earth.

hydrothermal solution.   Hot waters originating within the earth carrying mineral substances in solution.

hydrous.   Containing water chemically combined as in hydrates and hydroxides.

hydroxide.   A designation given for basic compounds containing OH^- radical. When these substances are dissolved in water, they increase the pH of the solution.

hydroxyl. Chemical radical OH⁻ composed of one atom of oxygen and one of hydrogen. It bears a negative charge in the ionic state; characterizing part of all bases and of all alkaline reactions and marked by great chemical activity.

hygrometer. An instrument designed to determine the percentage of moisture in a gas.

hypabyssal. General term applied to minor, usually shallow intrusions as sills and dikes, and to rocks of which they are made to distinguish them from volcanic rocks and formations on the one hand and "plutonic" rocks and major intrusions as batholiths on the other.

hyperpycnal flow. Sediment laden waters flowing down the side of a basin and then along the bottom as a turbidity current, with vertical mixing inhibited because the dense aqueous suspension seeks to remain at the lowest possible level.

I

ice.   Solid phase of water; occurs naturally as glaciers and elsewhere on or within the surface of the earth at O°C or below at standard conditions. When naturally occurring, ice is classified as a mineral or monomineralic rock.

ice cap.   Perennial cover of ice and snow over an extensive portion of the earth's land surface, as Antarctic and Greenland.

ice rafting.   Transportation of sediments and rock fragments of all sizes by floating ice, usually applies to sea ice and icebergs.

ichthyol.   An oil obtained by distillation and sulphonation of bituminous shales, afterwards neutralized with ammonia and salts; soluble in water.

ideal gas.   A hypothetical gas obeying exactly the laws of thermodynamics and the gas laws. All gases approach an ideal condition as their interatomic or intermolecular separation increases as pressure decreases.

igneous.   An adjective pertaining to molten earth materials (magma and lava); rocks forming from crystallization of magma or lava and the processes involved in the generation of magma, lava and igneous rocks.

igneous rocks.   Rocks solidified from silicate melt named magma (to knead). Granite and basalt are common igneous rocks.

ignimbrite.   A volcanic rock deposited by a glowing ash-flow eruption (nuée ardente) and composed of ash, crystals, and rock fragments. Deposit may remain hot for long periods allowing the constituent particles to weld together.

imbricate.   (1) A deposit of flat pebbles or cobbles stacked parallel to each other like shingles or a deck of cards. This condition is produced by strong stream or surf-related currents where the direction of slope of the rocks (imbrication) reflect the current direction. (2) A set of fault planes, most often thrust faults closely spaced and parallel to each other.

immiscible.   Liquids not forming a homogenous mixture, as oil and water.

impedance (geoph).   Opposition to current flow related to induction and capacitance generated by the current itself as well as the inherent resistance of the conductor; applied to circuitry of geophysical equipment or to phenomena occurring in rocks investigated using electrical or electromagnetic geophysical techniques.

impermeable.   Not allowing the passage of fluid. A formation may be porous, yet impermeable, owing to the absence of connecting passages between the voids.

impervious.   Impassable; applied to strata as clays, shales, etc. not permitting the penetration of water, petroleum or natural gas; often used interchangeable with impermiable.

impregnated.   Rocks or other bodies with pores more or less filled with extraneous materials as oil or tar.

impression.   Shape, usually concave or somewhat hollow left when a solid object is deposited on the surface of a soft sediment and then removed either physically or by chemical deterioration or solution. Impressions are normally of plants or animals but may be of crystals or other inorganic objects.

impression block (drill).   A block with wax or lead on the bottom run into a well and allowed to rest on a tool or other object lost in the well. In this manner, an idea of the size, shape, and position of the fish is obtained from an examination of the impression left on the wax and appropriate fishing tools may be selected.

imprint.   Used almost interchangeable with impression. Some authors prefer to use imprint for very shallow impressions.

impulse (geoph).   A sharp shock or short period burst of energy emanating from a blast or other mechanical source of energy during seismic prospecting.

incise.   To cut down or cut in deeply. Streams may be incised into plateaus and submarine canyons into continents and shelves.

incised meander.   A stream meander cut deeply into hard rock or resistant sedimentary rock; the meandering course having been acquired at an earlier time.

inclination (dip) (geoph).   In terrestrial magnetism, the angle at which the total magnetic field vector makes with its horizontal component.

inclined plane.   Sloping surface forming an angle with a horizontal surface.

inclinometer (drill).   A well surveying instrument for determining the direction and deviation from the vertical of a crooked hole.

inclusions.   (1) Pieces of older and generally very different rocks embedded in the rock of interest; called xenolith in igneous rocks and xenoblasts in metamorphic rocks. (2) Foreign matter trapped in mineral crystals during their formation; may include solid fragments as well as gas and fluid bubbles.

incompetent bed.   A sedimentary bed of low rigidity tending to flow when subjected to stress rather than flex or fracture; used in a relative sense with regard to contiguous beds. There is no absolute criteria determining whether a bed is competent or incompetent.

incongruent melting.   Melting of a solid to produce a solid and a liquid, neither of which has the same composition as the original solid.

incrustation.   A hard crust or coating of one material on another or on the walls of a cavity inside another material; may show well developed crystals, smooth crystalline masses or be amorphous.

index fossils.   A fossil which is characteristic of a formation or other stratigraphic unit believed to be unique to that unit. Biostratigraphic units may be named for the index fossils they contain; sometimes used as a synonym for guide fossil.

index map.   A relatively small scale map showing the location of sampling sites or of large scale maps related to detailed investigations.

index of refraction.   Ratio obtained by dividing the velocity of light in air by the velocity of light in the transparent medium under investigation. For some applications, the velocity of light in a vacuum is substituted for the velocity of light in air but in geology, the difference between these velocities is not significant.

indicator.   Substances in acid-base titrations in solution, change color or become colorless as the hydrogen ion concentration reaches a definite value; values vary with the indicator. In titrations for chloride, hardness, and other determinations, substances change color at the end of the reaction. Common indicators are phenolphthalein, potassium chromate, etc.

induced radioactivity.   Radioactivity produced in a substance after bombardment with neutrons or other particles.

induction.   Creation of a force field in rock by other than natural processes. A magnetic field may be induced in a body by creating a magnetic field around it. A body may become electrically charged or "electrified" by establishing an electric or electromagnetic field around it.

induction log.   A method of logging wells drilled for petroleum or natural gas. A continuous record of the electrical conductivity of strata traversed by the well and plotted against depth below earth surface.

induration.   Hardening of sediments through cementation, pressure, heat, or other processes.

industrial aromatics.   Products as benzene, toluene, xylene and solvent naptha divided into more than one grade.

industrial solvents. Aromatics or blends of aromatics and straight run solvents made in several grades depending upon their solvent power (Kb value) and rate of evaporation.

inert gas. Group 0 of the periodic table. Gases as helium, neon, argon, krypton, xenon and radon, not reacting readily with any other chemical element.

inertia. Property of a body which keeps it in motion once started until an outside force as friction slows or halts it; also the property that prevents a body from changing direction of its motion without interference by an outside force.

inertial flow. A rapidly moving high density current of water will move through a quiet body of water for some distance by its inertia before the force of friction slows or halts it or before the water in the current mixes with surrounding water thereby decreasing its density and reducing inertia. The increased density of the current is usually owed to a suspended sediment load, as when a sediment ladened river flows into a lake, ocean or sea.

inferior (geol). Lying below, physically or stratigraphically; does not reflect a value judgment, not equivalent to poorer.

infiltration. Percolation of water through pores into soil and rock. Deposition of mineral matter by the permeation of water carrying it in solution.

influent. Flowing into, as a tributary into a river, or river water into the zone of saturation, underground.

inhibited mud (drill). A drilling fluid having an aqueous phase with a chemical composition tending to retard and even prevent, inhibit appreciable hydration, swelling or dispersion of formation clays and shales through chemical and/or physical means.

inhibitor. Essential function of inhibitors is to prevent or retard oxidation. In drilling and producing operations, usually refers to something preventing corrosion.

inhibitor, chemical. An additive which arrests or prevents a chemical reaction by neutralizing or rendering harmless a reactive tendency in a product, as gum inhibitor, retards gum formation in gasoline. Generally, inhibitors are oil soluble compounds classed as phosphorus, sulfur, phosphorus-sulfur, and selenium compounds.

initial boiling point (IBP). Temperature in laboratory distillation when the first drop of distillate falls from the end of the condenser representing the boiling point of the lightest hydrocarbon in the series contained in any gas cut or fraction.

initial vapor pressure.   Vapor pressure of a liquid at a specified temperature and "zero per cent evaporation."

injected gas.   Gas put into an oil producing sand for the purpose of maintaining or restoring rock (formation) pressure, or for storage.

injection gneiss.   A metamorphic rock, a gneiss formed by injection metamorphism. A foliated intercalation, called lit-par-lit, of gneiss and granite; also called a migmatite, mixed rock.

injection, igneous.   Nearly synonymous with intrusion but usually restricted to intrusions with narrow feeding channels as dikes, sills, laccoliths, etc.

injection metamorphism.   Intimately intercalated sheets or layers of metamorphic and igneous rock, products of injection metamorphism. Term originally coined when all such occurrences were thought to originate by the injection of magma into country rock while undergoing metamorphism; now realized that a significant proportion of these occurrences developed by partial melting, in situ. In any case, the process takes place in zones, or regions on the borderline between pressure and temperature conditions of igneous activity and metamorphism producing rocks correctly called "migmatite" or "mixed rock" regardless of whether they were formed by injection or partial melting. *See* injection gneiss.

injection wells.   Wells in which fluid is pumped down the hole pushing reservoir fluids to a producing well.

inlier.   Area of rock outcrop surrounded by outcrop of younger rocks.

innage.   Refers either to the volume of liquid or the measured height of liquid in a tank or container as measured from the bottom of the tank.

inorganic compound.   Compound containing no carbons as clay or glass; exceptions are carbonates, considered inorganic, chemically.

inorganic filler.   Substance secured from a mineral source to enhance the properties of a grease is a mineral filler. Powdered mica, asbestos floats, talc, zinc oxides, powdered copper, are inorganic fillers employed in grease making.

in place.   Refers to a rock, mineral, fossil, exposure or structure not moved from its place of origin.

input gas.   Gas compressed and returned to a wellbore used for gas lift or as injection gas for pressure maintenance.

in situ.   In its natural position or original place.

in situ combustion.   Setting afire of some portion of the reservoir in order that the gases produced by combustion will drive oil ahead of it to the producing wells.

insolation (meteorology).   Sun's energy incident upon the surface of the earth or the direct solar radiation received per unit of horizontal surface.

insoluble.   Incapable of being dissolved in a liquid. A solid may be soluble in one liquid and insoluble in another.

insoluble residue.   Portion of a limestone not dissolved readily in hydrochloric acid. Common constituents are fragments of chert, shale, quartz and other siliceous materials as tests of radiolaria or diatoms. Carbonaceous material is occasionally present. A common technique applied to the study of well cuttings.

inspissation.   Evaporation of the lighter constituents of petroleum leaving the heavier residue behind, as in the formation of asphalt rock.

integrating meter (prod).   A meter that calculates, records and gives instantaneous readout of throughout volumes or rates. In many cases, the integrating, calculating portion will eliminate need for hand calculation or correction.

intensity (seis).   Degree of earthquake induced ground shaking experienced at the site of observation and not necessarily related to the epicenter of the earthquake. Measured on an arbitrary scale as the Modified Mercalli Scale of 1931.

interbedded.   Beds of different composition or lithology alternating closely with each other. Interbeds rarely exceed a few inches in thickness; interstratified.

intercalacted.   Pertaining to material interbedded with another kind of material.

interface.   Surface separating two media, across which there is a discontinuity of some property, as density, velocity, etc. or of some derivative of one of these properties in a direction normal to the interface.

interfacial tension.   Force required to break the surface between immiscible liquids. The lower the interfacial tension between the two phases of an emulsion, the greater the ease of emulsification. When the values approach zero, emulsion formation is spontaneous.

interfingering.   Laterally contiguous rock units with extensions of each projecting into the other so as to make a zigzag boundary between them.

interformational conglomerate.   Consolidated pebble to cobble bed within a formation whose constituents have a source outside the formation. Distinctly different from, but often confused with, intraformational conglomerate.

intermediate casing string.   String of casing set in a well after the surface casing. Its purpose is to keep the hole from caving, sometimes to afford a strong string of pipe for attachment of blowout preventers; also called protection casing.

intermediate fraction (refin).   One of the middle or intermediate products recovered from the crude during distillation; also known as intermediate cuts.

intermediate rock.   Rock, intermediate between silica rich, granite and silica poor, gabbro; containing between 52 per cent and 66 per cent silica.

intermediate gas-lift.   Pressurized gas, injected at intervals, into tubing down hole at a depth below top of fluid in tubing to lift slugs of fluid to surface.

intermittent stream.   A stream not flowing continuously, only during times of rainfall and high runoff. Stream course may be dry for periods of months or years.

intermontane.   Refers to structures, topographic features, glaciers or other geologic features lying or occurring between mountains.

internal fluid friction.   Friction caused by the property of cohesion within the oil itself, commonly referred to as fluid friction. A drop of heavy oil stops a watch largely because the power developed by the watch is not sufficient to overcome the fluid friction of the heavy oil.

internal preventer (drill).   An inside blowout preventer consisting of a check valve in the drill string circulating down the hole but preventing backflow.

internal-upset (drill).   An extra thick wall provided on the end of a drill pipe at the point where the pipe is threaded in order to compensate for the metal removed in the threading. Conventional drill pipe has the extra thickness on the outside. Internal upset drill pipe has the extra thickness on the inside with uniform straight wall on the outside. It is usually referred to as "internal-upset, external flush" pipe.

internal waves.   Waves of energy traveling through a fluid which exhibits a variation in density with depth as the ocean; observed most clearly at density boundaries as oceanic thermocline.

interstadial.   Interglacial; may refer to time or deposits.

interpreter (geoph).   A person who extracts information of geological significance from geophysical data.

interstitial.   Occurring or located in the openings or interstices between the grains of a rock or sediment.

interstitial water. Water contained in the pore spaces between grains in rock and sediment.

interval. (1) Distance measured perpendicular to the bedding, between the corresponding parts of two strata in a sedimentary formation; also called the stratigraphic interval. (2) Vertical distance between two adjacent topographic contour lines; also true for other varieties of contour maps.

intraformational. Formed or occurred within a formation; typical of materials or processes in the interior of a formation.

intrusion. (1) Forcing of masses of molten rock into or between other rocks. Diagram of three types of forceful intrusion; the laccolith and dome are concordant, and the diapir is largely discordant. (2) Invasion of a fresh water aquifer by salt water when too much ground water is pumped out along coastal areas or during times of drought; called salt water intrusion.

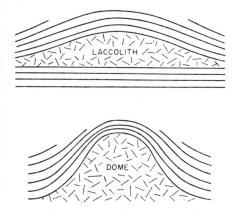
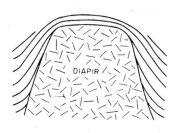

intrusive rocks. Igneous rocks which while molten penetrated into and between other rocks but solidified before reaching the surface of the earth or the sea floor-water interface. They may either have penetrated fissures in the original rocks or absorbed and replaced the original rocks. They are almost universally phaneritic (grain size large enough to be observed with the unaided eye).

inversion. Folding back of strata upon itself as when overturning a fold so succession appears to reverse.

invert oil-emulsion mud (drilling). A water-in-oil emulsion where fresh or salt water is the dispersed phase and diesel, crude, or some other oil is the continuous phase. Water increases the viscosity and oil reduces the viscosity.

invertebrate paleontology.   Study of fossil animals without backbone or spinal column; include fossil protozoans (tiny one-celled animals), snails, clams, starfish, crabs, sponges and worms.

iodine.   A nonmetallic element of the halogen group; its symbol is I. Its atomic number is 53 and its atomic weight is 126.91.

iodine monochloride.   Solution used for determining the iodine value of an oil.

iodine number.   Amount of iodine absorbed by oils, fats, and waxes; giving a measure of the unsaturated linkage present.

ion.   An atom with an unbalanced number of electrons and protons taking on an electrostatic charge. Acids, bases and salts (electrolytes) when dissolved in solvents, especially water, are usually more or less dissociated into electrically charged ions or charged molecule fragments due to loss or gain of one or more electrons. Loss of electrons result in positive charge producing a cation. A gain of electrons results in the formation of an anion with negative charge. The valence of an ion is equal to the number of electrons gained or lost.

ion exchange.   A chemical process involving the reversible interchange of ions between a solution and a particular solid material as an ion exchange resin.

ionic bond.   Chemical bond forming when ions of opposite electrical sign are attracted to each other by electrostatic forces.

ion pair.   A positive ion, cation, and a negative ion, anion; having charges of the same magnitude formed from a neutral atom or molecule by the action of radiation. An ion may be a simple electron.

ionization.   Any process by which a neutral atom or molecule loses or gains electrons, thereby acquiring a net charge; process of producing ions or electrostatically charged molecules.

ionization chamber.   Radiation measuring device dependent on the measurement of ionization created by the passage of radiation.

ionization gauge.   A vacuum gauge with a means of ionizing the gas molecules and a means of correlating the number and type of ions produced with the pressure of the gas.

ionization of gases.   Separation of gas atoms into charged particles, usually as the result of forcible collision with other particles.

ionizing particle.   A particle directly producing ion pairs in its passage through a substance. A charged particle having considerably greater kinetic energy than the ionizing energy appropriate to the medium.

isanomal contour (geoph).   A line connecting points of equal variations from a normal value.

island arc.   An arc-shaped chain of islands lying near a continental mass and bounded by a deep trench on the ocean side.

iso-.   A prefix designating similarity; many organic bodies although being composed of exactly the same number of the same atoms, appear in two or three, or more varieties, or isomers, which differ widely in physical and chemical properties; turpentine and tartaric are examples.

isobutane.   Colorless gas, with characteristic natural-gas odor. A member of the gas family at normal temperatures having four carbon and ten hydrogen atoms. One of the gases present in LPG; also used to make alkylate (100 octane gasoline) by a catalytic process combining it with butane.

isochron (geoph).   A contour line on seismic geophysical maps connecting points of equal arrival times or equal elapsed time between the seismic "shot" and arrivals from a reflecting horizon. Allows the construction of structure contour-like maps when vertical distances are unknown because seismic velocities are unknown or uncertain.

isoclinal.   Dipping equally in some direction, as limbs of isoclinal fold; a fold with parallel limbs.

isoclinal fold.   A fold in sedimentary rocks whose limbs have parallel dips; it may be an anticline or a syncline, or may be either vertical, overturned or forced over into an oblique position or recumbent (lying on its side). Diagram of some isoclinal folds in which the limbs of alternate anticlines and synclines are compressed into very steeply dipping or overturned strata.

isocline.   A series of strata dipping parallel to each other; a fold (anticline or syncline) so closely folded that the rock beds of the two sides or limbs have the same dip.

isoclinic line.   A line drawn through all points on the earth's surface having the same magnetic inclination. The particular isoclinic line drawn through points of zero inclination is given the special name of aclinic line.

isocracking (refin).   To convert a wide range of hydrocarbon feedstocks to lighter, cleaner and more valuable products; a wide flexibility to make different product slates that can emphasize high octane gasoline blendstocks, jet fuel, low pour point, low-sulfur diesel. LPG or low-sulfur fuel oil blendstocks. Employs a fixed bed catalyst system in an environment of recycle hydrogen under elevated pressure.

isoformate process (refin).   Olefinic naptha contacted with an alumina catalyst at high temperatures and low pressures to produce olefin isomers of higher octane number than those in the feed.

isogal.   A contour line of equal gravity values on the surface of the earth.

isogonic lines.   Imaginary lines joining places on the earth's surface at which the variation of the magnetic needle from the meridian or true north is the same.

isomer.   One of two or more compounds having identical elemental compositions but differing from each other in the arrangement of their constituent atoms or structure and perhaps in the nature of the interatomic bonds. The differences produce changes in physical properties as melting points, from one isomer to another; see polymorph.

isomerization process (refin).   A process changing one isomer into another by rearranging the atoms in a molecule. A product containing isomers is an isomate. At the refinery, this process can be employed to change normal butane, a member found in abundance, into isobutane, one not so plentiful, and which can be used to make alkylate and other effective blending agents; also for producing members which can be used to make synthetic rubber and many other synthetic products.

isomorphous.   Minerals exhibiting the same crystal form and atomic structure but of different elemental composition are isomorphous. Differences in chemical composition are limited to exchange or substitution of atoms of nearly identical size and identical or similar valence. When two minerals are isomorphous and grade in composition from one to another, they are said to be a solid solution series.

isooctane.   A liquid at normal temperature and a member of the gasoline family having 8 carbon and 18 hydrogen atoms in a spiderweb shape. Although found in crude oil, the principal source is from synthetic processes as the alkylation process.

isopach.   A contour line on a chart drawn through points of equal thickness of a sedimentary layer.

isopach maps.   An isopach map consists of lines drawn through points of equal interval between two stratigraphic horizons or points of equal thickness of some units. Isopach maps show the variations in the thickness of a reservoir producing oil and gas useful in planning the development of oil and gas fields and for estimating their reserves; also very valuable for changing the key horizons on which structure contours are drawn.

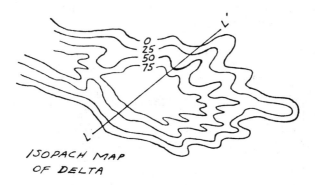

ISOPACH MAP
OF DELTA

isoparaffin.   Paraffin hydrocarbon of branched chain structure.

isopentane.   Colorless liquid obtained by fractional distillation from petroleum. A member of the gasoline family having five carbon and twelve hydrogen atoms.

isoprene.   A hydrocarbon C_5H_8 having the same chemical composition as natural rubber; produced synthetically from methane in the presence of a catalyst. It is a compound of considerable importance for the synthesis of rubber.

isoseismal.   Line connecting points of equal earthquake intensity; concentric to the epicenter.

isostasy.   A theoretical condition of density balance in the earth's crust as if it were floating in a homogenous fluid of higher density than any portion of the crust. As a result, segments of the crust stand topographically high because they are of lower density than surrounding low lying areas or because they have an extensive mass or root protruding downward into the fluid like an iceberg. When mass is eroded from high standing crustal masses as in mountains, the crustal mass tends to float higher in the fluid. As a result the loss of elevation of the mountain is less than the decrease of the mountain by erosion. When the crustal mass floats higher, high density fluid must move laterally to compensate for the upward movement. These movements, both of the crust and the high density fluid are called isostatic adjustment or isostatic compensation.

isostatic adjustment.   *See* isostasy.

isostatic compensation.   Departure from normal density of material in the lower part of a column of the earth's crust which balances or compensates landmasses topography above sea level and deficiency of mass in ocean waters and produces the condition of approximate equilibrium of the earth's crust known as isostasy.

isostatic correction (geoph).   Adjustment made to values of observed gravity at a station to account for the gravitational effects of adjacent mountains and their roots or adjacent low plains and the high density mantle more closely approaching the surface under such plains.

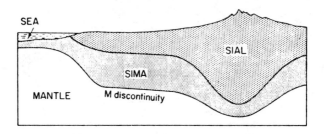

isothermal.   When a gas passes through a series of pressure and volume variations without change in temperature, the changes are isothermal. A line on a pressure-volume diagram representing these changes is an isothermal line. Equal or constant temperature with respect to either space or time.

isothermal compression.   A method of compressing a gas by which the temperature does not increase or decrease but remains constant during the compression cycle.

isotope.   Group of atoms, the nuclei containing the same number of protons but a different number of neutrons. The atomic numbers of isotopes are identical, but the mass numbers differ.

J

jack and circles (drill). Apparatus consisting of a powerful jack and a steel-toothed circle on which the jack moves; used to tighten the joints in a string of tools.

jackboard (drill). A device used to support the end of a length of pipe while another is being screwed on.

jack latch (drill). A fishing tool.

jack lines. Pull rod lines running from a central point or unit to a pumping jack.

jack pump (prod). A buildup pump used for pumping oil from gathering stations to a central dispatch tank.

jack-up drilling and production platform. A typical off-shore drilling and production platform having quarters for 50 men, a 90 foot by 90 foot heliport, cementing units, light drilling units, crane, runway and anchor winches. Diagram of the tilt-up/ jack-up drilling and production platform in position appears on p. 184.

Jacob's staff. A staff or rod of convenient and known height used in compass surveying for the same purpose as a tripod in transit surveying.

jamin action. A theoretical conception of the behavior of occluded gas in an oil sand reservoir. When the small gas bubbles are propelled through the sand because of the prevailing drop in pressure towards the well, they are distorted.

jars (drill). Two links, resembling chain links, hooked loosely together, fastened to the stem above the bit in a cable tool drill for the purpose of giving a sharp jerk to the bit.

jarring (drill). Using jars for releasing or tightening tools in the hole by hammerlike action of the jars.

jerk line (drill). A line connecting the band wheel crank to the drilling cable and which does the jerking to operate the spudding tool.

jerker line (prod). A line radiating from a common point of power to the jack of several wells, therefore several wells are pumped by one power unit.

jet bit (drill). A modified form of either a drag bit or roller bit utilizing the principle of the hydraulic jet to increase the drilling rate.

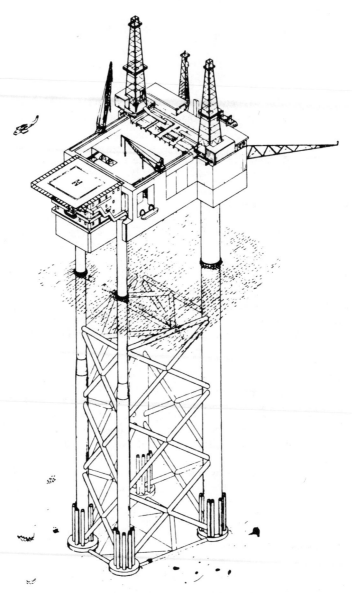

jack-up drilling and production platform

jet perforating.   An operation similar to gun perforating except that a shaped charge of high explosives is used to burn a hole through the casing instead of using the gun to fire a projectile.

jet pump.   A pump moving fluid, bringing it in contact with a rapidly moving stream of fluid of the same or different kind of motion imparted through friction. Injectors and aspirators are such pumps.

jetting out (drill).   An operation using the jet to clean out the cellar slush pit.

joint.   (1) A fracture in a rock, along which no appreciable movement has occurred. (2) A single piece of pipe or tubing usually about 22 feet in length.

joint system.   Two or more sets of joints commonly present in a rock mass; each set trending in a different direction.

jointing, columnar (igneous rocks).   Caused by the shrinking of magma as it cools and crystallizes, pulling in many directions, in many places, and through long periods of time. Under these strains, the rocks start to crack, and the cracks grow larger until they bring relief. As a result, the lava is crossed by many fractures or joints dividing it into columns and slabs. Joints differ with nature of rocks.

Joule's law.   No change of temperature when a gas expands without doing external work and without receiving or rejecting heat.

Joule-Thomson effect.   Cooling occurring when a highly compressed gas is allowed to expand so no external work is done. The cooling is inversely proportional to the square of the absolute temperature.

Jurassic Period.   Closely resembling the preceding Triassic, eastern parts of North America were still generally dry and elevated. Erosion of the ancesteral Appalachians was well on the way toward the formation of a peneplane. The Jurassic is perhaps best known for its unusual reptiles, dinosaurs were abundant and dominant. Ending in the Nevadian disturbance, when the Sierra Nevada and Coast Ranges on the west coast were formed. Associated igneous activity formed gold ores whose erosion led to the placer deposits in the streams of California.

juvenile water.   Water entering for the first time into the hydrologic cycle; released from the earth's interior through volcanic activity at a rate probably not exceeding one-tenth cubic kilometer per year.

K

K-Ar or K-A (geochronology). A pair of isotopes related to each other by radioactive decay. Potassium 40 (K) is the parent isotope while Argon 40 (Ar) is the daughter product. The pair serves as a popular means of determining absolute ages of the rocks containing it.

kame. Small, often cone-shaped hills of stratified sand and gravel or glacial drift; formed when streams fed by glacial melt water deposit their sediments in alluvial fan-like heaps at the icefront margins of the glaciers or at the bottom of circular "wells" or depressions in the glacier itself. Kames may occur superimposed on or cutting through terminal moraines, or on the level plains or gently hummocky topography on the proximal side of terminal moraines. When on the level plains, they are surrounded by outwash in the midst of hummocky topography they are resting on and/or surrounded by ground moraine.

kame terrace. A ridge of stratified glacial drift deposited by melt water between glacial ice and the glacial valley wall. Diagram of a condition leading to the devel-

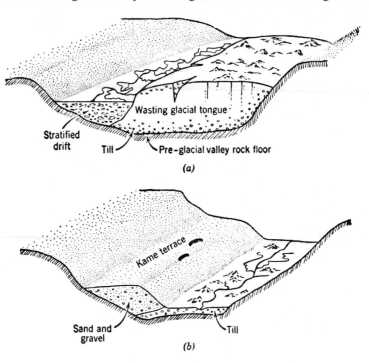

186

opment of a kame terrace. (a) The wasting ice tongue which formerly occupied the valley furnishes outwash to the zone between the valley wall and the ice margin. (b) After the melting of the ice, the belt of gravel is left as a terrace on the valley floor and locally rests upon till.

kaolinite.   A common clay mineral, hydrous aluminum silicate, formed by the replacement or alteration of feldspars or other minerals. Kaolin is a rock composed principally of kaolinite.

karst topography.   An irregular topography developing in most moist and humid climates by the action of surface and underground water on soluble rock, mainly limestone; sinkholes and caverns are characteristic. Rain water passes through sinks and circulates underground; consequently little or no integrated surface drainage develops. Limestone areas have many sinks separated by abrupt ridges or irregular hills; generally honeycombed below the surface, by tunnels and caves resulting from solution by ground water. Solution continues to enlarge underground openings below the watertable, which in turn leads to further collapse of the karst surface.

kauri-butanol value (solvent).   A measure of the aromatic properties of a solvent referred to as the K.B. value; number of cubic centimeters of test sample added to a definite quantity of a standard solution of kauri gum in normal butyl alcohol.

kelly (drill).   Heavy square of hexagonal steel member suspended from the swivel through the rotary table and connected to the drill pipe turning the drill string; has a bored passageway permitting fluid to be circulated from the swivel into the drill stem and up the annulus, or vice versa. The kelly transmits torque from the rotary to the drill string and permits free vertical movement for making hole.

kelly bushing (drill).   Device fitted to the rotary table through which the kelly passes, and by means of which the torque of the rotary table is transmitted to the kelly and to the drill.

Kelvin scale (absolute scale).   Named for Lord Kelvin, a British physicist. Temperature scale with the zero point at absolute zero; the freezing point of water, $273.16°K$; the boiling point of water, $373°K$.

kerogen.   A mineraloid of indefinite composition, consisting of a complex of macerated organic debris and forming the hydrocarbon content of kerogen "oil" shales. A stage in the formation of petroleum and natural gas requiring heat to complete the process.

kerosene.   A refined petroleum distillate; a fraction or cut between gasoline and gas oil with the temperature of distillation ranging between 105°F and 300°F depending upon specifications.

kerosene (color).   Color may be an indication of the amount of refinement and of the substance causing a smokey flame and wick deposits. Kerosene begins with water white, sometimes called super-fine. The next is slightly off-white, and called prime white, followed with the standard white which is the darkest shade, frequently called export white because it conforms generally with the color or exported kerosene.

kerosolene.   A mixture of hydrocarbons from coal or alberite oil, equivalent to petroleum ether, possessing a boiling point of 90°F.

kettle.   A block of stagnant ice becomes isolated from a receding glacier during wastage and partially or completely buried in drift before finally melting. When it disappears it leaves a kettle, pit, or depression in the drift. Larger varieties of kettles, tens of feet deep and hundreds of feet in diameter are called kettle holes by some authors but there is little consistent differentiation between kettle and kettle hole.

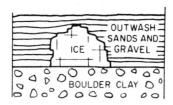
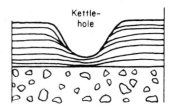

kettle topography.   As glacier ice wastes away, a topography of irregular knobs and hollows remains; also known as dead ice topography.

key (prod).   A hook-shaped wrench fitted to the square of a sucker-rod, used in pulling and running sucker rods of a pumping oil well.

keybed.   A well-defined and easily recognized bed facilitating correlation in geologic work; also applies to the horizon or bed on which elevations are taken or to which elevations are finally reduced in making a structure contour map.

kill (drill).   In drilling, to prevent the threatened blowout of a well by suitable preventive measures. In production, to stop a well from producing oil and gas so that reconditioning of the well may proceed. In both cases, usually accomplished by injection of dense fluid into the well.

killing a well (drill).   Bringing under control a well threatening to blow out; also applied to the procedure of circulating water and mud into a completed well before starting well-service operations.

kilometer.   Unit of distance in the metric system equal to 0.62 statute mile or 0.54

nautical mile. A statute mile equals 1.61 kilometers; a nautical mile equals 1.85 kilometers.

kinematic to Saybolt fural viscosity.   A formulated table for converting viscosity in centistokes to Saybolt fural viscosity as outlined under ASTM Designation.

kinematic to Saybolt universal viscosity.   A formula and table for converting kinematic viscosity in centistokes to Saybolt universal viscosity as outlined in ASTM Designation.

kinematic viscosity.   Ratio of the viscosity, centipoise in grams per millisecond, to the density, grams per cubic centimeter, using consistent units.

kinetic theory of gases.   Assumes that particles of a gas move in straight lines with high average velocity continually encountering one another and, therefore, changing their individual velocities and directions; pressure of the gas is due to the impact of the particles against the walls of the containing vessel.

knife-edge ridge.   When two cirques are formed on opposite sides of a peak, the divide between them may become extremely narrow and sharp; also called arete.

knockout.   A kind of tank or filter used to separate oil and water.

knockout coil (refin).   A water-cooled coil in the vapor line of stills for the distillation of the impure gasoline produced by the absorption process; designed to remove or "knock out" completely the heavy absorbent oil fraction and the water from the gasoline distillate.

knockout drum.   A drum of vessel constructed with baffles through which a mixture of gas and liquid is passed to disengage one from the other. As a mixture comes in contact with the baffles; the impact frees the gases and allows them to pass overhead, the heavier substance falls to the bottom of the drum.

knoll.   An elevation less than 500 fathoms (1,000 meters) from the sea floor and of limited extent across the summit.

Knox process (refin).   A vapor phase cracking process falling under the classification of a true vapor phase process.

knuckle post (drill).   A sand reel support in a derrick; used during cable drilling operations.

kogasin.   A German word for the oily fractions consisting of a mixture of straight and slightly branched-chain saturated and unsaturated hydrocarbons produced by the hydrogenation of carbon monoxide over iron, nickel or cobolt catalysts at low pressure. The part of the oil boiling in the gasoline range, 30 to 200°C, is kogasin I, and the high-boiling residue, over 200°C, is kogasin II.

L

laccolith. A concordant igneous intrusion, resembling a sill with a flat floor but an upward convex roof; mushroom-like in cross-section.

lacustrine. Related to the lake environment, inhabitants, processes and materials.

lacustrine plains. Lake plains formed by the emergence of a lake floor by either uplift and drainage or evaporation and sediment fill; also called lake-bottom plain.

ladder vein. Series of parallel ore-filled cracks in a dike; if the dike is vertical, the cracks are horizontal and resemble ladder rungs.

lag gravel. Coarse material remaining behind after the fines have been blown away. Some use the term synonymously with desert pavement, while others restrict lag gravels to accumulations of small extent developing from parent deposits of limited size.

lagoon. A body of water along a continental or island coast, having restricted or no connection with the sea; separated from the sea by coral reefs, longshore bars or bay barriers.

lamina. Thinnest discernible layer (unit layer) in a sedimentary sequence; less than one centimeter thick, separated from the material above and below, along readily discernible boundaries established by difference in grain size, mineralogy, and color.

laminar flow. Fluid elements flowing along fixed streamlines parallel to each other and to the walls of the channel of flow; moves with a differential velocity across the front varying from zero at the wall to a maximum toward the center of flow; the first state of flow in a Newtonian fluid; it is the second stage in a Bingham plastic fluid.

laminated. Thin layers of rock one upon the other.

lamination. Within sedimentary beds, unit layers less than one centimeter in thickness are laminae; a deposit of laminae. The laminae may be parallel to the bedding planes or at angle to them. In the latter case, the sediment is said to be cross-laminated or show cross-lamination.

lampblack. Product obtained directly from natural gas by burning it under plates or rolls with insufficient oxygen.

190

lamprophyre. Group of dike rocks of dark color and sugary texture, or sugary groundmasses. Most of them are more easily altered than basalt dikes and weather to rusty ferruginous outcrops.

land bridge. Dry land connection between two continents allowing migration of fauna and flora. The existence of some land bridges in the geologic past can be substantiated by physical evidence, but others are postulated on the basis of plant and animal distributions on now separated continents.

landslides. Varieties of mass wasting where the force of gravity exceeds the force of cohesion of a sloping earth surface resulting in failure or movement readily discernible and occurring at velocities of several to tens of miles per hour. Five examples of mass wasting commonly grouped under the term landslide: (1) A slump is the downward and outward movement of an unbroken block of earth. The mass rotates backward so that the slope of the upper surface of the block is diminished or reversed. (2) A debris-slide is a rapid downslope movement of a broken mass of unconsolidated material; may be of any size, expose bedrock in its course, and often produces landforms resembling morainal topography. (3) A debris-fall occurs where unconsolidated material drops from a vertical cliff. (4) Rockslides are a rapid downslope movement of newly detached bedrock occur-

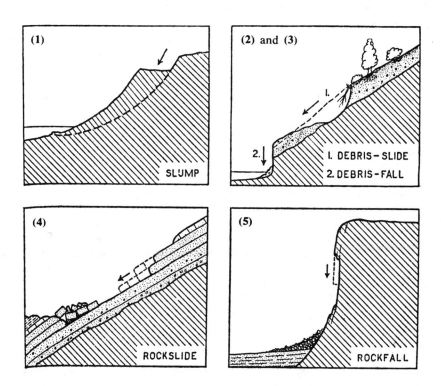

ring along bedding, joint, foliation surfaces, or any other planes or weakness. (5) Rockfalls are free falling portions of newly detached bedrock and are common in areas oversteepened by rivers, valley glaciers, or wave cutting.

lapilli (volcanic).   Volcanic ejecta or cinders ranging from four to 32 millimeters in diameter. Solid or porous fragments larger than ash, reaching the size of English walnuts. They are either angular or rounded by wear against each other during their violent ascent.

large calorie.   Amount of heat energy needed to raise the temperature of a kilogram of water, one degree centigrade.

latch jack (drill).   Fishing tool designed to catch the bail of a bailer.

latent heat.   Unit quantity of heat required for an isothermal change in the state of a unit mass of matter. The thermal equivalent of the energy expanded in melting a unit mass of a solid or vaporizing a unit mass of a liquid, or conversely the thermal equivalent of energy set free in the process of solidification or liquefaction.

latent heat of fusion.   Heat needed by a unit mass of a solid to convert it directly into a liquid without a change in temperature.

latent heat of vaporization.   Quantity of heat necessary to change one gram of liquid to vapor with no change in temperature and measured in calories per gram. When one pound of water is vaporized, the result is measured in Btus per pound.

lateral moraine.   Moraine deposited along the edge of a valley glacier between ice and valley wall, composed of angular rock fragments either fallen onto the glacier from the valley wall or plucked from the valley wall, of glacier; may resemble a kame terrace but unstratified.

laterite.   A red to red-brown soil which has undergone extreme weathering under humid tropical or subtropical conditions. Principle constituents are iron and aluminum oxides and hydroxides sometimes in sufficient concentrations to serve as ores.

lattice.   (1) Structure of the central part of a nuclear reactor, usually consisting of uranium fuel rods held in a moderator as graphite. (2) A hypothetical network of lines with the same symmetry elements as a crystal and with the line intersections representing the locations of the crystal's constituent atoms.

lava.   General name for molten rock poured out on the surface of the earth by volcanoes and for the same material cooled and solidified as solid rock.

lava cone.   Broad, gently sloping volcanic cone composed chiefly of solidified basaltic lava very fluid at the time of eruption; intermediate in size between a spatter cone and a shield volcano.

law of superposition.   General law stating that if undisturbed, any sequence of sedimentary rocks will have the oldest beds at the base and the youngest at the top.

layer.   A bed or stratum of rock.

layer depth effect (seis).   Weakening of the sound beam or seismic pulse owing to abnormal spreading as it passes from a positive gradient layer to an underlying negative layer.

leaching.   Process where the more soluble mineral compounds are removed in solution by percolating ground waters.

lead acetate, sugar of lead.   A poisonous white crystalline salt of lead used in disulphurizing.

lead base grease.   A mixture soap and mineral oil. This soap is generally prepared by the reaction of lead oxide as the alkali and fatty acid as the fat; has some extreme pressure properties and among the first to be used in the lubrication of gears.

lead naphthenate.   Lead soaps of naphthenic acids occurring naturally in crude oil; soluble in mineral oils and imparting film strength.

lead oleate.   A soap made by heating oleic acid with lead oxide.

lead tong (drill).   Pipe tong suspended in the derrick and operated by means of a wire line connected to the automatic cathead; sometimes called breakout tong.

lead-uranium ratio.   Proportion of lead and uranium in rock, mineral, or amounts of isotopes of these elements; used in radioactive age determinations.

lean gas (refin).   Gas from the absorber when making natural gasoline by the absorption process. Consists of the remains of wet gas after the condensable gasoline has been removed by the absorption oil.

lean oil (refin).   Absorption oil returned to the absorber after being separated from the gasoline vapors when making natural gasoline by the absorption process.

lease.   A legal document executed between the land owner or lessor and another party as lessee, granting the right to exploit the premises for minerals or other valuable substances; also applies to a tract of land undergoing mineral development on which a lease has been obtained.

leaseblock method.   Lease on oil land for which the royalty is arranged on a sliding scale basis, a fixed rate of royalty being paid on all oil produced to a certain output, and a lower rate for all production in excess of this amount.

lease condensate.   A natural gas liquid recovered from gas well gas (associated and non-associated) in lease separators of field facilities.

lease period method.   Method in which the royalty rate is based on the production of the well per unit of time, usually in barrels per day and the rate changed when the well produces less than the stipulated amount.

lease separator.   A facility located at the surface for the purpose of (1) separating casing head gas from produced crude oil and water at the temperature and pressure conditions of the separator, and (2) separating gas from that portion of associated and non-associated gas liquefying at temperature and pressure conditions of the separator.

left-lateral fault.   A strike-slip fault in which the ground opposite the observer appears to have moved to the left.

lens.   An ore body more or less eliptical in outline, thickest at the center and thinning out toward the edges; lenticular in shape.

lensing.   A thinning of the bed in several directions simultaneously causing a variation in the stratigraphic interval.

lens, sand.   A body of sand, thick in the center, thinning to the edges. A significant but subordinate part of a formation or other stratigraphic unit; may serve as a stratigraphic trap for petroleum or natural gas.

Leptometer.   An especially constructed viscosimeter invented by Lepenau; with two metallic cylinders placed side by side in a water bath and terminates below in stop cocks. Three sets of interchangeable jets are provided for attachment to the stopcocks. Standard oil is placed in one cylinder and oil to be tested in the other.

levee.   An elevated bank, natural or artificial, confining a stream or river and limiting its opportunities to innundate the flood plain. Natural levees develop gradually after repeated flooding. Velocity of sediment ladened water decreases markedly once it leaves the main channel in flood stages. The coarsest sediment is deposited immediately along the river course. In time, these dike-like deposits or levees of coarser sediment build up enough to be effective in flood control.

level, additive.   Refers to the total percentage of all additives in an oil.

leveling.   A technique for gathering data for the construction of topographic, structure contour and other maps; utilizes a surveying instrument, a level allowing the surveyor to establish the location of points of equal elevation.

lift (prod). Difference in elevation between the surface of the liquid being pumped to the elevation at which it is discharged as in gas lift pumping.

ligarine. A saturated naphtha boiling in the range of 68°F to 275°F (20°C to 135°C); used as an anesthetic and general laboratory use.

ligate tar. Destructive distillation of lignite produces a soft tar used directly in a diesel engine or redistilled to obtain products somewhat similar to the distillate of coal tar. This tar usually contains large quantities of paraffin, and unsaturated hydrocarbons and, upon exposure, absorbs oxygen from the air.

light ends. Lower boiling components of a mixture of hydrocarbons.

light fractions (refin). Among the first liquid products recovered from the crude during distillation.

light hydrocarbon distillation (refin). Gaseous and more volatile liquid hydrocarbons produced in a refinery collectively known as the light hydrocarbons or light ends. Light ends are produced in relatively small quantities from crude petroleum and in large quantities when gasoline components are manufactured by cracking and reforming. When a naphtha or gasoline component at the time of its manufacture is passed through a condenser most of the light ends do not condense and are withdrawn and handled as a gas. A considerable part of the light ends, however, remain dissolved in the condensate thus forming a liquid with a high vapor pressure.

light minerals. Detrital minerals in rock or sediment having a specific gravity of less than 2.8 and usually are light in color as quartz, feldspar, calcite.

light oils. Term generally used for all products distilled or processed from the crude up to, but not including, the first lube oil distillate. The products beginning with and following the first oil distillate are classed as heavy oils.

lignite oils. Heavy lignite oils pass a specific gravity of 0.89 to 0.97, a faint odor of creosote and usually a low viscosity. They dissolve to the extent of 20 to 60 per cent when shaken with two volumes of cold ethyl alcohol.

lignosulfonates (drill). Organic drilling-fluid additives derived from by-products of sulfite paper manufacturing process from coniferous woods. Some of the common salts, as ferrochrome, chrome, calcium and sodium, are used as universal dispersants while others are used selectively in calcium treated systems. In large quantities, ferrochrome and chrome salts are used for fluid-loss control and shale inhibition.

ligroine. Name given to a special distillate obtained after benzene, similar to kerosene, but with a slightly higher boiling point. A saturated petroleum naphtha boiling in the range of 68°F to 275°F (20°C to 135°C).

limb.   One of the two parts of an anticline, on either side of the axis.

lime.   *See* lime, quicklime.

lime base grease.   Grease made from lime soap; does not have high melting points, but is outstanding in its resistance to water and is not washed away or dissolved by it.

lime, hydrated.   A fine powder obtained by "slacking" lime by the addition of water.

lime, quicklime.   A white or gray substance obtained by calcining or roasting a natural rock known as limestone. It consists mainly of calcium oxide with which magnesium oxide is also generally found. Develops great heat when treated with water, with which it reacts to form slaked lime, $Ca(OH)_2$. It has strong caustic and alkaline properties and may be used to neutralize acids in petroleum refineries as in the case where the acidity of a distillate is to be eliminated after the acid treatment.

lime-treated muds (drill).   Commonly referred to as "lime-base" muds. These high pH systems contain most of the conventional fresh water additives to which slaked lime has been added to impart special properties. The alkalinities and lime contents vary from low to high.

lime treatment (refin).   Process of introducing lime into the still during distillation in order to reduce the acidity of the distillate. The alkaline characteristics of lime tend to neutralize sulfur acids, forming salts not vaporized and broken down by heat to form substances of non-acidic properties.

limestone.   A general term for a class of rocks containing at least 80 percent of the carbonates of calcium or magnesium. Varieties of limestones take their names from the source material, as algal limestone, reef limestone, coquina, crinoidal limestone, etc.

limonite boxwork.   Residual iron oxide or limonite derived from the weathering of an iron-rich sulfide and displaying a characteristic meshwork.

lineament.   A significant alignment of topographic features, ridges, lakes, streams, valleys, troughs, etc. usually related to a major fault.

linear amplifier (seis).   A pulse amplifier in which the output pulse height is proportional to an input pulse height for a given pulse shape up to a point at which the amplifier overloads.

linear transducer (seis).   A transducer for which the pertinent measures of all sound waves concerned are linearly related.

lineation. Parallel orientation of structural features of rocks expressed by lines rather than planes, as long dimensions of minerals or pebbles, striae of slickensides, intersection of bedding with cleavage, etc.

line hydrophone (seis). A directional hydrophone consisting of a single straight-line element, or any array of contiguous or spaced electroacoustic transducing elements, disposed on a straight line, or the acoustic equivalent of such an array.

liner. Any string of casing whose top is located below the surface. A liner may serve as the oil string, extending from the producing interval up to the next string of casing. A blank liner has no perforations. A perforated liner is perforated or slotted before being placed in the hole. A screen liner is perforated and then arranged with a wire wrapping which acts as a sieve to prevent the entry of sand particles.

lineshaft. A relatively long shaft with mounted pulleys used to transmit power from power producing equipment to pulleys, directly or through jackshafts or countershafts in a mill or factory.

liquefaction of gases. A gas may be liquefied by subjecting it to pressure and then cooling it to its critical temperature. This is done commercially with ammonia gas, sulphur dioxide and carbon dioxide in refrigerating plants, also extensively used in refineries and natural gas plants.

liquefied gas. Gas may be converted to a liquid by combining the effect of compression with cooling; in the manufacture of natural gasoline from casing head gas, the gas is generally compressed and then cooled while under compression, causing condensation and therefore resulting in liquefaction.

liquefied petroleum gas (LPG). Hydrocarbons, gaseous at normal temperature and pressure, liquefied by pressure for storage and transport, as butane and propane gases used for domestic and industrial purposes; known also as LNG (liquefied natural gas).

liquid. A state of matter in which the molecules are relatively free to change their positions with respect to each other but restricted by cohesive forces so as to maintain a relatively fixed volume.

liquid grease (calcium). Light and medium grade lubricating oil thickened with small quantities of calcium soap, while the heavy grades include a fixed oil or fat.

liquid immiscibility. Property causing some dissimilar liquids to remain discrete fractions and not mix or dissolve in each other.

liquid inclusion. Inclusion of liquid in solid crystals.

liquid phase cracking (refin).   Charging stock is heated in a pipe still to cracking temperature, and then passed to the reaction chamber as a liquid where further cracking takes place. The product leaves the reaction chamber as a vapor, passing successively through the evaporator, fractionator and stabilizer. The cracked gasoline comes from the bottom of the stabilizer and gases from the top. The residue from the bottom of the evaporator and fractionating towers may be recycled through the plant or run to storage.

lithification.   Process of induration, cementation, petrification, consolidation, and crystallization converting newly deposited sediment into rock. The process by which unconsolidated rock-forming materials are converted into sedimentary rock. *See* diagenesis.

lithofacies.   A geological data set comprising the rock record of the physical and organic aspects of any given sedimentary environment.

lithofacies maps.   Map expressing areal variations in the environmentally significant lithologic aspects of a sedimentary unit. The unit may be any convenient stratigraphic interval, as a formation or a member, or the strata of a given geological period. Most lithofacies maps show variations which can be expressed numerically, such as the ratios of the thickness of two lithologic types to each other or to the total thickness. Lithofacies maps may show where general conditions are favorable for the accumulation of oil and gas accumulation and help find stratigraphic traps.

lithology.   Study of the physical character of a rock, usually by macroscopic techniques; included are study and description of mineral composition, structure, texture, fabric, etc.

lithosphere.   Outer, solid portion of the earth; crust of the earth. Portion of the earth above the asthenosphere.

lithospheric plates.   *See* plates and plate tectonics.

lithotope.   An area and environment of uniform sedimentation. The layers of deposits of uniform or uniformly heterogenous composition and texture produced in a lithotope is a lithostrome.

lit-par-lit.   (*pronounced lee-par-lee*) Intimate intercalation of granite and gneiss brought about either by injection metamorphism or partial melting. Migmatites regularly exhibit lit-par-lit structure.

littoral.   Of or pertaining to a shore or coastal region. In the strictest sense, pertains to processes, materials, structure, and inhabitants in the area between low and high tide.

littoral zone.   In general, the coastal zone including both the land and nearshore water. In the strictest sense, the intertidal zone between high and low tide levels.

live oil.   Crude oil containing gas and not stabilized or weathered.

liver.   Intermediate layer of dark-colored oily material, insoluble in weak acid and in oil; formed when acid sludge is hydrolyzed by steaming.

LNG.   Liquefied natural gas.

load.   Quantity of sediment transported by a current of water, wind or glacial ice. Water- and wind-borne load may be carried in three ways: (1) Suspension in the medium of transport. (2) Bed load, rolling and sliding along the earth's surface and (3) Saltation, bouncing along spending part of the time in bed load and part of the time in suspension. Glacial load is carried either on the surface of the glacier or frozen into its interior.

load (drill).   In mechanics, the weight or pressure being placed upon an object. The load on the bit refers to the amount of the weight of the drill collars allowed to rest on the bit; the weight on the bit.

location.   A spot or place where a well is to be drilled. An engineer or land surveyor is generally employed to stake a location, and tie the location to some property line or fixed point by measurement kept on record.

lock box (refin).   A box with glass windows built in the run down lines from stills, so arranged that the streams of oil coming from the condenser coils may be watched at all times and samples for tests are drawn.

locke level.   A hand level for determining differences in elevation.

lode.   A mineral deposit filling a fissure in country rock; any ore deposit occurring within definite boundaries, separating it from adjoining rock; contrasted with placer or disseminated deposits.

loess.   Nonstratified, yellowish silt deposited primarily by the wind; consists of

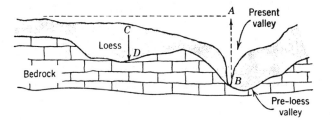

fresh, sharp-cornered particles of quartz, feldspar, calcite, and numerous other minerals mingled with some clay. Often associated with glacial deposits and thought to be the result of eolin reworking of moraine and outwash.

log.   A systematic recording of data made either while a well is being drilled, during temporary cessations in drilling or after a well is completed, as driller's log, electric log, and radioactivity log.

log, acoustic velocity.   Records the time required for an acoustic compression wave to travel through one foot of rock. Value of time or interval transit time is related to porosity. Log is usually recorded simultaneously with gamma ray log and a caliper log.

log, caliper.   Borehole caliper, a device used to determine the variations in well diameter from bottom to top. The caliper consists of three or four collapsible arms evenly spaced around a steel shank. The tool is lowered to the bottom of the hole, where the arms are released. It s then raised to the surface at a rate of about 100 feet per minute. During the trip, individual springs free each arm pressed against the sides of the hole and the deviations from the bit diameter are automatically and continuously plotted with the depths by an electrically operated recorder mounted on a service truck.

log, compensated acoustic velocity.   Similar to acoustic velocity but some of the extraneous responses related to irregular borehole geometry are removed.

log, compensated density.   Similar to density log, but compensated to minimize the effects of mud cake thickness variations in the tool response.

log, contact.   A record of two resistivity measurements made by two very shallow investigation pad mounted electrode systems. "Positive separation" of the two curves usually is an indication that mud cake is present between the pad and the rock.

log, density.   Records the value of formation bulk density. Tool emits a beam of gamma rays and the number of those gamma rays that reach the detector is related to formation density. Log is recorded with gamma ray and caliper.

log, electric.   In making an electric log of a well, an electric current is fed into the ground near the wellhead. An electric contact is lowered to the bottom of the well. This contact receives the current that passes through the earth and the mud in the well and delivers it to a device where the strength of the current or the resistance or resistivity of the rock is measured and recorded. As the contact moves up the well, the resistivity varies depending on the nature of the rock and its contents of oil and water. In this way, the location, thickness and some of the characteristics of the subterranean rock strata can be determined.

log, gamma ray.   In gamma ray logging, the differences in the radioactivity of different kinds of rocks are used. All rocks give off radioactive particles and gamma rays. The latter can easily penetrate several layers of steel casing and therefore are used as the indicating force in gamma ray logging. Dolomites, limestones and sandstones give off little gamma radiation, while shales, particularly those containing marine organic material give off considerably more gamma radiation. The gamma rays pass through a tube containing an ionizable gas becoming an increasingly better conductor of electricity, the more radiation it receives. The current flow is proportional to the intensity of radiation.

log, induction electric. Usually an SP curve, 16 inches to 18 inches (normal) resistivity curve, induction conductivity curve (~40 inches spacing), and the induction resistivity curve. In beds of ~6 inches or greater thickness, the induction resistivity value is usually very close to the value of formation resistivity. Induction log. (See Fig. p. 202).

log, long drilling time. Drilling time, or rate-of-penetration logs, consists of a curve plotted on a time-depth basis. The slope of the curve designates the speed of penetration. An abrupt break shows contact between rocks of unequal penetrability.

log, mud analysis. Mud analysis allows the detection of minute amounts of oil and gas in the drilling cuttings. The log constructed from the data consists of three curves and is usually combined with the lithologic and/or drilling time log. One curve shows the oil content, another the content of hydrocarbon gases other than methane and ethane and a third, the total gas content.

log, neutron. Similar to gamma ray but increases the gamma ray radiation from the rock strata by bombardment with neutrons. Depending on the kind of rock and its water or oil content, the neutrons penetrate to different distances before they are stopped by an atom. Since hydrogen is particularly effective in stopping neutrons, the presence of water and oil both containing hydrogen is readily determined. Schematic of radioactivity logging components. *(a)* components of neutron logging equipment. *(b)* components of gamma ray logging equipment. (See Fig. p. 203).

longitudinal waves (seis). Body waves moving by alternate compression and dilatation of earth's constituent atoms; primary or ''P'' waves are longitudinal.

long shore bar. A sand bar or ridge elongated parallel to shore. These bars are usually submerged at all times but occasionally they are briefly emergent at low tide.

longshore drift. Transportation of sediment along the coast by wave action.

long string. Last string of casing set in a well; the string of casing set through the producing zone, often called the oil string.

loran. Long-range navigation system using transmitters at fixed locations. A

202

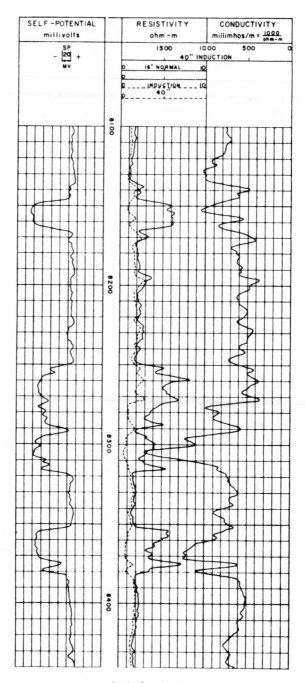

An induction log.

Schematic of radioactivity logging components.

(A)   (B)

highly popular technique in recent decades but now being displaced by satellite, minirange and other navigational systems.

lose returns (drill).   To encounter an interruption in the circulation of drilling fluid due to the fluid entering into a porous formation underground rather than returning to the surface.

lost circulation.   Loss of quantities of whole mud to a formation, usually cavernous, fissured, or coarsely permeable beds, evidenced by the complete or partial loss of drilling mud return.

Lovibond tintometer.   An instrument used for determining the color of refined petroleum oils, reading in Lovibond "color" numbers.

low grade metamorphism.   Metamorphism accomplished at low pressures and temperatures as in the zeolite facies; may overlap with sedimentary diagenesis on occasion.

LPG.   Industry initials for certain liquefied petroleum gases. (*See* liquefied petroleum gas)

lopolith.   A concordant intrusion associated with structural basins. In the simplest and ideal case the sediments above and below the lopolith dip inward toward a common center.

lube cut (refin).   A refinery term including part of the crude oil from which lubricating oils are made. When resulting from distillation, it is the lube fraction of overhead lube stock, or lube distillate.

lube oil distillate (refin).   General term for that part distilled from the crude oil an used in the manufacture of lubricating oil.

lube stock.   Fractions of crude petroleum of suitabl boiling range and viscosity to yield motor or other lubricating oils when completely refined.

lubex.   An aromatic extract from lubricating oil.

lubricants.   Terms applied primarily to greases, semi-solid compositions, and gear oils, even though some of the material are as much liquid oils as solids. Most generally greases may be grouped into three classes: (1) admixtures of mineral oil and solid lubricants, (2) blends of residuum, waxes, uncombined fats, rosin oils and pitches, and (3) soap thickened mineral oils. Common thickeners are sodium, calcium aluminum, lithium, and lead soap.

lubricating film.   Thin layer of oil or greases between rubbing surfaces. This film may be considered as subdivided into a series of thin sheets or laminations parallel to the surface, each lamination shearing or sliding over the other when movement between surfaces takes place. The film clings to both surfaces, hence, opposite sides of the film are dragged in opposite directions, shearing the film at an almost infinite number of planes.

lubricator (drill).   An extension of casing or tubing above a valve on top of the casing or tubing head. Lubricators are supplied with a pack-off, or pressure sealing device at the upper end to afford a seal on the wireline or other connection, attached to tools run into a well.

luster.   General appearance of a mineral's surface, particularly how it reflects light. Luster is described in subjective terms as metallic, pearly, silky or glassy.

lutite.   A fine-grained sedimentary rock consisting primarily of clay-sized material, most of which consists of clay minerals.

lux text.   A test designed to determine the amount of fixed oil or oils present in a liquid.

lye treating (refin).   Removes hydrogen sulphide and light mercaptans; sometimes used as a pretreat for sweetening and other processes. Process consists of mixing a water solution of lye, sodium hydroxide or caustic soda with a petroleum fraction. The treatment is removed as soon as possible after the petroleum fraction is distilled since contact with air will form free sulphur.

M

MCIF. 1,000 cubic feet of gas.

MMCF. 1,000,000 cubic feet of gas.

MMSCF. 1,000,000 cubic feet of gas at standard or base conditions of temperature and pressure according to established base condition in area of measurement.

MSCF. 1,000 standard cubic feet of gas.

MWD (Measurement While Drilling) (drill). A system of measuring parameters at the bottom of the hole and transmitting that information virtually instantaneously to the surface. Just about any sensor now in use can be interfaced with MWD. MWD will have wide impact in directional drilling and eventually optimize drilling to lower cost per foot gaining valuable information as formations are penetrated, plus being an invaluable safety tool. There are several systems being developed: (1) Pressure Pulse System: mud pulse system for MWD transmits signals by restricting mud flow inside the drill pipe and creating an increase in standpipe pressure. Pressure pulses are decoded at the surface and yield data from downhole sensors. (2) Acoustic Methods: transmission of sound waves through the drill pipe or through the mud stream. Acoustic transmission is similar to mud-pulse telemetry. The advantages of the acoustic method are simplicity, cost, and perhaps a higher data rate than mud impulse. (3) Hardwire Systems: requires special drill pipe or inside-the-string tools either with or without a jumper cable. Hardware systems offer the advantage in data rate and two way power and data transmissions. The number of downhole sensors permissible with a hardware system is unlimited. (4) Electromagnetic Methods: transmission of electromagnetic signals both through the earth and the drill pipe for underground measurements. The advantage of EM system includes high data rate, low cost, and ease of handling.

MacMichael's viscosimeter. An instrument of the torsion type with disc suspended in a cup of fluid while rotated, the torsion produced on the disc.

macrofossil. A fossil large enough to be studied with the unaided eye.

mafic. A term contrived from the "ma" of magnesium and "f" for iron and referring to dark colored silica deficient igneous rocks made up of femic mineral.

magma. Mobile, usually molten rock material generated at high temperature and pressure within the earth from which igneous rock is derived by solidification; when extruded onto the earth surface, lava.

magmatic. Related to bodies of molten rock within the earth.

magmatic differentiation.   Process where an originally homogeneous magma is segretated during the crystallization process into rocks of different composition. Commonly, early crystallized mafic minerals separate by gravitational settling or filter pressing, leaving the melt enriched in silica, sodium and potassium.

magmatic segregation.   *See* magmatic differentiation.

magmatic stopping.   A process of igneous intrusion whereby a magma gradually eats its way upward by breaking off blocks of country rock. Upward movement of magma by melting and prying off blocks of overlying rocks, thereby occupying their space.

magmatic water.   Water existing in, or derived from magma.

magnetic anomaly.   A distortion of the regular pattern of the earth's magnetic field due to local concentrations of ferromagnetic minerals.

magnetic declination.   At any point, the angle between the direction of the horizontal component of the earth's magnetic field and true north.

magnetic dip.   Angle between the horizontal and the direction of a line of force of the earth's magnetic field at any point.

magnetic field.   Magnetic lines of force surrounding the poles of a magnet. These lines of force are put out in all directions from the North Pole and are directed through space toward the South Pole. The space filled with these lines of force is called a magnetic field.

magnetic inclination.   Angle of the magnetic needle with the surface of the earth.

magnetic meridian.   Magnetic north. The north trending horizontal component of the earth's magnetic lines of force.

magnetic north.   Direction north at any point as determined by the earth's magnetic lines of force.

magnetic pole.   North magnetic pole is the point on the earth's surface where magnetic meridians cross and where the north-seeking end of a magnetic needle points directly down. At the south magnetic pole the same end of a magnetic needle points directly up. These poles are also known as dip poles.

magnetic variometer.   An instrument for measuring the difference in a magnetic field with respect to space or time.

magnetometer.   An instrument for measuring the earth's magnetic field. Early varieties and still in use utilize the torque imposed on a wire used to suspend a magnet as a means of determining the earth's field strength. Modern varieties utilize electromagnetic fields (flux gate magnetometer) or the precession of magnetic nuclei (proton precession magnetometer) to evaluate the earth's magnetic field strength.

magnetostriction. That property of certain ferromagnetic metals as nickel, iron, cobolt and manganese alloys causing them to shrink or expand when placed in a magnetic field.

magnitude, earthquake. Measure of energy released by an earthquake based on a logarithmic scale of the ratio of amplitude to period of earthquake waves; independent of location of recording seismograph. Scale extends from zero to approximately nine. There is no maximum limit to the scale, but it is unlikely that the earth's crust is strong enough to store stress sufficient to cause an earthquake greater than nine upon the release of stress. The scale progresses such that an earthquake of magnitude five is ten times greater than an earthquake of magnitude four and one hundred times greater than an earthquake of magnitude three.

main sill (drill). In a rig front, the main sill extends from a point under the derrick floor where it rests upon the nose sill, back to a point beyond the location of the sand reel. It lays at a slight angle to the center line of the rig, rests upon four mud sills, supports sampson, and the acts as the mainstay tying the rig front together.

make a connection (drill). To attach a joint of drill pipe on to the drill stem suspended in the well bore. The addition of this joint permits deepening of the well bore by the length of the joint to the drill stem.

making hole. Refers to progress being made at a given time when rotating the bit and deepening the well bore while drilling.

making a trip (drill). Hoisting the drill pipe to the surface and returning it to the bottom of a well bore for the purpose of changing bits, preparing to take a core, and other reasons.

making a joint (drill). Screwing a length of pipe into another length of pipe.

malthenes. Constituents of bitumen or pyrobitumen which are soluble in volatile aromatic-free petroleum spirits.

manganese nodule. Irregularly shaped mass precipitated on the ocean floor. Nodules vary in size from a few millimeters to nearly half a meter and show a concentrically laminated structure. They are generally some shade of brown, resembling the color of rust and found in areas of slow sedimentary deposition. Principal constituents are oxides and salts of manganese and iron. Accessory base and transition elements may be present in economically important amounts.

manifold.   A piping arrangement allowing for one stream of liquid or gas to be divided into two or more streams.

manometer.   A pressure gauge consisting of a glass tube shaped somewhat like a letter U with a scale at its back graduated or read in inches and decimal fractions of inches.

mantle.   (geoph) Part of the earth between the core and the mohorovicic discontinuity. (2) (geology) Loose material at or near the surface, above bedrock; includes soil and regolith.

manus tester.   An instrument for determining the flashing point of petroleum.

map.   (1) *n.* A two dimensional or flat representation of some portion of the earth. A map may contain two dimensional information as location of culture, sites of human construction, or geological exposure along any given plane, real or imaginary. A map may present three dimensional information by using contour lines as a topographic or isopach map. Maps may present information about the topography, geology and structure of the earth representing relationships at or beneath the earth's surface. Almost without exception, maps contain information allowing the reader to determine the geographic location of the area represented by the map, showing a north arrow to permit proper geographic orientation and includes a scale to allow the reader to relate map measurement to actual distances, thicknesses, on or in the earth. (2) *vb.* Process of gathering information necessary for the construction of a map or chart.

marble.   Metamorphosed limestone; a fine or coarse grained crystalline rock composed almost exclusively of calcite or dolomite.

marginal conglomerate.   A narrow, elongated sedimentary deposit of pebbles and/or cobbles paralleling a shore line and grading into finer sedimentary deposits off-shore. It moves laterally to reflect changes, with time, in the shore line or sea level and therefore time transgressive. When deposited as sea level is rising or land level, sinking, it may be called a basal conglomerate. Marginal conglomerates are useful in establishing the locations of ancient shorelines and whether the depositional environment was transgressive or regressive.

marginal junction.   A lithospheric plate boundary along which the two plates move in parallel but opposite direction slipping past each other; also called translational plate boundary, transform fault or transform boundary.

marine abrasion.   Erosion of a bedrock surface by wave controlled corrasion or abrasion resulting in a marine terrace.

marine geology.   Principles, practices, processes and materials of geology as they pertain to the modern marine environment. Marine geology and geology differ in the way data are collected. Classical field and sampling techniques must be modified greatly or be supplemented by remote sensing.

marine terrace. A flat, horizontal surface or platform with a thin veneer of marine sediment; found close to shore line. Usually emergent, above sea level but occasionally submerged. Wave action plays an integral role in marine terrace formation hence their presence reflect changes in sea level or land level with time. Older marine terraces may lose their planar/horizontal character because of tectonic warping or disruptions since the formation of the terrace.

marker bed. (1) In stratigraphy, a bedrock surface or stratigraphic level readily discernable or easily found and widespread in areal occurrence; used as a reference bed, surface or level when measuring a stratigraphic section or constructing a geological outcrop map. (2) In geophysics, an extensive seismic reflecting horizon yielding a clearly discernible signal with recognizably unique character on a seismogram; also a horizon whose velocity contrast with contiguous beds is such that a characteristic variation in the refraction time-travel curve is uniquely recognizable over large distances.

Marl. Porous masses of shells and shell fragments accumulating on the bottoms of many fresh-water lakes. Large amounts are formed by the lime secreting alga *Chara*. Designates certain marine sediments, soils in which clay and calcium carbonate are present in about equal amounts. An old term with no single accepted definition.

marsh funnel (drill). A calibrated funnel commonly used in field tests to determine the viscosity of drilling mud.

marsh gas. A volatile odorless and colorless gas coming from marshes and swamps where vegetation is in process of decay; when found in petroleum called methane, when purified, known as beninum; the simplest paraffin hydrocarbon.

mass number. Number of protons plus the number of neutrons in the nucleus of an atom; approximately equal to the atomic weight.

mass wasting. Downslope movement of material under the influence of the force of gravity and without the direct benefit of another agent of erosion as rivers, wind, glacier, etc. A process or agent of erosion occurring when the force of gravity exceeds the forces of cohesion within a sediment or rock mass resulting in slope failure. Mass wasting includes landslides, rockfalls, creep, slump, mud flows and other related or similar occurrences. Mass wasting events may be triggered when slopes are undercut or oversteepened by stream, glacial, wave or wind erosion.

mast (drill). A portable derrick, capable of being erected as a unit, distinguished from a standard derrick which cannot be raised from a working position as a unit, but is of bolted construction and must be assembled part by part.

matrix.   Fine grained material filling the interstices between larger grains or completely surrounding larger grains or fossils. Fine grained material surrounding phenocrysts in an igneous porphyry is usually called ground mass but may be called matrix. The sediment in which a fossil is embedded is called a matrix. The fine grained, silt and clay sized particles material surrounding or between sand grains in a sandstone is called matrix.

matterhorn.   An alpine peak carved by the headward erosion of alpine glaciers and resembling the Swiss mountain peak of the same name.

mature coast.   A coast straightened by cutting back of headlands and bridging of bays by sand bars. Sandy beaches are developed and stable. Coastal lagoons and marshes are greatly diminished or absent.

mature stream.   A stream eroded to grade and efficient in carrying sediment delivered to it by tributaries. Being at grade, or nearly so the mature stream accomplishes little or no downcutting but is capable of extensive lateral cutting provided the material through which it flows is not too hard or resistant. Mature streams characteristically meander across broad flood plains created by lateral cutting.

mature topography.   A stage in the development of the topography of a region identified by large, well defined stream drainage systems with closely spaced tributaries and a resulting destruction of preexisting upland erosional surfaces and maximum slope development. Flat land is restricted to occasional flood plain growth along limited portions of the major streams.

meander/meandering stream.   A single broad loop typical of loops or bends exhibited by a meandering stream. Meander develops by lateral cutting of a mature stream and migrates across the flood plain of the meandering stream by cutting on the convex side of the stream course and filling on the concave side. Meandering streams frequently erode to cut off a meander thereby creating an oxbow lake.

meander scars.   Low lying arcuate ridges and depressions resulting from the cutting and filling action of a meandering stream as its meanders migrate.

measure (drill).   An accurate measurement of the depth reached in a well by measuring the drill pipe or tubing as it is run into the well.

measure out (drill).   Measurement of drill pipe or tubing as it is pulled from the hole, usually to determine the depth of the well or the depth to which the pipe or tubing was run.

mechanical dewaxing (refin).   First step in the process of dewaxing is to add naphtha to prevent the stock from becoming a frozen mass at low temperatures. The chilled mass is run through a wax press to separate the wax. Propane is used as a solvent to deasphalt the stock and to act as the refrigerating medium for dewaxing. The deasphalted oil and propane from the deasphalter, under pressure, go to the dewaxer where at the atmospheric pressure the liquid propane vaporizes. The refrigerating effect accompanying vaporization chills the lube stock. While chilled, the stock is forced through wax presses separating the frozen particles of wax from the oil.

mechanical sediment.   Sediment consisting of the broken particles of preexisting rock transported mechanically by water, wind or ice; synonymous with clastic sediment.

mechanical weathering.   Breakdown or weathering of the earth's surface by mechanical means as frost wedging and root wedging; synonymous with physical weathering.

medial moraine.   Ridge of drift or moraine formed by the joining of adjacent lateral moraines below the junction of two valley glaciers. When two valley glaciers flow together, the lateral moraines along the sides of the glaciers which join, coalesce and are incorporated into the interior of the new glacier showing as a dark line of moraine parallel to the direction of ice flow in an active glacier or as a morainal ridge parallel to the walls of an abandoned glacial valley.

medium-curing cut back.   An asphalt cement thinned with a kerosene type of distillate.

megabreccia.   A rock consisting of randomly oriented clasts with sharp edges and of very large size, in excess of 100 meters in horizontal dimension.

megafossil.   *See* macrofossil.

megagroup.   A rock stratigraphic unit made up of stratigraphic groups and representing a major event in geologic history.

megaripple.   A ripple of great size developing in fluvial, tidal or marine environments. Opinion varies on the size of the ripple to be called a megaripple but in general they are several tens of centimeters to a meter high and exhibit a wavelength of from one to 100 meters.

melange.   A mixture of materials; a mappable unit of material intensely sheared and consisting of a confused assortment of poorly sorted blocks, slabs and fragments.

melt.   A liquid resulting from the fusion of a rock or a liquid remaining after the partial crystallization of a magma or lava.

melting point.   Temperature at which a solid substance melts or fuses at a given pressure. For asphalt, the melting point is defined as the temperature at which the asphalt is soft enough to permit a steelball to drop through a disc of asphalt supported in a ring suspended in water (ring and ball method). Grease melting point is determined by placing a small amount of grease on a bulb of a thermometer and heating in hot air until the grease begins to run off. Mineral melting points are determined by observing when a crystal loses its shape or the structure deteriorates.

meniscus. Curved upper surface of a liquid column, concave when the containing walls are wetted by the liquid and convex, when not. To read the gravity of a liquid in a glass cylinder, the meniscus is noticeable in the hydrometer stem and must be taken into consideration for an accurate reading.

menstrum. Any substance dissolving a solid body; a solvent.

mer (maximum efficiency rate). Highest rate of production from an oil well consistent with proper maintenance of underground pressures, gas oil ratios and other good production practices.

Mercalli Scale. Scale of earthquake intensity. A means of recording the strength of ground movement resulting from an earthquake based primarily on varying degrees of disruption and damage from I (shaking detected only by seismograph) to XII (almost total destruction). Subdivisions are arbitrary and based on observations made without regard to location of the earthquake focus or epicenter.

Mercapfining (refin). Sweetens various petroleum-derived hydrocarbon streams. Mercapfining is a fixed-bed, liquid-phase catalytic process for converting mercaptans to disulfides, remaining dissolved in the hydrocarbon stream. Conversion is accomplished by oxidation of the mercaptans, with air, in the presence of a fixed bed catalyst.

mercaptans (refin). Organic compounds of carbon, hydrogen, and sulfur, giving gasoline an offensive garlic odor; also make necessary larger quantities of anti-knock compounds to increase anti-knock qualities to predetermined levels but generally are not corrosive. They may be removed or converted to the less objectionable desulphide by the doctor treatment.

mercaptides (refin). Compounds of metals with mercaptans. Learmercaptides, formed in sweetening have the general formula R_1S PbS R_2, in which R_1 and R_2 are alkyl, aryl, or arkalkyl radicals.

mercurization test. A test with mercuric acetate determining the presence of unsaturated hydrocarbons in a petroleum oil.

mercury freezing test. A test determining the contamination of propane by higher-boiling hydrocarbons.

Merifining (refin). Treatment of catalytic cracked hydrocarbons with caustic to remove organic acids, mercaptans and other sulfur compounds.

mesa.   A low, flat-topped mountain or tableland bound on at least one side by a steep cliff.

mesh.   A measure of fineness of a woven material, screen or sieve; as a 200 mesh screen has 200 openings per linear inch. A 200 mesh screen with a wire diameter of 0.0533 mm (0.0021 inch) has an opening of 0.0074 mm or will pass a sperical particle of 74 microns.

meson.   A charged or neutral particle having about 200 times the mass of an electron; mesons are released from the nuclei of atoms when they are struck by very high energy radiation. In the classification of subatomic by mass, the second lightest of such particles, intermediate between that of the lepton and the nucleon.

mesothermal.   Hydrothermal deposits formed at intermediate temperatures and pressure.

Mesozoic Era.   Mesozoic (meso, middle + zoic, life) Era lasted about 170 million years, consisting of three periods, Triassic, Jurassic, and Cretaceous; an era that saw the development and extinction of dinosaurs, giant flying reptiles, and strange toothed birds. In this era, flowering plants developed, and hardwood trees rose to challenge the dominance of evergreens in the primitive forests. It was a time of mild climates and the first appearance on earth of the mammals, highest class in the animal kingdom. During the entire Mesozoic Era in North America, most of the submergence, marine deposition, and mountain building took place in the western part of the continent. The eastern half of the continent was above sea level for the most part and underwent steady erosion. The name means middle life and indicates an era of transition between the time of "ancient life" and Cenozoic. The environmental changes taking place during the Appalachians revolution, wiped out many groups of animals and plants, those surviving became the ancestors of many thousands of new forms.

metal deactivators.   Organic compounds suppressing the catalytic action of metal compounds contained in Hydrocarbon distillates as cracked gasoline and promoting the formation of gum. These metal compounds are usually either copper compounds retained in the gasoline as a result of copper sweetening or other catalytic metals with which the gasoline came into contact during the course of its handling and utilization.

metallic naphthenate.   A compound of naphthenic acid obtained from petroleum and other sources used as a wood preservative, paint dryer, etc.

metalloid.   A borderline element with properties characteristic of metals and non-metals, as carbon, arsenic and silicon.

metamorphic.   Pertaining to or resulting from the process of metamorphism.

metamorphic facies.   An assemblage of minerals reaching equilibrium during the metamorphism under a specific range of temperature and pressure and from which the range of temperature and pressure of formation may be deduced.

metamorphic grade.   A general measure of the difference between a parent rock and its metamorphic derivatives in relation to temperature and pressure conditions of metamorphism; extent of metamorphism.

metamorphic rocks.   Rocks produced by the action of heat, pressure, and chemical activity operating in the solid state upon rocks within the earth, usually through long periods of time. These factors operate to produce recrystallization, either partial or complete, of the minerals of the rock. New minerals appear and may develop a wholly new texture and fabric in the rock, as slate, true marble, gneiss and schist. Occasionally metamorphic rocks form at or near the earth's surface and in a short period of time as under a lava flow because of the intense heat. Metamorphic rocks are most often found associated with mountain building and convergent plate boundaries.

metamorphism.   Change in the form, texture, fabric and mineralogy of a rock in response to a change in its environmental pressure, temperature and chemistry from that of its formation. Metamorphism is exclusive of changes classified as sedimentary or diagenesis and igneous or melting. Metamorphism is commonly associated with mountain building and convergent plate boundaries.

metasomatism.   Interaction between solid rock and its chemically active pore fluids during metamorphism. When this term was first introduced into the geologic literature, it was defined as a related but alternative process to metamorphism. Now it is viewed as a usually necessary part of the metamorphic process enabling many of the textural and mineralogical changes.

metastable.   A phase or system thermodynamically unstable but for which the reaction threshold energy is too great for a change to occur spontaneously. Such systems can withstand small disturbances without reaction but cannot withstand large disturbances. The concept is convenient for geochemical applications to disequilibrium mineral assemblages remaining unchanged for long periods of geologic time.

meteoric.   Pertaining to the earth's atmosphere; ground water of recent atmospheric origin may be called meteoric water.

meter.   (1) An instrument for measuring and recording the volumes of gases and liquid. (2) Fundamental unit of length in the metric system, equal to 39.37079 inches or 3,2808 feet.

meter, proportional gas.   Meter automatically measuring a proportional part of the volume flowing through it.

meter, venturi.   Meter employing a venturi tube to detemine the factor used in calculating the flow of fluid.

methane (CH_4).   Light, odorless, flammable gas, the chief constituent of natural gas; also produced by the partial decay of plants in swamps, so that its occurrence is not uncommonly misinterpreted as an indication of the presence of petroleum. The first member of the gas family composed of one atom of carbon and four of hydrogen.

methyl-butyl-ketone process (refin).   A solvent dewaxing process chilling the mixture of oil, solvent and wax and then removing the wax by means of a filter.

methyl-ethyl-ketone.   A solvent used in dewaxing process; sometimes referred to as MEK process.

mica (drill).   A naturally occurring flake material of varying size used in combating lost circulation; chemically an alkali aluminum silicate.

micelles (drill).   Organic and inorganic molecular aggregates occurring in colloidal solutions. Long chains of individual structural units chemically joined to one another and laid side by side to form bundles. When bentonite hydrates, certain sodium or other metallic ions go into solution, the clay particles plus its atmosphere of ions is technically known as a micelle.

micrite.   Lithified calcareous ooze. An entire rock may consist of this material or may occur as matrix in limestone. Micrite is microcrystalline, crystal diameter less than four microns; chemically precipitated calcareous mud.

microbreccia.   A rock consisting of randomly oriented, angular sand grains which may or may not contain matrix in the interstices. Some authors consider graywacke to be a microbreccia.

microcoquina.   Partially cemented, sand-sized shell fragments.

microcrystalline.   Rocks composed of minute crystals visible only with a microscope.

microcrystalline wax.   Plastic high melting point petroleum wax secured by removing most of the oil from petroleum by solvents or other means; has a finer and less apparent crystalline structure than paraffin wax.

microfossil.   A fossil too small to be seen or studied effectively with the unaided eye. Fossilized remains of minute flora and fauna or minute fragments of larger organisms studied best with a microscope.

microlog.   An electrical resistivity borehole log indirectly measuring properties of porosity or reservoir permeability by means of very closely spaced (two to three centimeter) electrodes.

micron. μ = mu. A unit of length equal to one millionth part of a meter, or one thousandth part of a millimeter.

micropaleontology. Study of fossils so small that they are best studied under a microscope. These tiny remains are called microfossils and usually represent the shells or fragments of minute plants or animals; includes identification, classification, description and observation of occurrence.

micropore. A pore so small that capillary forces holds water in the pore against the force of gravity.

microseism. A feeble oscillatory disturbance of the earth's crust, detectable only with a very sensitive seismograph; usually associated with strong wind and/or ocean surf.

mid-boiling point (refin). Point at which 50 percent by volume of a petroleum fraction having a symmetrical distillation curve distills.

Mid-Continent crude. Oil produced principally in Oklahoma, Kansas and North Texas.

mid-oceanic ridge. A long submarine mountain chain running through the Atlantic, Indian and South Pacific oceans and associated with divergent plate boundaries.

migration. (1) Movement of oil or gas through the pores of rock; the rate of movement varying with the permeability of the rock, the viscosity of the oil, and other factors. (2) In geophysics, a geometric procedure whereby a horizon, as detected from seismic reflection events, is mapped in its true spacial orientation and location.

millidarcy. Unit of measurement of reservoir or aquifer permeability; one thousandth part of a darcy.

milliliter (ML). A metric system unit for the measure of volume; 1/1000th of a liter.

mineral. (1) In general, a material composed of, or pertaining to inorganic substance, not pertaining to vegetable or animal matter. (2) In geology, a naturally occurring homogeneous inorganic, crystalline solid with unique chemical composition and/or atomic structure.

mineral additive (grease). An additive having a mineral source and used as a filter in grease manufacturing, particularly in heavy duty grease. Among the mineral additives used for this purpose are talc, asbestos, mica, molybdenum and graphite. They are used to impart definite characteristics to the grease to meet specific operating problems.

mineral cleavage.   Cleavage of a mineral is its tendency to split easily, leaving a smooth flat surface. These surfaces are parallel to planes of atoms weakly bonded to each other.

mineral hardness.   A mineral's ability to resist scratching. If a mineral is scratched by a knife, it is softer than a knife. If it cannot be scratched by the knife, the two are of equal hardness or the mineral is the harder. Degrees of hardness are related to a simple standard scale known as the Mohs scale which ranges from 1, hardness of talc to 10, hardness of diamond.

mineral luster.   Appearance of the mineral in ordinary light due to the light reflected from its surface. If the mineral looks like metal as do galena and pyrite, its luster is said to be metallic. If a mineral looks glassy, like quartz, its luster is vitreous or glassy.

mineral oil.   Refers to a wide range of products derived from petroleum and within the viscosity range of products spoken of as oils. The term mineral implies that the oil is not derived from living animals or plants.

mineral oil, naphtha.   A limpid or yellowish liquid, lighter than water consisting of hydrocarbons. Petroleum is heavier than naphtha, and dark greenish in color when crude. Both exude from the rocks but naphtha can be distilled from petroleum.

mineral pitch.   Asphaltum.

mineral resin.   Any one of certain mineral hydrocarbons as asphalt and bitumin. A term applied to the solid bitumins.

mineral rights.   Rights of ownership, conveyed by a deed of gas, oil, and other minerals beneath the surface. In the United States, the mineral rights are the property of the surface owner unless disposed of separately.

mineral salts and water (crude petroleum).   Crude petroleum as received at the refinery usually has some saltwater mixed with it. The salts in the water are generally the chlorides of sodium, calcium and magnesium; the amounts are expressed as pounds per thousand barrels of crude (PTB). During distillation, these salts decompose and form hydrochloric acid which corrodes condensing equipment. Special metals are used in such equipment to resist, the action of the acid; lime or ammonia is added during distillation to neutralize the acid.

mineral seal oil.   Cut between kerosene distillate and gas oil, widely used as a solvent oil in gasoline absorption processes.

mineral, specific gravity.   Ratio of the weight of a mineral and the weight of an equal volume of water, thus a cubic inch of quartz weighs 2.65 times as much as a cubic inch of water. The specific gravity of quartz is 2.65. Nearly all minerals are heavier than water, so their specific gravity numbers are larger than one.

mineral streak.   Color of the powdery stripe made when a mineral specimen is drawn across a plate of unglazed porcelain. For many minerals, it is not the same as the color of the solid lump. Iron pyrite crystals are yellow, but their streak is greenish-black. Hematite is red or black, but its streak is always red.

mineralization. Process of change or metamorphism by which minerals are secondarily developed in a rock, especially the introduction of ore minerals into previously existing rock masses.

mineraloid.   Noncrystalline mineral-like substances.

minseed oil.   A bloomless petroleum product, used in connection with linseed oil for cheapening purposes or as a substitute.

minute pressure.   Capacity of a gas well is sometimes measured by quickly shutting the gate valve and noting the gage pressure each minute; the pressure at the end of the first minute is used to estimate the volume of the gas and is termed the minute pressure.

miogeosyncline.   A term in wide usage before plate tectonics gained acceptance. Attempts have been made to modify the definition to suit plate tectonic theory but many geologists advocate abandoning the term. In general, it refers to the thick shallow water, sedimentary deposits of the continental shelf as they extend along the craton.

mist drilling.   A method of rotary drilling whereby water and/or oil is dispersed in air and/or gas as the drilling fluid.

mixed aniline point.   A test for minimum solution temperature of a mixture of aniline, heptane, and a hydrocarbon; used to indicate the aromaticity of the hydrocarbon.

mixed base grease.   Utilizes a soap having two or more different greases.

mixing.   A technique used in seismic prospecting for oil. Electronic combination of the energy of different channels to enhance the signal to noise ratio. Electronic mixing became possible with the development of magnetic tape recording devices.

mixture.   Combination of two or more substances united in such a way that each retain its original properties.

moderator.   Material, as graphite or heavy water used in a nuclear reactor to slow the neutrons without absorption by repeated collisions making them more likely to produce fission in the uranium fuel.

Mohorovicic discontinuity (Moho or "M" discontinuity).   Sharp discontinuity in composition or crystalline phase between the outer layer of the earth, the crust, and the next inner layer, the mantle; discovered by Mohorovicic from seismograms.

Mohs Scale.   A scale for abrasive or scratch measurement of mineral hardness based on the hardness of the following minerals: (1) talc; (2) gypsum; (3) calcite; (4) fluorite; (5) apatite; (6) orthoclase, (7) quartz, (8) topaz, (9) corundum, (10) diamond.

molar.   A strength of solution involving one gram molecular weight of solute per liter of solution.

mold.   A void in a rock left by the dissolution of a fossil's original skeletal material. A mold may be either internal or external as impressions left in the surrounding rock by shells or other organic structure.

molecular weight.   Sum of the atomic weights of all the constituent atoms in the molecule of an element or compound.

molecule.   Smallest unit into which a compound can be divided without losing its characteristic properties.

mond gas.   A variety of semi water-gas. A producer gas made from coal; when used as a fuel it is termed mond fuel gas.

monoclinal ridge.   A ridge formed by a simple, steplike bend in horizontal beds of rocks.

monocline.   (1) Strata inclined in a single direction, especially a steplike fold,

(1)

(2)

produced by local downbending of a nearly horizontal strata or locally steepened dip of gently inclined beds. Oil is obtained from monoclines, often in great quantities. (2) Some authors equate monocline to homocline, a succession of beds dipping in one direction. Diagrams of monocline and homocline.

monomer.   A single molecule of a compound of relatively low molecular weight. A chemical compound containing only one structural unit per molecule; capable of forming polymers under certain conditions and temperatures.

montmorillonite (drill).   A clay mineral commonly used as an additive to drilling muds. Sodium montmorillonite is the main constituent in bentonite. The structure of montmorillonite is characterized by a form consisting of a thin plate-type sheet with the width and breadth indefinite, and thickness, that of the molecule. The unit thickness of the molecule consists of three layers. Attached to the surface are replaceable ions. Calcium montmorillonite is the main constituent in low yield clays.

moraine.   Unstratified deposits of unsorted glacial sediments; unstratified glacial drift. Deposits take the form of ridges, blankets and mounds left either by alpine or continental glaciers.

Mother Hubbard bit (drill).   Used for working in hole mudding up easily, or where the rock is hard and contains wide seams. The bit is almost as wide at the top as at the cutting point and the steel is exceptionally thick, resulting in the bit nearly filling the hole making it less likely to slip off into the slanting openings. The water course is wide, and rounding. The sharp shoulders in the bit causes it to cut its way through the mud when pulling out. Cable tool bit used for drilling shale or slate.

mouse hole (drill).   A hole drilled under the derrick floor and cased with a length of drill pipe temporarily suspended for later connection to the drill string.

mouse hole connection.   Procedure of adding a length of drill pipe or tubing to the active string. The length of pipe is placed in the mouse hole, made up to the kelly, then pulled out of the mouse hole and subsequently added to the string. Diagram of a mousehole connection for adding a joint of pipe as hole is drilled. (See Fig. p. 221)

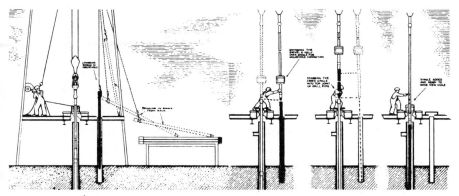

MOUSE   HOLE   CONNECTION

mouse trap (drill).   A fishing tool, cylindrical in form, with the bottom and open end fitted with an inward opening valve; tool is designed to fish out small metal parts lost in a well.

mud.   (1) A non-specific term denoting any soft, sticky, mixture of water and silt or clay-sized particles. Organic constituents are often present but abundance is highly variable. Mud may be marine or terrigenous. (2) In drilling, a water- or oil-base drilling fluid whose properties have been altered by solids, commercial and/ or native, dissolved and/or suspended. Used for circulating out cuttings and many other functions while drilling a well. Mud is the term most commonly given to drilling fluids.

mud balance (drill).   A beam balance consisting of a cup and a graduated arm carrying a sliding weight and resting on a fulcrum; used for determining the density or weight of drilling mud.

mud cake.   Coating of drilling mud on the walls of a well drilled by rotary methods. Mud or clay minerals accumulate as water escapes into the wall rock from the drilling fluid.

mud circulation (drill).   Pumping mud downward through the drill pipe and returning it to the surface through the annulus is normal circulation. In some cases the mud is pumped downward through the annulus and returned through the drill pipe, a procedure called reverse circulation.

mud conditioning (drill).   Treatment and control of drilling mud to insure the proper gel strength, viscosity, density, etc. Treatment may include the use of additives, the removal of sand or other solids, the removal of gas, the addition of water, and other measures that prepare the mud for conditions encountered in the well.

mud cracks.   Mud cracks are a result of a deposit of silt or clay drying out and shrinking. The cracks outline roughly polygonal areas, making the surface of the deposit look like a cross-section of a large honeycomb. Eventually, another deposit may come along to bury the first. If the deposits are later lithified, the outlines of the cracks may be accurately preserved for millions of years. When the rock is split along the bedding plane between the two deposits, the cracks will be found much as they appeared when they were first formed, providing evidence that the original deposit underwent alternating flooding and drying. Mudcracks are useful in determining the tops and bottoms of deformed sedimentary beds.

muddy.   Refers to water containing a mixture of sand, silt and clay-sized material or to a sediment consisting of sand, silt and clay-sized material in unspecified proportions.

mud flat.   A relatively level depositional surface along a shore line, frequently in a bay or estuary, emergent at low tide and submerged at high tide. Mud flats consist of silt, clay and some sand-sized material; soft and sticky when wet and hard when dry.

mud flow.   A variety of mass wasting where predominately silt and clay-sized sediment is suspended in water to create a viscous fluid mass capable of flowing down hill. In arid and semi-arid climates, flash floods may grade into mud flows as the water run-off picks up an increasingly large load of sediment. Mud flows slow to a stop as water is lost by evaporation or sinking into the ground. Immobilized mud flows show distinctive flow structure including a lobate front of flow. Mud flows may grade into earth flows as viscosity of flow increases, causing flow to change from fluid to plastic in nature.

mud hog (drill).   Pump circulating mud laden fluid through the drill stem and the well when drilling by the rotary method.

mud log.   Record of the continuous examination of drilling mud and drill cuttings to detect gas, oil or water and to gain stratigraphic and/or paleontologic information from the formations being penetrated.

mud pit (drill).   Reservoir or tank circulating the drilling mud to allow sand and fine sediments to settle out, where additives are mixed with mud, and a temporary storage for the fluid before being pumped back into the well.

mud, salt-water (drill).   A drilling mud with the water saturated or nearly saturated with salt. Muds containing salt concentrations of more than one percent are generally considered to be salt-water muds. Freshwater mud is often converted to salt-water mud when thick beds of salt are to be drilled.

mud sock (drill).   A device used in drilling tools to clean mud or sand out of a well.

mud still (drill).   An instrument used to distill oil, water, and other volatile materials in a mud and determine oil, water, and the total solid contents in volume percentage.

mud stone.   Lithified mud. Blocky or massive and well indurated sedimentary rock, similar in all respects to shale except lacking shale's fissility and laminations.

multiple completion wells.   A well equipped to produce oil and/or gas separately from more than one reservoir.

multiple reflection (geophysics).   In reflection seismology, a pulse emanating from the surface travels downward until reflected by a particular horizon and returned to the earth's surface to be recorded. Occasionally the pulse will not return directly to the surface but will be bounced or reflected back down to be reflected again off the original horizon. Pulses traveling directly by the shortest time path will be recorded as the principal reflection or first arrival off a seismic reflecting horizon. Pulses reflected up and down once or several times are recorded as multiples of the principal reflection.

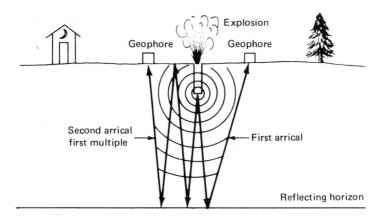

multiple zone well completion.   Method used on a single well to produce from several different formations.

nannofossil. Fossils of organisms as discoasters and coccoliths close to the lower limit of size resolved by an optical microscope. Advances in electron microscopy have been of great benefit to those who study very small life forms and fossils.

nannoplankton. Ocean surface floating or drifting organisms, plankton ranging in size from 5 to 60 microns.

Nansen bottle. An oceanographic device used in a series of several bottles suspended from a submerged cable in order to take a water sample. Bottles are lowered into the water while attached to the cable or hydrowire in an upside down and open position. When it is time to take samples a heavy weight (messenger) is sent sliding down the cable until it reaches the first Nansen bottle, causing it to turn to an upright position, close and release another messenger. The newly released messenger slides down the cable until it encounters the second Nansen bottle and the process is repeated. In this way the bottles are closed sequentially until all are secured. The cable is next pulled to the surface where the water samples are decanted and analyzed or stored. It is customary to attach a reversing thermometer to each Nansen bottle thereby obtaining an ocean temperature profile simultaneously with the water samples.

Nansen cast. The process of lowering or casting a series of Nansen bottles over the ship's side into the ocean and then recovering them. *See* Nansen bottle.

naphtha. General term for a refined, partly refined or unrefined product, including liquid products obtained from natural gas falling within the following distillation range. At least ten percent distill below 347°F and at least 95 percent distill below 464°F when distilled by the method described in ASTM designation. An artificial, volatile, colorless liquid is obtained from petroleum utilized in two grades, the light used as a solvent in the manufacture of paint and varnish and as a cleaning fluid, and the heavy, used for blending with casing head gasoline to produce motor gasoline and for turpentine substitute.

naptha bottoms (refin). Residue left in a steam still after distilling naphthas.

naphtha gas. Illuminating gas charged with the decomposed vapor of naphtha.

naphtha scrubber (refin). Tower in which gas is washed with mineral oil to absorb naphtha. The scrubbing consists of simply spraying the mineral oil down through an ascending current of gas under pressure.

224

naphtha treating plant (refin). A series of connected vertical and horizontal tanks with naphtha distillates treated continuously with acid and alkali.

naphthalene. An aromatic hydrocarbon containing a double benzene nucleus; one of the principal constituents of coal tar, forming brilliant white, plate-like crystals of peculiar odor.

naphthalene oil. A liquid aromatic hydrocarbon obtained by the distillation of coal tar, containing about 40 percent solid naphthalene.

naphthene. A group of cyclic hydrocarbons, also termed cycloparaffins or cycloalkanes. Polycyclic members are also found in the high boiling fractions. One of a class of hydrocarbons, isomeric with olefins but differ from them by behaving as saturated compounds.

naphthene base crude. Crude oil as produced in the Gulf Coast district, commonly called asphalt base crude. Depending upon their source, also called coastal crude.

naphthene base oil. Lubricating oil distilled or processed from crude oil classified as a naphthene base crude, frequently and improperly called asphalt base crude.

naphthene (Cycloparaffin) Series. Naphthene series of hydrocarbons, also known as the cycloparaffin series, is a saturated, single covalent bonds, homologous, closed ring series, with the general formula CnH_{2n}. Isomerous with the olefin, alkene series of the same composition, but members of the olefin series are structurally open chain and unsaturated.

naphthene hydrocarbons. Unsaturated hydrocarbons, similar to paraffins in their properties but the carbon atoms in each molecule arranged in a close ring. Therefore, cycloparaffins are named cyclopentane, cyclohexane, etc. They occur in naphthas and higher boiling fractions. In gasoline, they have medium octane numbers and high chemical stability. In lubricating oil, they occur as complexes of several naphthene rings associated with paraffins and aromatics. Naphthene can be changed into higher octane aromatics by catalytic dehydrogenation.

naphthene series. A hydrocarbon series having the general formula CnH_2. These are ring or cycle compounds very seldom found in the lighter fractions of petroleum but frequently found in the heavier fractions. The naphthenes contain compounds as cyclopropane, cyclopentane, cyclohexane, etc.

naphthenic acid. General term applied to the whole family of organic acids occurring naturally in crude petroleum principally of the monocyclic structure. Valuable materials are present in considerable quantities in certain crudes and secured from the gas oil cut. They are complex organic acids and are frequently used to make soap employed in the manufacture of grease, paint dryers, emulsifying or demulsifying agents, etc.

naphthenic crudes.   Crude oil containing a relatively large percentage of naphthenic-type hydrocarbons. Lubricating oils made from these crudes are normally distinguished from similar oils made from paraffinic crudes, both oils equally well refined by lower gravity, lower carbon content and pour point, and a lower rating in viscosity index.

naphthenic hydrocarbons.   Saturated hydrocarbons with at least one closed carbon ring; with one ring only the general formula is $CnH_{2}n$.

narrows.   A steep-walled narrow part of a usually wider stream valley; generally occurs when stream flows from region of soft bed rock to region of hard bed rock.

native element.   Designates the uncombined state of an element occurring in nature commonly applied to metals, as native silver, copper, gold, platinum or bismuth.

natural bridge.   An arch of rock completely spanning an open space; formed naturally by action of stream or ocean wave erosion or by the partial collapse of the limestone roof of a cave.

natural gas.   A highly compressible and expansible mixture of hydrocarbons having a definite specific gravity and occurring naturally in a gaseous form. The principal component of natural gas, along with the approximate percentages are methane 80 per cent, ethane seven percent, propane six percent, isobutane 1.5 percent, butane 2.5 percent, pentane 3.0 per cent. In addition, natural gas may contain appreciable quantities of nitrogen, helium, carbon dioxide and contaminants as hydrogen sulfide and water vapor.

natural gas, commercial.   Propane, butane and iso-butane in natural gas is largely removed and used to make LPG and for other uses. The remaining methane and ethane are principally used for heating and constitute the commercial grade of natural gas.

natural gas liquids.   Portions of reservoir gas liquefied at the surface in lease separators, field facilities, or gas processing plants.

natural gasoline.   A gasoline made from wet gas vaporizing at a relatively low temperature and having high anti-knock value. For these reasons, natural gasoline is used as a blending stock to provide the starting fractions of the gasoline and to improve anti-knock quality. It may be made by (1) compression process, (2) absorption process, or (3) adsorption process.

natural gasoline (stable grade).   Gasoline stabilized and practically free of dissolved gases evaporating at a much faster rate than the liquid members.

natural gasoline plant.   A plant for the extraction of fluid hydrocarbons as gasoline and LPG from natural gas.

natural gasoline (unstable grade).   Gasoline not stabilized to remove dissolved gases evaporating at a much faster rate than the liquid members. On evaporation, these dissolved gases may carry away a large portion of the volatile liquid members causing high storage and handling losses.

natural glass.   Vitreous, amorphous, siliceous substance forming when lava cools and solidifies too quickly for crystals to form. Obsidian is perhaps the most common and well known example of natural glass. The presence of natural glass shards in continental margin sediments can be used to indicate and possibly measure proximity to an igneous volcanic arc associated with convergent plate boundaries or subduction.

natural levee.   A parallel pair of natural dike-like embankments containing a stream as it flows, usually meanders, across its flood plain. The embankment or levee is formed during flood stages when the stream spills out of its banks, slows and deposits coarse, sand and silt-size sediment in a ridge immediately beside the stream channel. The ridge, embankment or natural levee builds higher with each successive flood until it is possible for the bed of the stream to be elevated above the surrounding flood plain. When the natural levee has attained sufficient height to elevate the stream above its flood plain, the situation is ripe for a disastrous flood if the natural levee is ruptured.

naumeme number.   Results of a test indicating the amount of fixed oils and fats contained in a compound oil.

nearshore.   A widely used but poorly specified term denoting a zone beginning at the low tide line and continuing offshore for an indefinite but generally short distance. The breaker zone, longshore drift and wave induced long shore currents are found in the nearshore region.

neck.   (1) Igneous. A vertical, usually shallow igneous intrusion with undergone erosion and modification of its original shape. Some necks represent the frozen lava left behind in the throat of a volcano on the event of its last eruption and exposed when subsequent erosion removed the softer surrounding layer of cinders and softer lava. Other necks may be remnants of laccoliths whose thicker center portion survived the erosional processes which first stripped away the overlying material and then cut into the thinner, weaker margins of the laccolith. Necks may also represent solidified feeder conduits to laccoliths, volcanos or similar features. Most writers agree that a neck must be isolated by erosion and stand as a butte as with the Devils Tower. (2) Stream. A narrow strip of land between two bends of a meandering stream. (3) Coast. A narrow peninsula, isthmus, cape or promontory along a shore.

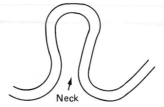

neck cutoff. A geomorphic feature occurring when a stream erodes a meander neck, cutting off the meander from the main channel of a stream and transforming the meander into an oxbow lake.

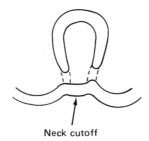

Neck cutoff

needle.   A prominent, sharp, rock pinnacle or spire in mountainous topography. A product of differential erosion, along closely spaced joint sets.

nektonic.   A free and actively swimming pelagic organism.

Neogene.   A system and period of late Cenozoic age, comprising the Miocene, Pliocene and Pleistocene series or epochs.

neohexane.   Colorless volatile liquid, characterized by a very high octane rating in internal combustion engines made by the thermal catalytic union or alkylation of ethylene and isobutane, each recovered from refinery gases resulting from the cracking of petroleum.

nepheloid zone.   A suspension of fine organic matter and clay-sized sediment particles in sea water forming a zone about 200 to 1,000 meters thick near the bottom of the continental slopes and rises of the world's oceans.

neritic.   Portion of the marine environment extending from the low tide level to the continental shelf break or from shore to water depth of 100 fathoms.

network.   In surveying and geophysical prospecting (gravity, magnetic and seismic) a pattern or configuration of stations, often arranged to provide a check on the consistency of the measured values.

neutral.   Between or halfway; a chemically neutral substance contains equal concentrations of H^+ and OH^- and is neither acid nor alkaline; may be derived by fractional distillation.

neutral oil.   Designation for light grades lubricating oil secured by distillation and, as a rule, prepared without chemical treatment by acid or alkali, but refined by simple filtration. Neutral oils are made in a range of viscosities under 200 Saybolt viscosity at 100°F. Stocks are quoted in wholesale as 150 neutral, meaning a lubricant oil having a Saybolt universal viscosity of 150 to 100°F.

neutral flame.   A flame where complete combustion occurs, neither an oxidizing flame or a reducing flame.

neutralization.   A reaction in which the hydrogen ion of an acid and the hydroxyl ion of a base unite to form water and a salt.

neutralization number.   Weight in milligrams of potassium hydroxide needed to neutralize the acid of one gram of oil; an indication of acidity of an oil. The result of a test is used to determine the acid and base constituents in petroleum products including Diesel fuels, burning oils, illuminating oils, both new and use.

neutrino.   Particles with no charge and almost no mass, released from the nucleus of atoms along with the emission of beta rays. A subatomic particle of zero or near zero, rest mass, having no electric charge, postulated by Fermi (1934) in order to explain apparent contradictions to the law of conservation of energy in beta particle emissions.

neutron.   An elementary nuclear particle with a mass approximately the same as that of a proton and electrically neutral, its mass is 1.008982 mass units. Neutrons are commonly divided into sub-classifications according to their energies as follows: thermal, around 0.025 electron volts; epithermal, 0.1 to 100 electron volts; slow, less than 100 electron volts; intermediate 10^2 to 10^5 electron volts; fast, greater than 0.1 million electron volts.

neutron diffraction.   The study of the structure of crystalline materials by the scattering of neutrons; very closely analogous to the study of crystals with x-rays in use for many years.

neutron flux.   Sum of the distance travelled by all the neutrons in one cubic centimeter in one second.

neutron lifetime.   Length of time typically about 0.001 second between formation of a neutron in a reactor and its disappearance occurring in various ways.

neutron number.   Number of neutrons in a nucleus; equals the difference between the mass number for that nuclide and the atomic number.

newton.   Unit force in the MKSA system; that force which gives to a mass of one kilogram an acceleration of one meter per second squared.

Newton's friction law.   Tangible force as in the force in the direction of the flow per unit area acting at an arbitrary level within a fluid contained between two rigid horizontal plates, one of which is in steady motion, is proportional to the shear of the fluid motion at that level.

Newton's law of cooling.   Rate of cooling of a body under given conditions is proportional to the temperature difference between the body and its surroundings.

Newtonian fluid.   Basic and simplest fluids from the standpoint of viscosity consideration in which the shear force is directly proportional to the shear rate. These fluids will immediately begin to move when a pressure or force in excess of zero is applied.

n-heptane.   A hydrocarbon used to rate the octane number of gasoline.

nipple.   A short pipe threaded on both ends used to connect up fittings in a convenient manner.

nipple chaser (drill).   A material man whose duty is to procure and deliver, to the drilling rig, the necessary tools and equipment to carry on the work.

nitrate.   A soluble mineral containing nitrogen and oxygen and frequently associated with another element as sodium or potassium.

nitric acid.   A powerful active acid combining with nearly all metals and having a strong oxidation effect on organic materials; strong corrosive mineral acid of formula HNO_3.

nitrile.   An organic cyanide, characterized by the univalent group CH, which on hydrolysis yields an acid with elimination of ammonia.

nitro-benzene.   A compound made by treating benzene with nitric acid.

nitrobenzene process.   A single solvent refining process for lubricating oils. *See* solvent refining.

nitrogen.   A gaseous non-metallic element, having little chemical affinity for other elements. It forms many unstable compounds, the most noted is dynamite. In itself, it is nearly inert and does not enter the process of combustion although it constitutes about 77 percent of the atmosphere. Many natural gas wells produce great volumes of nitrogen gas, detrimental because it lowers the heat content of the natural gas by displacing the methane and ethane.

nitrogen base. Compounds such as amine, which may be considered a substitution product of ammonia (NH_3): a compound containing trivalent nitrogen, capable like ammonia, of combining with acids to form salts containing pentavelent nitrogen.

nitrogen compounds (crude oil). Crude petroleum contains nitrogen compounds which, like the sulphur compounds were formed form the original organic tissue that produced the hydrocarbons. Most crude petroleum contains less than 0.1 percent nitrogen. Crudes with high sulfur and asphalt contents usually have high nitrogen contents. Nitrogen compounds are found in all fractions of crude petroleum but are more plentiful in the higher boiling fraction.

nitromethane. A solvent secured by processing methane and an organic compound.

nitrogen trioxide N_2O_3). A brown gas, blue solid or liquid with a specific gravity of 1.477 and melting point of $-102°F$.

nodule. An irregular, knobby surface body of mineral matter differing in composition from the sediment or sedimentary rock in which it has formed. If flattened, it usually lies parallel to the bedding planes of the enclosing rock and sometimes adjoining nodules coalesce to form a continuous bed. Nodules vary greatly in size but rarely exceed one foot in maximum dimension. Silica in the form of chert or flint comprise the most common nodules which are thought to form when silica replaces some of the materials in the original deposit, or when silica from the tests of radiolaria or diatoms is concentrated. Concretionary lumps of manganese, cobalt, iron and nickel are found widely scattered on the ocean floor. *See* maganese nodules.

nonane. A hydrocarbon of the paraffin series contained in gasoline of boiling point 105°C and formula C_9H_{20} .

non-carbon oil. Applies to (1) an oil containing little or no free carbon in suspension, (2) an oil containing a minimum of these compounds, or (3) one which when decomposed contributes little so-called carbon deposits.

non-combustible. A substance which will not combine with oxygen to generate heat.

non-condensable gas (refin). Includes the various refinery gases not condensing into a liquid while passing through a condenser. Unless collected these gases are classed as a refinery loss. A gas classed as non-condensable may become a condensable gas if the temperature in the same condenser is lowered or the pressure increased.

non-conductive mud (drill). Any drilling fluid, usually oil-base or invert-emulsion muds, whose continuous phase does not conduct electricity as in oil. The spontaneous potential (SP) and normal resistivity cannot be logged, although other logs as the induction, acoustic velocity, etc, can be run.

non-soap grease. A grease-like lubricant not containing soap; consists of heavy residual stock and mineral oil.

non-viscous neutral oils. A neutral oil having a viscosity at or below 135 Universal Saybolt seconds at 100°F.

Norma-Hoffman stability test. A static, accelerated oxidation test in a bomb measuring the rate at which oxygen is absorbed while oxidizing the readily oxidizable constituents in petrolum products, fixed oils, and greases.

normal (chem). Denoting a situation or solution of such strength that one liter contains one gram atom of replaceable hydrogen or its equivalent. In organic chemistry, pertaining to or designating aliphatic hydrocarbons or hydrocarbon derivatives in which no carbon atom is united with more than two other carbon atoms.

normal fault. *See* fault, normal.

normal heptane. Hydrocarbon compounds with an octane rating of zero used as a reference fuel ingredient in making motor fuel octane number tests.

normal paraffin. Paraffin hydrocarbon of straight chain structure.

normal solution. A concentrated solution containing one gram-equivalent of a substance per liter of solution.

nose sill (drill). Short piece of timber placed under the end of the main sill of a standard rig front of a well rig.

notch. A deep, narrow cut in the base of a sea cliff made by breaking waves.

novaculite. An extremely fine grained, highly indurated chert. It is white or pale gray, semitranslucent and composed of microcrystalline quartz with some fine-grained chalcedony.

nuclear fission. Division of a heavy nucleus into two approximately equal parts. For the heaviest nuclei, the reaction is highly exothermic, the release of energy being about 170 million electron volts per fission. A well known example is the fission of the compound nucleus formed when U^{235} captures a slow neutron.

nuclear fusion.   Coalescing two or more atomic nuclei.

nuclear precession magnetometer.   A magnetometer that utilizes the precessional characteristics of hydrogen nuclei when in an ambient field. The data of this instrument is in the form of a frequency measurement, which in turn is proportional to the magnetic field intensity.

nuclear radiation.   Corpuscular emission as alpha and beta particles or electromagnetic radiation as gamma rays, originating in the nucleus of an atom.

nuclear reaction.   An induced nuclear disintegration; a process occurring when a nucleus comes into contact with a photon, an elementary particle, or another nucleus.

nucleus.   Central core of an atom, composed largely of protons and neutrons. Although very tiny compared to the size of the entire atom, the nucleus contains most of the atom's mass. An atom consists of a nucleus surrounded by electrons in number equal to the protons in the nucleus.

nuclide.   A special atom characterized by the constitution of its nucleus. The nuclear constitution is specified by the number of protons, number of neutrons, and energy content, or alternatively by the atomic number, mass number, and atomic mass. To be regarded as a distinct nuclide, the atom must be capable of existing for a measurable lifetime, generally greater than 10^{-10} second.

nuée ardente (hot cloud).   A French term applied to a highly heated mass of gas-charged lava ejected more or less horizontally from a vent or pocket at the summit of a volcano, onto an outer slope down which it moves swiftly, however slight the incline, because of its extreme mobility. Products of cooling, degassing and solidification of nuee ardente are usually welded tuffs.

O

observed gravity. Uncorrected value of gravity at a station measured with a gravity meter.

obsidian. Homogeneous natural glass with low percentage of water. The word is of classic and ancient origin but is now used for silica rich volcanic glasses with a prefix to indicate the composition as in rhyolite-obsidian, etc. *See* natural glass.

occlude. Take in and retain in pores or other openings; to absorb; used particularly with respect to absorption of gases by certain substances without losing their characteristic properties.

oceanic basalt. Basalt occurring in the ocean basins of the world. The lava originates in either the asthenosphere or upper mantle. Oceanic basalts are found associated with mantle hot spots or with divergent plate boundaries (mid-ocean ridges). Oceanic basalts range in composition from tholeiitic to alkali.

oceanic crust. Mass of basaltic material approximately five kilometers thick lying under the ocean bottom and covered by a thin sedimentary veneer.

octane. Any of the group of isomeric hydrocarbons (C_8H_{18}) of the paraffin series; specifically, normal octane, a colorless liquid boiling at 124.6°C found in petroleum.

octane number. Term numerically indicating the relative antiknock value of a gasoline, based upon a comparison with reference fuels isooctane (100 octane number) and normal heptane (0 octane number). The octane number of an unknown fuel is the percent of volume of isooctane, with normal heptane which matches the unknown fuel in knocking tendencies under specific conditions.

octane value. Blending agent having two octane numbers or values. One of these, the actual value is the octane number when tested alone. The other is the blending value, the octane number they impart when blended in a fuel.

odor test (refin). A test to determine whether a fuel is "sweet" or "sour" during or after refining.

offlap. Arrangement of sedimentary units in a depositional basin whereby the shoreward edge of each succeeding younger unit is farther offshore than the unit on which it lies. The bottom most oldest unit was unconformably deposited on the underlying rock. Diagram shows an offlap resulting from progressive emergence of a land area either by tectonic uplift or a progressive recession of the sea. (See Fig. p. 235)

234

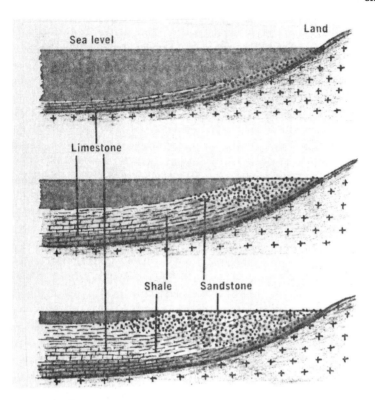

offset well. Well drilled opposite to a well on an adjoining property. An offset well is located at the same distance from the property line as the well which it offsets.

offshore drilling. Drilling for oil in the waters of the sea or large lake. Drilling platforms for offshore operations may be mobile floating units with submersible bases, or permanent structures used as production platforms when drilling is completed.

offshore platform. Two typical offshore platforms (a) steel structure (b) concrete gravity platform. (See Fig. p. 236)

oil. General term for a water insoluble viscous liquid possessing lubricating properties. There are three principal classes of oil: **(1)** Mineral oils derived from petroleum, coal and shale, consisting of hydrocarbons; **(2)** Fixed or fatty, oils from animal, vegetable and marine sources, consisting chiefly of glycerides and esters of fatty acids; **(3)** Essential oils, with characteristic odors, perfumes, volatile products, mainly hydrocarbons derived from certain plants. An unctuous combustible substance, liquid or at least easily liquefiable on warming and soluble in ether but not in water.

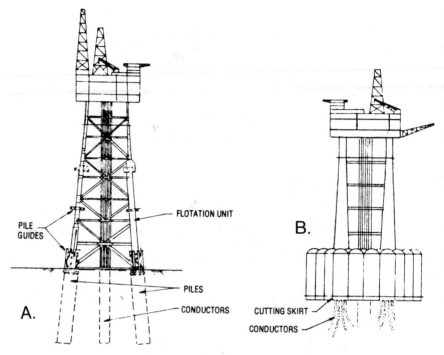

PILE GUIDES

FLOTATION UNIT

B.

A.

PILES

CONDUCTORS

CUTTING SKIRT

CONDUCTORS

offshore platform

oil base mud.   Drilling mud with clay particles suspended in an oil rather than wa-ter. The term "oil-base mud" is applied to a special type drilling fluid where oil is the continuous phase and water, the dispersed phase. Oil-base mud contains blown asphalt and usually one to five percent water emulsified into the system with caustic soda or quick lime and an organic acid.

oil bases.   Crude oils are classified into paraffin base oils and asphaltic base oils according to the nature of the predominating solid residuals. Mixed base oils are those containing both paraffin and asphaltum.

oil bloom.   An irridescent or opalescent reflection from the surface of petroleum oils and resin oil, an indication of the base of the oil, whether paraffin or asphaltic.

oil cold test.   Test made to determine the point at which clouding or coagulation sets in.

oil column.   Difference in elevation between the highest and lowest portions of various producing zones.

oil cut.   A mixture of oil and drilling mud recovered in testing.

oil dehydrating plant.   Plant for ridding oil of excess water or of water held in emulsion; may employ a simple heating process, an electrolyitic process, or a chemical treatment.

oil emulsion.   Mixture of water and oil in which the oil is more or less permanently suspended in the water in the form of very small droplets. When prepared as a soluble oil, the emulsion becomes a lubricant fluid.

oil-emulsion water (milk emulsions) (drill).   A drilling fluid in which the oil content is usually kept between three to seven percent and seldom over ten percent, can be considerably higher. The oil is emulsified into fresh or salt water with a chemical emulsifier. Sometimes sodium carboxymethylcellulose (CMC), starch, or gum may be added to the fresh- and salt-water systems.

oil field emulsion.   Most crude oils when reaching the surface contain water, the percentage varying from a few to fifty percent or more. Chemicals used to break up these emulsions are usually of the water-in-oil type and differ from the oil-in-water type encountered in the course of petroleum refining.

oil field water analysis.   Water analyses are used to solve a variety of problems connected with the exploration and development of oil and gas pools. Probably the most important geological use of oil field water analysis is its application to the interpretation of electric well logging. Reservoirs in multipay fields frequently may be separated or correlated by their water analysis. This is particularly useful in lenticular sand reservoirs. Radical changes in the abundance or chemistry of water when passing from a shallow to a deeper formation indicate different geologic environments. In the absence of mechanical leaks, a sudden reduction in the abundance of the brine suggests the crossing of an unconformity. Water analysis may tell whether the water produced with the oil comes from the bottom of the well and is part of the formation water of the reservoir rock, or whether it is a shallower formation water entering the well because of improper cementing of the casing or leaks due to breaks in the casing.

oil field water (connate water).   Original seawater in which marine sediments were deposited, presumably originally filling all the pores. Most reservoir waters are quite different in chemical composition from seawater; they have undoubtedly circulated and moved and, in fact, have probably been completely replaced since the sediments were deposited. Most oil field waters are brines, characterized by an abundance of chloride, especially sodium chloride, and they often have concentrations of dissolved solids many times greater than that of modern seawater.

oil field water (free water).   Most oil and gas pools occur in aquifers, that is, in water-saturated, permeable rocks. The water may be meteoric, connate, or mixed in origin. The water confined in the interconnected pores of a reservoir rock may be considered a continuous and interconnected body of water. Such water is confined, as in the water system of a city and is ready to flow toward any point of pressure release. It is called free water to distinguish it from the interstitial or attached water.

oil field water (meteoric water).   Water that has fallen as rain and has filled up the porous and permeable shallow rocks, and percolated through them along bedding planes, fractures, and permeable layers. Water of this type contains dissolved oxygen and carbon dioxide and is carried into the ground from the vadose zone above the ground water table, where the oxygen may react with sulfides to produce sulfate and the carbon dioxide may react to produce carbonates and bicarbonates. The presence of carbonates, bicarbonates, and sulfates in oil-field water suggests that at least some of the water has come from the surface.

oil field water (mixed water).   Characterized by both a chloride and a sulfate-carbonate content suggesting a multiple origin; presumably meteoric water mixed with or partially displaced by connate water of the rock. Mixed waters may occur near the present ground surface or may be found below unconformities.

oil from gas.   A process combining members of the gas family to make members of the lubricating oil in a manner somewhat similar to the combination of these members to make polymerized gasoline.

oil gas.   Obtained by the destructive distillation of petroleum. The raw gas contains carbon, naphthalene and sulfur compounds. The carbon and naphthalene are recovered and the gas is cleaned of sulfur compounds by the use of a scrubber.

oil gas tar.   Produced by cracking oil vapors at high temperatures in the manufacture of oil gas.

oil horizon.   Upper surface of oil in a well or the stratum in which it is located.

oil hydrometer.   Device used for measuring the gravity of oil; consists of a weighted closed glass tube floating at various heights corresponding to the gravity of the oil.

oil insoluble (sludge).   Parts or ingredients in sludge not dissolved in a petroleum product as naphtha. Asphaltene for example, is not soluble in a petroleum product and therefore is an oil insoluble.

oil-in-water emulsion mud (drill).   Commonly called "emulsion mud." Any conventional or special water-base mud to which oil has been added. The oil becomes the dispersed phase and may be emulsified into the mud either mechanically or chemically.

oil lenses.   Small irregular patches of oil bearing sands having a cross section similar in appearance to an optical lens.

oil liver.   Gelatinous or jelly-like mass formed in lubricating oils.

oil log.   Record kept during the drilling and life of a well preserving all data about the geology, depth, drilling difficulties and production.

oil, nuetral.   An oil obtained by fractional distillation of crude oil and not requiring treatment with sulphuric acid.

oil, non-viscous.   Neutral oils below 135 seconds viscosity; Saybolt viscosity at 100° F.

oil of paraffin.   Colorless to yellowish, limpid oil having a specific gravity of about 0.880 and not boiling below 360°F.

oil pool.   Term applied to a producing oil field or the accumulation of petroleum underground. In reality, however, oil does not exist underground in pools or lakes. It is distributed throughout the interstices of oil sands or sedimentary formations.

oil pulp.   A soap used to make nonfluid oils and greases. Consists of a salt of the fatty acids as aluminum stearate and aluminum palmitate, dissolved in oil to act as a thickener or stiffener.

oil pumping.   Production of crude oil from a well by a plunger pump, the sucker rods of which work inside the petroleum tubing. Sometimes referred to as on-the-beam pumping.

oil reclaiming (filtration).   Oil passed through a filter as it comes from equipment and then returned for reuse after filtration, in the same manner that crank case oil is cleaned by an engine filter.

oil reclaiming (gravity).   Method employed to remove solids from oil placed in settling tanks. The lighter the body of the oil, the higher will be the rate of settling; also the higher the temperatures, the more rapidly the solids will settle from the oil. However, when heated it must be accomplished without causing circulation in the body of the oil keeping the small solids in a state of motion, as when the oil is heated from the bottom or at only one area.

oil regenerator.   Device for economizing heat during the process of refining. The heat absorbed in cooling of one product is used for heating another oil in the course of manufacture so that the demand for fuel is decreased; sometimes known as "heat exchanger."

oil reserve.   Oil remaining underground. The quantity is known to a high degree of probability and its recovery is commercially feasible at present day prices and costs.

oil rights.   Ownership of oil under a given tract of ground is transferred from one owner to another by lease or purchase. In leasing or purchasing oil property, it is more usual to lease or purchase both oil and gas rights, though in many cases the two substances are disposed of separately.

oil sand.   Crude oil is located not in large pools, as the term oil pool would indicate, but in one or many layers of porous oil saturated rock, each of which is called an oil sand. These formations as a rule, are overlaid with an impervious cap rock preventing the oil and gas from escaping while salt water usually occupies the lower parts of the sands.

oil saver (drill).   Device attached to the top of the casing, casing head or control head to prevent waste of oil and gas around the drilling cable.

oil separator.   Tank, usually a sunken concrete construction adjoining a battery of stills, into which waste oils and surface sewage are run for separation. Oil is recovered for further refining by skimming the surface with a swing suction pump. Oil thus removed is termed "separator slop."

oil shale.   Properly called kerogen shale because the hydrocarbon content is pyrobituminous and insoluble in ordinary petroleum solvents. Oil shales are organic shales that yield petroleum hydrocarbons upon destructive distillation. An oil-bearing sedimentary formation containing crude shale oil or kerogen which may be extracted in a retort by destructive distillation. It is dark green to brownish in color and like petroleum crude, distilled and refined into many products.

oil solubles.   Products dissolved in a petroleum product as gasoline gum, certain forms of tar, etc.

oil still.   Device for heating and vaporizing crude oil or partly finished oils where various hydrocarbons are separated by difference in their boiling points.

oil string.   A string of casing used to keep the oil well open through the rock formation, down to or through the producing formation.

oil structure.   A term used by the geologist when referring to a geologic configuration proven to possess oil or indicating the possibility of possessing oil.

oil, topped.   Oil with the light naphtha, gasoline and kerosene fractions removed by straight distillation at low pressure and temperature.

oil traps (refin).   Separators used in steam and ammonia pipe lines for the separation of oil and greases from the steam or ammonia gas.

oil traps.   Diagrams show places where oil and gas accumulates. *(a)* anticline; *(b)* salt dome; *(c)* unconformity. The oil-pools are shown in black. (See Fig. p. 241)

oil water interface.   Oil will float on top of water due to the difference in specific gravity. The bottom level of the oil column and top level of the water column are at a common level called the interface.

oil zone.   Formation or horizon of a well producing oil. In any given reservoir,

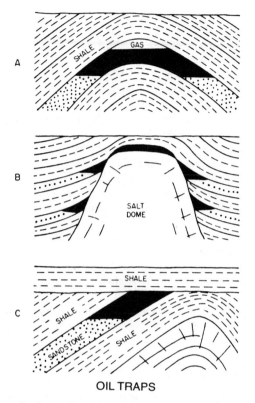

OIL TRAPS

the oil zone is usually immediately under the gas zone and on top of the water zone if all three fluids are present and segregated.

oiliness. Characteristic of a liquid responsible for the degree of friction between two surfaces accounted for, on the basis of viscosity alone. The ability of a lubricating oil to orient itself on bearing surfaces so as to form new surfaces with a low coefficient of static friction.

oleates. Salts produced by the action of oleic acid on the various metals.

olefiant. Forming an oil or an oil forming substance.

olefiant gas. Ethylene, C_2H_4 sometimes found in natural gas, is the most illuminative constituent in coal gas; heavy carbureted hydrogen.

olefins. Open chain hydrocarbons having one or more double bonds per molecule; specifically, hydrocarbons belonging to the monoolefins or ethylene series $(C^{n2}H_2{}^n)$ having one double bond per molecule as compar ed with diolefin, triolefin, etc. Examples are ethylene, propylene, isobutene and n-butene. The main source of the olefins containing two to four carbon atoms in the molecule, ethylene, propylene, and the butylenes are the gases obtained from cracking operations. The most important of these gases is ethylene for its conversion to chemicals as ethylene glycol and in the manufacture of plastics. Higher molecular weight olefins of six or more carbon atoms are obtained by thermal cracking of the waxes produced during the refining of lubricating oils. They are used for the manufacture of detergents, lubricating oil additives and higher alcohols.

olefin hydrocarbons. Unsaturated hydrocarbons. Olefins are almost entirely formed by cracking, a considerable portion of cracked gasoline consists of monoolefins. Diolefins are also preserved, but these are so unstable chemically that they react with one another to form gum, therefore they must be removed from gasoline. The monoolefins have names corresponding to those of the paraffins from which they are derived but have the ending -ene. Diolefins also have names corresponding to paraffins but end in -diene as in butadiene.

olefin polymer oil. Synthetic lubricating oil consisting of a variety of polymers and made by a catalytic process.

oleic. Pertaining to; derived from; or contained in an oil.

oleic acid. An organic acid obtained principally from animal oils and fats, and olive oils. It forms salts known as the "oleates" used in the preparation of some lubricants.

oleoginous. Property of oiliness found in petroleum lubricants.

oleo oil. A fixed oil containing oleic acid giving it a red cast or color; sometimes called suet oil.

oleum. A strong sulfuric acid containing an excess of sulfur trioxide; used in the chemical refining of special oils and to secure oils which are odorless, colorless and tasteless; also known as fuming sulfuric acid.

oleum liver. Made from sludge produced in manufacturing white oils with fuming acid; source of green acid and the green acid soaps used extensively for breaking emulsions; also sold for fat splitting.

oleum spirits. A petroleum cut, boiling between 300° and 400°F.

olisostrome. A deposit of throroughly mixed sedimentary materials occurring as the result of submarine gravity sliding or slumping. It may consist of mud and blocks of sediment in a chaotic mass of sufficient thickness and extent to be mappable. Usually found as lensoid layers inter calated with normal sedimentary beds of the continental slope and rise environment.

omnidirectional hydrophone (seis). A hydrophone whose response is essentially independent of angle of arrival of the incident sound wave.

once run oil.   Second and last cut of the crude shale oil when first distilled.

oncolite.   A rounded, concentrically laminated calcareous nodule or structure formed in the same way as a stromatolite by secretions from blue-green algae but much smaller, rarely exceeding 10 centimeters in the largest dimension.

onlap.   Arrangement of sedimentary units in a depositional basin whereby the shoreward edge of each succeeding younger unit is closer to shore than the unit on which it lies. The sequence of units is laid down unconformably on the preexisting rock. Onlap results from the progressive submergence of a land either by tectonic subsidence or a progressive raising of sea level.

on the line.   Refers to a tank when it is being emptied into a pipe line.

on the pump.   Flow of an oil well cannot be induced naturally making it necessary to install pumping equipment.

oolite.   Oolitic limestone formed of round grains of calcite, ooliths which look like fish eggs. This resemblance accounts for the rock's Greek name meaning "eggstone."

oolith.   A small rounded accretionary grain generally 0.5 to one millimeter in diameter usually occurring in great numbers so as to give the appearance of fish eggs. Most ooliths show fine concentric laminations surrounding a grain of sand, a fossil fragment or an organic pellet. Most ooliths consist of calcium carbonate but occurances are known where they are composed of dolomite, silica, iron oxide, iron sulfide and other materials. Most geologists interpret them to be indicative of inorganic precipitation in a shallow, warm, wave agitated environment.

oolitic.   Sediment or sedimentary rock with ooliths as its principal constituent. A rock made up of calcareous ooliths is an oolitic limestone.

ooze.   Fine grained deep-sea deposits containing more than thirty percent of organic residues; soft mud or slime.

ooze, diatomaceous.   A soft mud consisting predominantly of the siliceous cell walls of single-celled marine algae known as diatoms. The greatest concentration of diatomaceous ooze occurs in the North Pacific and the Antarctic Ocean.

ooze, globigerina.   A soft mud consisting predominantly of the limy shells of minute one-celled animals called Globigerina; covers large portions of the floors of Atlantic, Pacific and Indian Oceans.

ooze, radiolarian.   A soft mud consisting predominantly of the delicate and complex siliceous hard parts of minute marine protozoa called radiolaria. Radiolarian ooze predominates in a long east-west belt in the Pacific Ocean just north of the equator.

opalescence.   A term generally applied to an oil, particularly lubricant oil containing visible wax; means milkiness or cloudiness.

opalescent.   Reflecting a milky or pearly light from the interior of the fluid or having an opaline play of colors of varying shades.

open cup tester.   Instrument for determining the flash point of an oil in which the flame is exposed to atmosphere.

open flow (prod).   Maximum delivery rate at the surface for a gas well with a back pressure at the sand face of one atmosphere plus tubing friction plus weight of a column of gas from surface to formation.

open flow test.   A means of determining the unrestricted volume of gas flowing from a well during a given period of time.

open hole (drill)   Unprotected hole below the shoes of the last landed string of casing when drilling for oil; uncased part of the well.

open sand.   Sandstone, with sufficient porosity and permeability to provide good storage for oil.

optical activity.   Most petroleum has optical activity, the power to rotate the plane of polarization of polarized light. This is measured with a polariscope in degrees per millimeter and the average range is from zero to one point two degrees. If the plane is rotated to the right, the substance is said to be dextrorotary; if to the left, levorotary. All crudes either are optically active or contain distillation fractions, particularly in the intermediate range, 250 to 300°C at 12 millimeters mercury, that are optically active.

optimum separation point.   Pressure and temperature conditions where the maximum point of condensation occurs in the separators under field operations; applied to natural gasoline recovery.

ore.   Rock, sediment or other naturally occurring material from which it is profitable to extract economically valuable minerals. In most cases, the ore minerals are valuable for their metal content. The word ore is derived from an Anglo-Saxon word meaning a lump of metal.

ore shoot.   Concentration of primary ore along certain parts of a rocky opening.

organic.   A substance or compound pertaining to or derived from vegetable or animal matter, including carbon artificially produced from materials originating from these sources.

organic acidity (lubricating oil).   When properly refined lubricating oil is free from mineral or inorganic acids and alkalies. However, all contain natural organic substances, either weak acids or upon oxidation, form acidic compounds. The proportion of these acidic compounds present in an oil is commonly referred to as its organic strong acidity. Organic acidity is represented by the difference between total acid number and strong acid number. Additive type oils may have an initial acid or alkaline reaction, distinctive of the type of additive in the oil.

organic acids.   Acids formed in organic matter as the fatty acids, etc. also occur in small quantities in petroleum. Any acids containing the carboxyl group (COOH).

organic filler (grease).   A product secured from animal or vegetable sources as wool, cotton, hemp, horse hair, goat hair, ground cork, and sawdust used as organic fillers in grease making.

organic grease.   Applies to any grease containing a fixed oil, fixed fat or derivative of these products; may also include any grease containing a petroleum oil.

organic rock.   Rocks consisting of the fossilized remains of plant or animal life as coal, limestone, diatomaceous earth, etc.

organic sediments.   Sediment consisting of the remains of plants and animals as shells, ferns, etc.

organic solvent.   Has an organic source as alcohols, ketones, esters, etc. When used as a lacquer solvent, it is known as an organic lacquer solvent.

orifice meter.   Instrument used to measure fluid flow by recording differential pressure across a restriction placed in the flow stream and the static or actual pressure on the system.

orifice well tester.   Device for measuring gas. A precision drilled nipple attached to the open end of a line with a plate having small opening attached to the end of the nipple and a connection to measure gage pressure; designed for low pressure and low volume testing.

orogenic.   Pertaining to or designating an orogeny or mountain building disturbance; contrasts with epeirogenic.

orogeny.   Large scale regional mountain building processes, including folding, faulting, and igneous activity; usually associated with convergent plate boundaries.

orsat apparatus.   A portable device for the analysis of gases, consisting of a glass measuring tube surrounded by a glass jacket filled with water and a set of three glass absorption pipettes.

ortho-. A chemical prefix denoting what may be called a normal or ordinary compound distinguishing it from others which are not normal, or to denote two adjacent positions in the benzene ring and occasionally in other rings.

Orthoflow, fluid catalytic cracking (refin). To convert petroleum products into valuable light olefins, gasoline, and furnace oil. The products are high octane gasoline, furnace or diesel oil blend stocks, carbon black feedstock, light olefins and isobutane for alkylation and dry gas.

orthogneiss. Name of a gneissose rock derived from rocks of igneous origin.

orthoquartzite. A clastic sedimentary sandstone consisting almost exclusively of quartz grains. Matrix is absent or rare. Frequently interpreted to represent a beach sand. This term is old but still in use by some authors but most authors prefer to use the equivalent term, quartz arenite.

outage. Difference in volume between the standard temperature at 60°F and the volume in the same tank at temperatures below 60°F. In cases where the volume has increased due to temperatures greater than 60°F, the difference is called overage or innage. The difference between the full or the rated capacity and the actual content of a barrel, tank or tank car.

outcrop. Naturally protruding or erosionally exposed or uncovered portion of a rock bed or formation. That part of a rock formation appearing at the surface.

outwash. Great bulk of stratified drift generally intimately associated with end moraines. Outwash is the product of redistribution by water, of rock fragments carried to the end-moraine. Ideally, outwash forms a large apron spreading outward from the glacial terminous and decreasing in over-all particle size grade with increased distance from the ice until fine sand and silt dominate the deposits. In valley glaciers, the outwash is confined by the valley walls whereas in continental ice sheets the distribution of an individual outwash apron is depended upon the localities of major streams draining the ice. An outwash fill is often called a valley train. Many are pitted with kettle holes, meaning that the deposit was built by receeding ice.

outwash deposits. Coarse glacial material rolled or carried by streams flowing from melting ice fronts and finally deposited.

outwash plains. Plains formed by deposition of materials washed out from edges of a glacier.

overage. Increase in volume of a liquid due to temperature increases.

overburden. Waste overlying the good stone in a quarry or open pit mine; worthless surface material covering a body of useful mineral.

overhead (refin).   In a distilling operation, portion of the charge vaporized.

overhead drums (refin).   Part of the condensing equipment on certain types of stills, generally consist of large drumlike chambers in which the vapors may expand and condense so that only the lighter vapors tend to escape from the drum and continue along the vapor line. The condensate rises back into the still. This feature of condensing equipment is now obsolete, being replaced by more elaborate fractioning towers.

overhead products (refin).   All products taken off from an oil still in a vapor form.

overlap.   (1) In stratigraphy, the extension of one formation beyond another or over the beveled edge of another. Lateral extension of a rock unit beyond the limits of the next underlying unit. (2) A thrust fault in which the shifted strata double back over themselves.

overpoint (refin).   Initial boiling point in the process of distillation; temperature at which the first drop falls from the tip of the condenser into the condensate flask.

override.   A royalty payment or percentage of the gross income from production deducted from the working interest.

overshot (drill).   A fishing tool with a bell shaped end, designed for the purpose of bringing into line and connecting onto a piece of lost pipe or casing in a deep well.

overthrust.   Low-angle reverse fault with the hanging wall pushed over the foot wall for a considerable distance.

overthrust fault.   *See* overthrust.

overturned.   Having been tilted past the vertical and hence inverted in outcrop; said of folded strata and of the folds themselves.

overturned fold.   In geology, a fold with both limbs inclined in the same direction, but one more steeply inclined than the other.

oxbow lake.   A crescent shaped lake found on the flood plain of a meandering stream. When a meandering stream straightens its course by cutting off a meander, the disconnected meander becomes an oxbow lake.

oxidation oven.   A laboratory apparatus for determining the rate at which atmospheric oxygen combines with an oil.

oxidation, rock.   Rocks decompose or decay when their component minerals are altered chemically by oxidation. In the process of oxidation, oxygen is added to the rocks, especially to the iron compounds. Ferrous silicates as pyroxenes, amphiboles, and olivine convert their ferrous iron to ferric oxide with accompanying color changes from green or black to red, yellow, or brown. The oxidation of pyrite composed of iron and sulfur, leads to the formation of sulfuric acid and ferric oxide.

oxidize.   Combination with oxygen by direct contact or contact with some oxidizing agent. Also to lose electrons in an oxidation reaction.

oxidized mineral.   Mineral formed by the direct union of an element with oxygen as with corundum, hematite, magnetite, cassiterite.

oxidized zone.   Portion of an ore deposit subjected to the action of surface waters carrying oxygen, carbon dioxide, etc.; zone where sulphides have been removed or altered to oxides, hydroxides, and carbonates.

oxo process (refin).   A two stage process for producing alcohols by treating olefin mixtures with carbon monoxides and hydrogen to give aldehydes, reduced by hydrogen.

oxygen.   Gaseous element occurring free as about 21 percent of the atmosphere Chemically very active and supports combustion, the combination of oxygen with other substances generally produces heat; taken from Greek words meaning acid producing.

oxygen compounds (crude oil).   Compounds containing oxygen also exist in crude oil. These compounds are most organic acids but include phenols, and complex oxygen-nitrogen, and sulphur containing asphaltenes and resins. The oxygen content of a crude oil is indicated by its acid content usually naphthenic in composition, and may contain up to two per cent acids. The acids may be present in any petroleum fraction but reach their highest concentration in the gas oil fractions. Organic acids are reached or readily removed from petroleum products by treatment with a lye solution.

oxygenated oil.   Oils partly oxidized as with blown oils.

ozocerite.   A crude natural paraffin.

ozokerite (earth wax).   From Greek meaning "odoriferous wax." A naturally occurring mineral wax usually dark-brown in color containing mineral matter and oil. On purification it yields a white to yellow microcrystalline wax which may contain small quantities of oil. Fully refined ozokerite is a hard, white microcrystalline wax substantially free from oil which was formerly and still occasionally known as deresin.

P

packed tower (refin). Fractionating or absorber tower filled with small objects to effect an intimate contact between rising vapors and falling liquid.

packer fluid (drill). Any fluid placed in the annulus between the tubing and casing above a packer. Along with other functions, the hydrostatic pressure of the packer fluid is utilized to reduce the pressure differentials between the formation and the inside of the casing and across the packer itself.

packstone. A carbonate rock associated with reefs. Fragments of organic skeletal material are packed closely together. Intergranular pores are usually filled with a fine grained calcareous matrix but the strength of the rock is dependent upon intergranular contact rather than the matrix.

pahoehoe. A Hawaiian term for basaltic lava flows with a billowy or ropy surface.

paleo. Prefix indicating great geologic age or existence early in geologic time as in paleobotany, the study of fossil plants.

paleocurrent. An ancient current, usually of water, existing during the deposition of a sediment and presumably influencing the sedimentary depositional environment. The existence and direction of flow of the current is inferred from structure as crossbedding, ripples and other sole marks preserved in the sedimentary rock.

paleogeologic map. A map showing the distribution of outcrops of various geological formations as they were at some period in the geologic past. These maps are usually described by affixing the prefix ''pre'' to the age of the oldest existing strata deposited after the period represented by the map. They are used for expressing the geologic relationships which existed on an old erosion surface prior to its burial and consequent involvement in an unconformity.

paleomagnetism. Natural remnant magnetism of rocks reflecting the direction of the earth's magnetic field at the time of the rock's formation. Most frequently studied in igneous extrusive rocks but also studied less effectively in sedimentary rocks. Used to determine the geographic positions of crustal plates at various times in geologic history as well as to measure spreading rates of divergent crustal plates and to detect magnetic polar reversals in geologic history.

paleontology. From Greek palaios, ancient plus ontos, a being plus logos, word or discourse. The study of fossils preserved from past geologic ages, including classification, phylogeny and chronology of earth's history.

paleosol.   A soil formed in the geologic past and preserved totally or partially in the stratigraphic position of its formation. Burial by younger sediments protect paleosols from destruction by erosion. May be difficult to recognize and interpret if it has never been buried or if exhumed subsequent to burial; useful in the study of climates and climatic changes in geologic history.

palimpsest.   Describes a circumstance where a recent process produces features which partially obscure, destroy or reconstruct patterns created by an older process. The term was used originally in regard to metamorphic structures or textures and superimposed drainage patterns. The term is now coming into wide use in sedimentology to describe autochthonous sediment deposits which continue to exhibit some of the attributes of the source sediment. Sediments which were deposited originally on continental shelves during low sea stands of Pleistocene time and subsequently have been reworked by modern shelf processes are relict and modern at the same time; such sediments are called palimpsest.

palinspastic map.   A map showing paleogeographic or paleotectonic features as they appeared prior to being distorted by faulting or folding. A map of an area modified by straightening folds and returning faults to prerupture position in order to restore ancient configurations.

paludal.   Pertaining to swamps or marshes and to material deposited in a swamp environment.

pan.   (1) A natural shallow depression often occupied by a lake. If the lake waters evaporate in time, the depression will be lined with salt crystals and called a salt pan. (2) The upper portion of the b-zone of soils in temperate climates may be hardened or cemented by precipitation of salts previously leached from the a-zone. The hardened zone is called pan or hardpan.

Pangaea.   A super continent formed during Mesozoic time when all continents were joined. The northern portion, named Laurasia, was sometimes partially and sometimes completely separated from the southern portion, named Gondwanaland, by a large sea called Tethys.

parabolic dune.   A dune with a long, scoop-shaped form when perfectly developed, exhibiting a parabolic shape with the horns pointing upwind.

paraconformity.   An obscure or uncertain unconformity above and below which the beds are parallel and there is little physical evidence of a long lapse in deposition.

paraffin.   Broad term embracing many grades of wax. A hydrocarbon belonging to the methane series, obtained from paraffin base petroleum, first obtained from dry distillation of wood. A white, waxy, odorless, tasteless substance, harder than tallow, softer than wax, with a specific gravity of 0.890; melting point is variable depending somewhat upon its origin ranging between 109°F and 151°F (43°C and 65°C). An ultimate analysis yields on the average, carbon 85 percent, and hydrogen 15 percent. It is insoluble in water, indifferent to the most powerful acids, alkalies and chlorine, and can be distilled unchanged with strong sulphuric acid. Warm alcohol, ether, oil of turpentine, olive oil, benzol, chloroform, and carbon disulphide dissolves it readily.

paraffin acid.   Acid formed by oxidizing paraffin with a concentrated solution of nitric acid.

paraffin asphalt petroleum.   "Mixed-base" crude oils of the Midcontinent and southern fields containing both paraffin and asphaltum.

paraffin base crudes.   Crude petroleum containing solid paraffin hydrocarbons but no asphalt.

paraffin base oils.   Paraffin base crude oils are so-called because they carry from two to six percent of paraffin wax in solution. They carry a fairly high percentage of gasoline and kerosene and only a little sulphur, oxygen or nitrogen and have a very low specific gravity.

paraffin base petroleum.   Crude oil carrying solid paraffin hydrocarbons and practically no asphalt.

paraffin butter.   A variety of native paraffin used in making candles; also one of the last products of petroleum distillation.

paraffin coal.   Grades of bituminous coal used in the production of burning oils and paraffin.

paraffin dirt.   A clay soil of rubbery or curdy appearance and texture resembling Art Gum in appearance; occurs in the upper few inches of a soil profile in the vicinity of gas seeps and is probably formed through the biodegradation of natural gas by low-rank organisms; found along the Gulf Coast of Texas and Louisiana. There is probably no true paraffin in it.

paraffin distillate.   At ordinary temperature, crystalline product ready for pressing, serving as a base for paraffin wax and paraffin oils. It should be distinguished from the wax distillate produced by the steam reduction of fuel oil since the latter contains wax in a rather amorphous form.

paraffin gas.   Gas with a name ending in -ane as methane, butane including saturated hydrocarbons gas.

paraffin hydrocarbons.   Paraffins are saturated hydrocarbons containing the maximum number of hydrogen atoms. For this reason they are chemically inactive under most conditions but do react chemically when broken apart. Their carbon atoms form chains. Those without branches or straight-chain structure are normal paraffins. Those with branches are paraffins; isobutane, for example will react to form the high octane material, aldylate, whereas normal butane will not. Paraffins with four or less carbon atoms are gaseous at ordinary temperatures; those with five to fifteen are liquids; those with over fifteen are waxes. The lighter gaseous paraffins, methane, ethane are components of natural gas. The heavier gaseous paraffins propane, butane, and isobutane are components of liquefied petroleum gas.

paraffin oil.   A near white, wax free oil secured by pressing wax distillate with a low lubricating value as white oil.

paraffin presses (refin).   Refinery equipment for the separation of paraffin oil and crystallizable paraffin wax contained in paraffin distillates; a combination filter and press.

paraffin scale (refin).   Crude paraffin wax remaining in the "pans" after sweating.

paraffin wax.   Colorless, more or less translucent mass, crystalline when separated from solutions, without odor and taste, slightly greasy to touch, consisting of a mixture of solid hydrogen carbons of the paraffin series. When found in petroleum, has a crystalline structure and is associated with the product distilled from petroleum.

paraffinicity.   A term describes paraffinic structure of products. Products made from paraffin base crudes have a higher paraffinicity than those made from naphthene base crudes.

paraffinic acid.   Obtained by oxidation of paraffin by concentrated nitric acid or by chromium trioxide mixture; easily oxidized by air when warm and exposed to light.

paraflow.   A condensation product of an aliphatic and aromatic hydrocarbon blended with a relatively high pour point lubricating oil causing a decided lowering of the pour point and improving its viscosity temperature and coefficient, and the viscosity index, without removal of wax.

paragenesis.   General term for the order of formation of associated minerals in time succession, one after another.

parallel evolution.   Similar development of organisms with related, yet different, ancestries under similar environmental conditions.

parasitic cone. A cinder cone on the side of a large volcano.

parathenes (refin). High grade oil hydrocarbons selected from lubricating oil stocks by the solvent process of refining.

parent rock. Original rock from which sediments were derived to form later rocks.

partial condenser (refin). An exchanger, located in or on the top of a tower condensing a part of the overhead vapors so that they drop into the tower as reflux.

partial melting. Selective melting of a rock, involving initial melting of felsic minerals followed by the melting of the more mafic minerals as temperature increases.

particle. A small unit of matter, often a single crystal, or part of a single crystal; may be angular or sharp, rounded or smooth and of any shape, spherical, platey, or other.

particle size. Some dimension of a particle useful in expressing its size such as volume, largest diameter, intermediate diameter, and diameter of an equivalent sphere.

particle size analysis. Techniques by means of which bulk measurements of the grain sizes of the constituent particles of a sediment sample are made. This may be accomplished by measuring settling velocities of grains in water or by sieving.

particle size distribution. Distribution of particles by weight or number in each of several size fractions. Particle size distribution reflects the erosional process transporting and depositing a sediment. By measuring the particle size distribution of a disaggregated sedimentary rock, a sedimentologist may identify many features or physical parameters of the environment of the rock's original deposition.

parting. Separation of crystals along planes that are not true cleavage planes.

partition coefficient. Ratio of the concentration of a substance in two different phases. The two phases may be two immiscible liquids or a liquid and a gas phase, as in gas chromatography.

Pascal's law. States that the same pressure exerted at any point upon a confined liquid is transmitted in all directions throughout the liquid.

pattern shooting (exploration seismology). Placing and firing explosives in a geometrical pattern designed to enhance the signal detected by geophones. Usual goals are to improve signal to noise ratio and to generate maximum useful energy while minimizing surface effects which may be destructive or disruptive to adjacent social activities.

pay sand. Portion of an oil or gas sand in which the oil or gas may be found in commercial quantity.

Pease's Electric Tester. An instrument in which the vapor of petroleum is ignited by an electric spark passing above the oil cup, resting in a water bath.

peat. Partially reduced plant material containing approximately 60 percent carbon and 30 percent oxygen. An intermediate material in the process of coal formation. Peat is an incomplete coal or an immature coal not sufficiently compacted or aged to produce true coal. It is the product of vegetation growing and being buried in peat bogs, and is of soft brittle nature. An inefficient fuel because it forms a considerable amount of ash and a great amount of water, and for that reason it is not used when other fuels are available.

peat tar. Tar obtained from the distillation of peat. The distillate obtained contains from two to six percent of tar.

pedalfer soils. An old term remaining in limited use, designating soils which develop under humid conditions and are enriched in aluminum and iron. These soils develop where the rain fall is 25 to 30 inches or more.

pedigree mud (drill). A descriptive term for high chemical content rotary mud including barium, sulfate, caustic soda, soda ash, sodium bicarbonate phosphates, etc.

pediment. Gently inclined planate erosion surface carved in bedrock and generally veneered with fluvial gravels; occurs between mountain fronts and valleys, or basin bottoms and commonly forms an extensive bedrock surface over which the erosion products from the retreating mountain fronts are transported to the basins.

pedocal. An old term remaining in limited use designating a soil characterized by an accumulation of calcium carbonate in its profile; characteristic of low rainfall.

pegmatite. A hypabyssal igneous rock which is very coarse grained; forms by crystallization from very fluid, water-rich magma and predominantly granitic in composition. When the term "pegmatite" is used alone it implies granite pegmatite. The list of pegmatite minerals include feldspar, quartz, micas, tourmaline, topaz, garnet, fluorite, beryl, spodumene and many others.

pelagic. A geologic term applied to any accumulation of deep sea sediments; also used to designate materials and organisms endemic to the marine environment as opposed to materials transported to the marine environment from terrestrial sources.

pelagic-abyssal sediments.   Deep sea sediments essentially free of terrestrial material. A small proportion of the clay-sized sediment fraction may be of terrigenous origin, transported to the sea primarily by the wind.

pelagic deposits.   Material formed in the deep ocean and deposited there as ooze.

pelagic limestone.   Rock formed principally of calcareous tests of planktonic foriminifera deposited in deep water.

Pele's Hair.   A special type of basalt pumic first described when drops of lava flew up from the boiling lava lake of Kilaua volcano in Hawaii. The drops stretched into thin threads of "hairs" of glass which the wind drifted into tangled masses.

pelite.   A very fine grained sedimentary rock principally argillaceous but also may contain minute particles of quartz and rock flour.

pelitic gneiss.   A gneissose rock derived from metamorphism of pelites or other argillaceous sedimentary rock.

pellet.   A small, approximately 0.2 mm diameter, well rounded, ovoid to irregularly shaped homogenous aggregate of clay-sized calcareous material, micrite. Most pellets are considered to be the feces of worms or mollusks.

penecontemporaneous deformation.   Folding, faulting, flow or slump of sediments during or immediately following deposition. Most frequently observed in sedimentary rocks derived from soft, clay mineral-rich sediments.

peneplane.   Pene, almost; plane, flat surface. Broad regional gently undulating erosion surface closely approaching stream base level. The surface presumably has been carved by subareal erosion (stream and mass wasting) over a long period of time.

penetration test.   A measure of the consistency of bitumen or grease by determining the depth to which a standard needle or cone penetrates under standard conditions.

penetrometer.   A device used to determine the consistency of asphalt and bituminous substances, by measuring the distance a standard needle will vertically penetrate a sample of asphalt when weighted with 100 grams for five seconds at 25°C.

penex (refin).   A process to improve the octane rating of pentane and/or hexane fractions from refinery naphthas and natural gasolines by isomerization over a specially prepared platinum-containing catalyst in the presence of hydrogen.

Pennsylvania crude. Oil produced in Pennsylvania, New York, West Virginia and parts of Ohio, average gravity 40 deg API; contains a high percentage of paraffin-base lubricating stock.

Pensky-Martens Tester. An instrument for determining the flash point and fire point of petroleum products. A modification of the Pensky-Martens tester devised by the U.S. Bureau of Mines is being used in place of the original instrument.

pentane. Any of three isomeric hydrocarbons; C_5H_{12} of the methane series; specifically the one of normal structure $CH_3(CH_2)_3CH_3$, three of which occur in petroleum. The others are 2,2-dimethylpropane, $CH_2(CH_2)_2CH_3$ and isopentane, $(CH_3)_2CHCH_2CH_3$

pentylene. A member of the unsaturated group of hydrocarbons known as the olefins or ethylenes seldom occurring naturally in crude oil but developed by distillation at high temperatures; formula C_5H_{10}

pepper (refin). Fine particles of sludge produced in acid treating; ordinarily the main body of sludge settles out quickly but in treating lubricating oils considerable "pepper" may be formed and remain in suspension.

peptize. A procedure used to deflocculate or disperse sediment samples into an aqueous suspension prior to particle size analysis. Peptizing methods include physical agitation, boiling, washing with distilled water to remove electrolytes promoting flocculation and adding of chemical reagents which inhibit flocculation.

peptized clay (drill). A clay to which an agent has been added to increase its initial yield as with soda ash, frequently added to calcium montmorillonite clay.

perched water table. Ground water lying above the regional ground water table and separated from it by impervious strata.

perco process (refin). A process for sweetening sour gasoline; commonly referred to as the copper sweetening process because it employs the use of a copper solution.

percolation. Flow of water or other natural fluids through the small openings of a porous and permeable sediment or rock; flow is usually laminar, slow and downward. Often used in reference to downward movement of water through the zone of aeration on its way to the water table.

percussion drilling. Principle of pounding rock into small fragments achieved by means of a chisel-shaped bit alternately lifted, dropped, and turned in the well.

The pulverized rock is removed periodically. Cable-tool drilling is percussion drilling.

perfect gas. A gas obeying Boyle's law. The nearest approach to perfect gas is to be found in helium, hydrogen, oxygen, and nitrogen.

perfect lubrication. Formation of a complete, unbroken film of liquid over each of two surfaces moving relatively one to the other with no metallic contact.

perforate. Pierce with holes, as is sometimes done to a well casing after it has been set in the well. There are several types of tools for perforating casing. Some employ expanding points forced against the interior of the casing wall by the operation of "jars", the others employ a series of high explosive cartridges with piercing bullets that are shot through the casing. When casing is set in unconsolidated sand, it is often necessary to perforate it.

Period. Fundamental unit of the standard geologic time scale during which a standard system of rocks was formed as the Devonian, Cretaceous and Tertiary periods.

Periodic Law. "The properties of the elements are periodic functions of their atomic weights." This law shows that a relationship exists between the atomic weights of certain elements, which on account of the similarity of their properties are usually grouped together to form a family. Thus, lithium, sodium and potassium have a very similar chemical property and are therefore in one class.

Periodic System. Classification of the chemical elements in nine groups demonstrating the fact that the physical and chemical properties of an element and its compounds vary periodically with the atomic number of the elements.

Periodic Table. A chart showing the orderly arrangement of the chemical elements and including the name and symbol of each element, its atomic number and weight and its electron distribution.

perma frost. A layer of soil or bedrock at a variable depth beneath the surface of the earth in which the temperature has been below freezing continuously from a few to several thousand years; permanently frozen subsoil.

permanganates. An oxidizing agent as potassium permanganate. Permanganates are salts of permanganic acid.

permeability. Ability of a rock to transmit fluid through pore spaces; a key influence on the subsurface rate of flow, movement and drainage as in groundwater and petroleum. There is no necessary relation between porosity and permeability. A rock may be highly porous and yet impermeable if there is no communication between pores. A highly porous sand is usually highly permeable.

permeability coefficient.   Factor relating steady-state rate of flow of gas through unit area and thickness of a solid barrier per unit pressure differential at a given temperature.

permineralization.   A process whereby the pore space in fossilized skeletal material fills with mineral matter of a somewhat different composition. In this manner, the fossil is made stronger and more resistant to destruction.

peroxide.   An oxide whose molecules contain two atoms of oxygen linked together as in peroxide of hydrogen. In chemistry of petroleum, peroxides are present in unburned gases ahead of the flame during knocking operation, and in general, the knock intensity varies with their concentration. The peroxide concentration is reduced when the lead tetraethyl is present, allowing it to slow down the reactions occurring prior to inflammation.

pervious rock.   Permeable stratum or a formation containing voids through which water will move under ordinary hydrostatic pressure.

petrifaction.   Process of petrifying or changing into stone; conversion of organic matter including shells, bones and the like, into stone or a substance of stony hardness. Original definition imposed an atom by atom replacement reaction on the process. In recent times, this has been shown to be incorrect and use of the term has changed to either mean fossilization in general, or to be synonymous with permineralization.

petrocene.   Greenish-yellow hydrocarbon with a pearly luster and needle-like crystals, obtained by the distillation of petroleum residue.

petrochemical.   A contraction of the words "petroleum" and "chemical," originally coined to designate chemicals of petroleum origin. It is now loosely used to cover a wide variety of products and cannot be defined specifically.

petrography.   Description and systematic classification of rocks.

petrol.   Variant for petroleum or its derivatives, particularly gasoline or motor spirits.

petrolatum.   A jelly-like product obtained from petroleum and having a microcrystalline structure, associated with residue remaining in the still after distillation. Before bleaching, it is called yellow petroleum, after bleaching it is called white petroleum. Residual lubricating oils as cylinder oils and bright stocks are made from paraffinic or mixed based crudes. These reduced crudes contain waxes of the microcrystalline type. Removal of these waxes from reduced crude produces petrolatum, a grease-like material most familiar in a refined form of petroleum jelly.

petrolatum album. Highly refined petrolatum of near "snow white"grade.

petrolatum liquid. Colorless to slightly yellowish, transparent liquid possessing a specific gravity of 0.840 to 0.940 at 25°C. It is soluble in either, chloroform, carbon disulphide, benzene, benzol or boiling alcohol, but is scarcely soluble in cold or warm alcohol, and is insoluble in water.

petrolatum wax. A high-boiling wax product obtained from cylinder stock, having a reddish brown color similar to that of the oil, but with further refining may be produced in any shade up to white; usually mixed with oils of proper color to make commercial "petrolatums."

petrolenes. Portion of asphalt soluble in hexane.

petroleum. Derived from the Latin peta, rock, and oleum, oil, and includes hydrocarbons found in the ground in various forms from solid bitumen, through the normal liquids to gases. Defined as a material occurring naturally in the earth which is predominantly composed of mixtures of chemical compounds of carbon and hydrogen with or without other non-metallic elements as sulfur, oxygen, nitrogen, etc.

petroleum asphalt. Commercial name of the asphaltic residue resulting from the distillation of asphaltic base crude oils. A pitch black substance of varying consistency used for asphalt paving, waterproofing, shingle impregnation, and many other purposes.

petroleum bloom. Fluorescence of iridescent cast seen in light reflected from the surface of certain petroleum hydrocarbons and resin oils.

petroleum coke. Solid carbon or coke remaining as a residue in tar stills after high temperature distillation. The solid carbon content or fixed carbon varies from 88 to 95 percent with an ash content varying from a trace of 0.3 percent, sulfur from 0.5 to 0.1 percent.

petroleum cuts. A "cut" is made during the process of distillation at the end of each commercial grade of distillate. When all of the hydrocarbons included under the commercial specifications of gasoline have been driven off, the next group under kerosene is collected by raising the temperature of the still to vaporize the compounds of a still high boiling point. Each cut is made at a specified temperature determined by the commercial demand for certain groups of compounds.

petroleum distillates. Generally includes the light and heavy oil distilled from crude oil and from cuts previously distilled.

petroleum fractions. A group of hydrocarbon compounds distilled from petroleum between a specified high and low temperature.

petroleum spirits.   A water-white petroleum distillate having a minimum flash point of 100°F and an end point of under 410°F; used as a paint thinner and solvent; also called *mineral spirits.*

petroleum sulfonate.   A petroleum stock treated with sulfuric acid, as technical white oil.

petroliferous.   Containing or yielding petroleum.

petroline.   Solid substance, analogous to paraffin, obtained in the distillation of Rangoon petroleum.

petrology.   Study of the mineral composition, texture and origin of rocks. One aspect is the study of drill cores and cuttings and geological investigations of sedimentary rocks in connection with oil exploration.

PGO hydrotreating (refin).   A process to upgrade the large quantities of pyrolysis liquid fractions boiling in the 400 to 650°F range resulting from pyrolysis operations utilizing heavy liquid feedstocks.

pH.   An abbreviation for potential hydrogen ion; numbers range from zero to fourteen, seven being neutral; indices of acidity, below seven or alkalinity, above seven of the fluid. The numbers are the log to the base ten with algebraic sign reversed of the gram-molecular hydrogen ion concentration.

phaneritic.   Designating igneous rocks with constituent grains coarse enough to be studied with the unaided eye.

phase (geochem).   A homogeneous, physically separable constituent of a geochemical system, as in a solidifying lava, the melt is one phase and each mineral is another.

phenocryst.   Relatively large and conspicuous crystals surrounded by much smaller grains in a porphyritic igneous rock.

phenol (carbolic acid).   White crystalline, deliquescent compound (C_6H_5OH) possessing a burning taste and odor resembling creosote. A caustic poison used as a solvent for refining lubricating oil.

phenol extraction (refin).   Improves viscosity index, color and oxidation resistance, and reduces the carbon and sludge-forming tendencies of lubricating oils. The products are stable, light colored neutrals and bright stocks meeting the most rigid viscosity index specifications.

phenol oil.   Product of coal tar distillation containing carbolic acid.

phenol process (refin).   A single solvent refining process using phenol as the selective solvent.

phi grade scale.   A logarithmic transformation of the Wentworth grade scale for size classification of sediment grains, based on the negative logarithm to the base two of the particle diameter.

Phi Grade Scale	Modified Grades (Millimeters)	Wentworth Sediment Class	
−8	>256. 0	Boulders, Rock_____	
−7	128. 0	Cobbles_____	
−6	64. 0		
−5	32. 0		
−4	16. 0	Pebbles_____	Gravel
−3	8. 0		
−2	4. 0		
−1	2. 0	Granules_____	
0	1. 0	Very Coarse_____	
+1	0. 5	Coarse_____	
+2	0. 25	Medium_____	Sand
+3	0. 125	Fine_____	
+4	0. 0625	Very Fine_____	
+5	0. 0313		
+6	0. 0156		Silt
+7	0. 0078		
+8	0. 0039		
+9	0. 00195		
+10	0. 00098		Clay
+11	0. 00049		
+12	0. 00024		
>+12	<0. 00024	Colloids_____	

phosphate (drill).   Certain complex phosphates, usually sodium tetraphosphate ($Na_6P_4O_{13}$) and sodium acid pyrophosphate (SAPP, $Na_2H_2P_2O_7$), are used either as mud thinners or for treatment of various forms of calcium and magnesium contamination.

phosphoric acid process (refin).   Polymerization process employing phosphoric acid as the catalyst.

phosphorite.   Sedimentary rock containing nodular and irregular masses of concretionary calcium fluorophosphate.

photo electron.   An electron ejected from its parent by the interaction between that atom and a high energy photon.

photo emission.   Emission of electrons from the surface of various substances, including sodium, potassium, lithium, rubidium and caesium when irradiated by light. Each substance has a minimum threshold frequency (q.v.) for the incident light, below which no emission takes place.

phylogeny.   Development or line of descent of a group of organisms; important in determining relationships between fossils in the geologic record.

physiography.   Branch of geology and/or geography dealing with the description of the earth's surface.

piauzite.   An asphaltoid substance, melting at 315°C, having a brownish or greenish black color and specific gravity of 1.220. After fusing, it burns with an aromatic odor and leaves about six percent ash; soluble in potassium hydroxide and in ether.

pick.   Selection of an event on a seismic record deemed significant in relating to seismic data to earth structure.

piedmont.   Foot of a mountain. A relatively gently sloping surface formed by a series of laterally overlapping alluvial fans or a pediment located along an actively eroding mountain front.

piedmont glaciers.   Glaciers formed by the coalescing and spreading of one or more valley glaciers on a plain adjacent to a mountain front.

piercement dome.   Salt plug that rises and penetrates rock formations to shallow depths; same as diapir.

pig.   A scrapping tool forced through a flow line or lines to clean out wax or other deposits.

pillow lavas.   Subaqueous lava flows that have spread out, torn up chunks of the muddy bottom, rolled and twisted them and then crystallized around them. Pillow lava is typical of divergent plate boundaries or mid-ocean ridges.

Pintisch gas.   A gas obtained by cracking gas oil.

Pintisch gas tar.   Tarry residue remaining from the cracking of gas oil in the manufacture of Pintisch gas.

pipe still (refin).   A still in which heat is applied to the oil while being pumped through a coil or pipe arranged in a suitable firebox. After leaving the heating zone, the pipe runs to a fractionator where a portion of the oil is taken off overhead as vapor and the liquid portion removed continuously.

pipe still distillation (refin).   A process during which the product being distilled is contained in pipes or tubes surrounded by flame or heat. The heated oil and vapors then enter the fractioning tower where the light volatile gasoline vapors are collected at the top of the tower, going to the condenser. The heavier products come from points lower down on the tower.

pipe thread protectors. Short threaded ring screwing into a pipe or into a coupling to protect the threads while the pipe is being handled or transported.

pipe tongs (drill). A heavy tool hung on a cable and used for screwing pipe and tool joints.

pitch. (1) The bottom product from the bubble tower in a refining operation. Pitch is actually asphalt and may be used directly as produced, as one grade of asphalt. It is sometimes processed into heavier grades, sometimes blended with gas oils to make a fuel oil. (2) Term used to designate the inclination of any line that lies in a plane, as lineation in a bedding plane, or slickensides in a fault plane.

pitch glance. Very hard and brittle bitumen with a low content of saturated hydrocarbons.

pitch, straight run. A petroleum pitch run directly during distillation to the desired consistency without compounding or after treatment.

pitman. A connecting rod, designed to convert reciprocating motion to rotative motion or vice versa; connects the walking beam with the main crank.

pitot tube. An impact tube, usually a piece of tubing bent 90 degrees and inserted in line with open end facing into flow stream. A flow measurement device.

pitted plain. A glacial outwash plain with numerous small kettle holes.

placer. A sedimentary deposit where alluvial processes have mechanically concentrated gold or other heavy minerals as platinum, cassiterite, etc. An alluvial or glacial deposit containing particles of gold or other valuable minerals.

placer mineral. A heavy mineral found in a placer deposit; usually gold, platinum, cassiterite or stream tin, magnetite, zircon, etc.

plane table. (1) A board mounted on a tripod, designed to support an alidade or other surveying instruments. (2) A procedure utilized to make topographic, geologic, or other maps using a plane table and alidade.

plankton. Free floating or weakly swimming organisms living in oceans or lakes. They are most densely concentrated in the upper zone of a water body, diminishing in abundance below the depth of penetration of sunlight.

plastic deformation. A permanent change in the shape of a solid occurring without fracturing or rupture. In most earth materials, caused by the application of stress over a long period of time, occurring by material flow rather than fracture.

plastic fluid.   A complex, non-Newtonian fluid in which the shear force is not proportional to the shear rate. A definite pressure is required to start and maintain movement of the fluid. Plug flow is the initial type of flow and only occurs in plastic fluids. Most drilling muds are plastic fluids.

plastic viscosity.   A measure of the internal resistance to fluid flow attributable to the amount, type, and size of solids present in a given fluid; expressed as the number of dynes per square centimeter of tangential shearing force in excess of the Bingham yield value inducing a unit rate of shear. The value, expressed in centipoises, is proportional to the slope of the consistency curve determined in the region of laminar flow for materials obeying Bingham's Law of Plastic Flow.

plasticity.   Property possessed by some solids, particularly clays and clay slurries, of changing shape or flowing under applied stress without developing shear planes or fractures. Such bodies have yield points and stress must be applied before movement begins. Beyond the yield point, the rate of movement is proportional to the stress applied but ceases when the stress is removed.

plate.   A large rigid segment of the earth's crust. Some plates consist of exclusively ocean crust as in the Pacific Plate, while others include both oceanic and continental crust as in the North American Plate. Thickness of plates is variable but measured in tens of kilometers. Plates are free to move with respect to each other at rates measured in centimeters per year. Plates interact in several ways along their boundaries. Plates may move toward each other creating convergent boundaries resulting in collision if both plate margins are made up of continental crust or in subduction, if two oceanic plate margins or oceanic and continental plate margins converge. Plates may move away from each other creating divergent plate boundaries which results in spreading or the creation of new oceanic crust in the gap left by the diverging plates. Plates may move parallel to each other creating translational or transform boundaries resulting in rift zones. *See* plate tectonics.

plate efficiency (refin).   Degree, expressed as a percentage with which an actual plate of a distillation column or counter-current tower extractor approaches the equilibrium effected by a perfect plate.

plateau.   A region of great extent, relatively flat and elevated appreciably above its surroundings. Usually used to name tectonically elevated regions, residual areas withstanding erosion to the lower levels of their surroundings or land built by successive eruptions of flood basalts. The term is frequently applied to continental features as the Colorado Plateau but may also be used in reference to submarine occurrences as the Blake Plateau.

plate tectonics.   A concept which came into popular acceptance by the geologic community in the late 1960's, stems from Weggener's early ideas about continental drift and more recent theories of sea floor spreading and polar wandering. The concept has revolutionized geologic thought concerning tectonics or mountain building and has supplemented or replaced the earlier geosynclinal concept. According to the plate tectonics theory, the crust of the earth is subdivided into several large rigid segments or plates free to move with respect to each other at rates measured in centimeters per year. According to plate tectonics, mountain building occurs when two plates converge. If the convergent plate boundaries both consist of continental crust, mountains as the Himalayan chain develop. If an oceanic plate boundary converges with a continental plate boundary, the oceanic plate is subducted by the continental plate and mountains as the Andean chain development. *See* plate.

platform.   (1) A level surface formed by erosion or deposition as a terrace, bench, or shelf. (2) A flat, elevated surface as a plateau or peneplain. (3) An extensive, flat region which has experienced broad uplifts; usually located in the interior of a continent. The surface usually truncates old rocks and structures and may or may not be covered by a thin veneer of sedimentary rocks or sediments.

platforming (refin).   A process to upgrade low octane naphthas to premium quality fuels; to produce high yields of high quality hexane to octane aromatics for petrochemical feedstocks and LPG from naphthas while upgrading them.

playa.   Flat-floored center of an undrained desert basin; dried up lake basin in an arid region.

plot.   A map, usually without detail, serving as a guide for more detailed geological or topographical work; to lay out to scale from field notes, a map to show locations, size, and form of the survey.

plug.   (1)To stop up a well by cementing a lock inside the casing or capping the well with a metal plate. (2)An intrusion with vertical walls and nearly circular outline; usually the plug consists of igneous rock as in the case of a volcanic neck or conduit. Occasionally low density sedimentary materials intrude overlying sediments and are called plugs, but more often they are given another name. Salt domes are occasionally referred to as salt plugs.

plug and abandon (drill).   An expression often abbreviated "P&A", describing the act of placing plugs in a dry hole, then abandonment.

plug back.   To place cement in the bottom of a well for the purpose of excluding bottom water, sidetracking, or producing from a formation already drilled through; also can be accomplished by means of a mechanical plug set by wire line, tubing, or drill pipe.

plug flow.   Movement of a material as a unit without shearing within the mass. Plug flow is the first type of flow exhibited by a plastic fluid after overcoming the initial force required to produce flow.

plunge.   Angle between any inclined line and a horizontal plane. The term is used to designate the inclination of the axis of an oreshoot, the axial line of a fold, the attitude of lineation.

plunger lift.   A method of lifting oil using a swab or free piston propelled by compressed gas from the lower end of the tubing string to the surface.

pluton.   A great body of intrusive igneous rock that formed beneath the surface by cooling and consolidation of magma.

plutonic.   Deep-seated igneous rocks, generally coarsely equigranular in texture; formed by solidification of a molten magma deep within the earth and crystalline throughout. Some authors call metamorphic rocks of deep seated origin, plutonic.

pneumeter.   An apparatus used for measuring the contents of tanks, reservoirs, standpipes, etc. It operates on the principle of measuring the hydrostatic head of the oil by means of the compressed air pressure necessary to equalize the pressure of the head.

pocket.   A cavity or a thin discontinuous bed of limited extent yielding a sudden but temporary flow of oil, gas or water, or containing a limited quantity of economically valuable minerals.

pod (or Podbielniak analysis).   A precision distillation procedure used to separate low-boiling hydrocarbon fractions quantitatively for analytical purposes.

podsol (podzol).   An ashy-gray or gray-brown soil of the pedalfer group. This highly bleached soil, low in iron and lime, is formed under a matting of organic or plant debris in cool, temperate climates.

poise.   Unit of absolute fluid viscosity and therefore, a unit of shear taken through a given cross sectional area of the film and at standard velocity. Numerically, the poise is the shearing force expressed in dynes per square centimeter per second.

polish rod.   A steel rod with a polished surface passing through the stuffing box of a pumping well; attached to the top of the sucker rods.

pollen.   Multi-celled microgametophytes of angiosperms and other seed plants. The tough, outer coverings of pollen grains fossilize readily. Because of its airborne distribution, fossil pollen is found distributed widely in sedimentary rocks deposited in a wide variety of environments, even in rocks normally not containing other kinds of fossils. Pollen is important in age dating and correlating sedimentary rocks and in determining the types of plants in existence at the time of preservation.

polmitin. A fatty acid obtained from palm oil and used in the manufacture of grease.

poly. From Greek, polys, a prefix used in the construction of compound words, meaning many. In chemistry, when several molecules are united to form a large molecule, they are said to be polymerized.

polyalkalyene glycol oil. A synthetic lubricating oil made by the reaction of an aliphatic alcohol with certain alkylene oxides.

poly butane oil. A synthetic lubricating oil produced from isobutenes and normal butenes by a catalytic process.

polycondensation. When two different chemicals react under the influence either of heat or of a catalyst, they build up large molecules and generally are involved in the elimination of water. The reaction is usually arrested at the stage at which polymers are thermoplastic, they soften under heat and before cross linking between the molecule chains takes place giving a thermoplastic polymer. In this "prepolymer" state, they are made as moulding powders or syrups, the final shape being achieved by heat and pressure causing interlocking between the molecule chains.

Polyco Process (polymerization). A process employing the principles of polymerization to make nonhexane by combining ethylene with isobutane.

polycycle. A term applied to the molecules containing more than one ring of atoms.

polycyclic. A term applied to compounds chiefly of carbon and hydrogen in which the carbon atoms are represented as grouped in two or more rings.

polyethylene. A plastic, chemically inert substance made by polymerizing ethylene under pressure; molecular weight, 1000 to 3000; melting point, 115°C to 125°C.

polyforming. A term used to cover a process charging methane and butane gases with naphtha or gas oil under thermal conditions to produce gasoline.

polymer. A compound, usually of high molecular weight, formed by the linking of simpler molecules or monomers. A substance in which the original molecules have been linked together to form giant molecules. Natural rubber is a polymer of isoprene.

polymerization.   Chemical combination of two or more molecules to form larger molecules called polymers. Polymerization process is employed in several catalytic processes to make gasoline and other petroleum products.

polymerize.   A technique employed to link original molecules together to form giant molecules. It is employed in the production of many products from petroleum as synthetic rubber, the conversion of gases to gasoline and the changing of certain liquids.

polymerized gasoline.   Gasoline made by catalytic process employing the principle of polymerization and frequently called polygas. It is a blending agent not a base stock having a high anti-knock quality and used to improve it.

polymorph.   A mineral sharing the same chemical composition with another physically distinct mineral, as diamonds and graphite are both polymorphs of carbon.

polymorphism.   Ability to exist in two or more crystalline forms; presence of two or more morphologic forms together in the same species.

pool.   Oil accumulation exploited or produced by a well or group of wells. Not a pool in the sense of an open body of fluid but rather is a relatively large region underground made up of highly porous and permeable rock saturated with petroleum.

pore space.   Open space or void between individual grains in a rock mass. Pores may vary greatly in size and shape. Porosity of oil sand ranges from 10 to 25 percent.

porosity.   Amount of void space in a formation or rock, usually expressed as percent voids per bulk volume. Absolute porosity refers to the total amount of pore space in a rock, regardless of whether or not that space is accessible to fluid penetration. Effective porosity refers to the amount of connected pore spaces, as in the space available to fluid penetration.

porous.   Cellular; having a high percentage of pore spaces.

porphyritic.   A textural term for igneous rock in which larger crystals called phenocrysts are set in a finer ground mass, crystalline or glassy, or both.

position isomerism.   Isomerism arising from a difference in the position occupied by one or more functional groups in the isomers.

positive segment.   Part of the earth's crust rising progressively during geologic time.

post still (refin). A still consisting of a closed vessel with condenser attached. The use of a fractionating column may be included but is not necessary. Post stills are only used for batch distillation.

potassium (drill). One of the alkali metal elements with a valence of one and an atomic weight of about 39. Potassium compounds, most commonly potassium hydroxide (KOH) are sometimes added to drilling fluids to impart special properties, usually inhibition.

Potassium-Argon Method (dating). Method of absolute dating using an isotope of potassium, K^{40}, accounting for one-one hundredth percent of all potassium. Although this is a small proportion, potassium unlike uranium, is a common element in the rocks and minerals of the earth's crust and the radioactivity of this isotope is responsible for a large part of the normal background radiation detected at the earth's surface. K^{40} decays ultimately to an isotope of argon, Ar^{40}. The half life of the disintegration to argon is 1.3 x 10^9 years.

potassium bichromate. A deep orange-red transparent salt formed by potassium and chromic acid; sometimes used in standard solutions for comparing oil colors and as a refining agent in a number of special procedures or processes.

potassium carbonate. K_2CO_3. A very active white alkaline substance coming in lumps, sticks, or granular powder, and is deliquescent potassium salt of carbonic acid; also called potash, pearlash, and salt of tartar.

potassium hydroxide. Variously known as potash, potassium hydrate, or caustic potash, a very active alkali rapidly absorbing both moisture and carbon dioxide gas; specific gravity 2.044, melting point 360.5°C; used in the manufacture of water soluble soaps and oil refining for neutralizing sulphuric acid left after washing distillates. A solution of the salt in water is used in gas analysis for absorbing carbon dioxide.

potential. A measure of the capacity of a well to produce oil or gas; when a well is completed, its productive capacity is predicted by an official test. The capacity, shown by this test is known as the well's potential.

pothole. Hole worn in the bedrock channel of a stream by pebbles and cobbles whirled by turbulent water and made to function like a drill. These holes are circular to elliptical in outline and a few inches to scores of feet in depth. They are most often observed in a stream channel during low water or along the bedrock walls where they have been left stranded after the stream has cut its channel downward. Potholes are most common in narrow valleys, but they may occur in broad valleys as well. They are also found along rocky sea coasts where they are drilled by rocks under the influence of the turbulence of breaking waves.

potrero. An accretionary ridge separated from the coast by a lagoon and barrier island.

pour point.   The lowest temperature at which an oil fails to flow under prescribed conditions; provides an indication of the presence of wax in gas oils, diesel oils, and fuel oils which might cause pumping difficulties.

pour point depressant. An additive employed to depress or reduce the pour point temperature at which lubricating oils become congealed, generally limited to a paraffin base oil.

pour stability. Ability of a pour depressant treated oil to maintain its original ASTM pour point when subjected to storage at low temperature approximating winter conditions.

powerforming (refin). A process to produce Powerformates of 85 to 102 or more, research octane number clear from low octane naphthas. Alternatively, the process may be operated to produce aviation blending stocks or to give high yields of benzene or other aromatics and LPG. Powerforming also produces large quantities of hydrogen used to desulfurize or improve qualities of other refining products.

power tools (drill). Equipment operated hydraulically or by compressed air for making up or breaking out drill pipe, casing, tubing, and rods.

ppm or parts per million. Unit weight of solute per million unit weights of solution, solute plus solvent, corresponding to weight-percent except that the basis is a million instead of a hundred.

Prandtl number. A dimensionless unit used in fluid mechanics; the ratio of the kinematic viscosity of a fluid to the thermal conductivity of the fluid.

precipitant. A substance added to a solution promotes the formation of a precipitate.

precipitate. A substance separating in solid form from a liquid as the result of some chemical or physical change; differentiated from a substance held only mechanically in suspension known as sediment.

precipitation number. A Number of millimeters of precipitate formed when 10 milliliters of lubricating oil are mixed with 19 milliliters of petroleum naphtha of a definite quality and centrifuged under definitely prescribed conditions. The precipitation number should indicate the amount of asphaltic bodies dissolved in the lubricating oil, although a certain amount of paraffin bodies may separate together with the asphaltic bodies.

press.   A device used in refining to separate paraffin wax from filtrate.

press drip.   Oil coming from presses after paraffin distillate is chilled, pressed and the wax completely separated.

press distillate.   Oil recovered when paraffin distillate is pressed to separate wax from the oil.

pressing (refin).   Operation of removing crystallized paraffin wax from chilled oil by pumping the mixture through a filter press where the wax is retained on canvas cloths and the oil passes through.

pressure.   A force or thrust distributed over a surface divided by the area of the surface normally indicated by a gage in pounds per square inch.

pressure base.   An absolute pressure agreed upon as a basis for converting the volume of gas metered to a correct volume.

pressure distillate.   Originates from a pressure or cracking still, as cracked gasoline, cracked gas oil, etc.; also known as cracked distillate. The light gasoline bearing distillate products from the pressure stills produced by cracking, as contrasted with virgin or straight run stock.

pressure drive.   Process occurring when water, compressed air, or high pressure gas is introduced into an oil producing horizon in order to force oil to move to a well where only small pressure is maintained and where the oil may be lifted to the surface by mechanical means.

pressure gauge (drill).   An instrument for measuring fluid pressure. A pressure gauge usually registers the difference between atmospheric pressure and the pressure of the fluid being measured by indicating the effect of such pressures upon a measuring element as a column of liquid, bourdon tube, a weighted piston, a diaphragm, or other pressure sensitive devices.

pressure gradient.   A scale of pressure difference wherein there is a uniform variation of pressure from point to point. For example, the hydrostatic pressure gradient of a column of water is about 0.433 psi per foot of vertical elevation (one kilogram per square centimeter per ten meters). The ''normal'' pressure gradient in a well is equivalent to the pressure exerted at any given depth by a column of 10 percent salt water extending from that depth to the surface, 0.465 psi per foot.

pressure head.   Pressure due to a column of fluid indicated in pounds at a given point of the column and converted into feet height. For a liquid, divide the pressure in pounds by the product .433 times the specific gravity of the liquid.

pressure regulator.   A valve controlling pressure in a pipe line; downstream from the valve.

pressure stills (refin).   Stills in which the liquid oil and oil vapors are held at pressures above atmosphere by release valves while the temperature is raised to a point where the oil will decompose or crack at a satisfactory rate, giving lower-boiling products; refers to a cracking still.

pressure viscosity.   Lubricating oils exhibit a peculiar property when subjected to pressure, they increase in viscosity. The exact reason for the behavior is not entirely understood, but laboratory experiments have proved that any oil will become progressively more viscous as the pressure increases. In the majority of oil firms, this property of rising viscosity is offset to a considerable extent by the rise in temperature due to fluid friction.

preheaters.   Any form of apparatus in which heat is applied to a material prior to its introduction into the main apparatus. The application of heat is usually affected by means of hot bodies cooled and whose heat would otherwise be wasted.

premium grade gasoline.   Highest grade of gasoline marketed for automobile use resulting in blending stable natural gasoline with correct quantities of straight run catalytic or cracked gasoline to produce a balanced gasoline.

prepared degras.   A blend of fixed oils used to make a compound oil having special properties.

pressed distillate.   A refinery term for lubricating stock after it has passed through a press to remove paraffin wax.

primary mineral.   A mineral crystallized from magma or deposited directly from solution, not transported nor produced by weathering.

primary recovery.   Crude produced from a well without the use of any secondary recovery methods or practices.

primary sedimentary structure.   A structure created during the deposition of a sedimentary layer and before lithifaction; includes structures related to current velocity and direction as well as sedimentation rates, consolidation and dewatering. Examples include bedding, cross bedding, ripple marks, sole marks and some structures related to slump and flow of the unconsolidated sediment. *See* penecontemporaneous deformation.

primary waves (p).   Compressional, longitudinal earthquake waves. These waves reach seismographs first and can travel through both liquids and solids. Earthquake wave with the highest velocity having a speed of 7.75 to 13.64 kilometers per second, one of two body waves.

prime city naphtha.   A refinery term for a petroleum solvent having an API gravity ranging between 68 and 73 degrees and commonly called PC cut or prime city cut.

prime white kerosene.   Has a color between water white kerosene and standard white kerosene; also called prime white oil.

prime white oil.   Kerosene of prime white color intermediate in color between water white and standard white.

prodelta.   Portion of a delta lying seaward of the delta front. The prodelta normally is found below the effective depth of wave erosion. Foreset beds are deposited as part of the prodelta in deltas.

produce gas.   A gas produced from low grade fuel as lignite for power purposes. A special producer furnace is used with air and steam mixed in the gas automatically.

producing horizon.   A formation, sequence of strata or level below the earth's surface which when penetrated by a well yields significant and economically important volumes of oil or natural gas.

production curves.   Curves plotted for individual wells, for pools or for fields, showing the rate of production at successive periods of time. Normally used for the purpose of studying the rate of decline of wells and pools.

production sand.   Sand or sedimentary strata in which crude oil is found.

production tubing.   Last string of pipe put in an oil well after the oil has ceased to flow naturally. A working barrel containing the pump is lowered near the bottom and then operated by sucker rods inside the tubing. The tubing is capped at the top with a stuffing box for the polished pump rod to pass through.

productivity test.   A test of a well's capacity to produce, usually conducted at different pumping rates or rates of flow.

profile.   A cross section to scale illustrating variations in data of geologic significance. Typical profiles are (1) stream profile, recording variations in elevation of the stream with regard to horizontal distance of flow, (2) structure profile, showing variations in dip and strike of strata and faults along a vertical plane, (3) geophysical profile, showing variations in gravity, electrical or magnetic field strength in a horizontal direction, (4) seismic profile, a compilation of seismic data received from a line of seismometers or of seismic data recorded from a series of shots on land or data recorded continuously by seismic equipment towed behind a ship. *See* profiling and sub-bottom profiling.

profiling. Geophysical prospecting technique, usually electrical on land or seismic at sea. Energy source and receivers are moved simultaneously across the structures of interest.

progradation. Expansion of a coastline, or portion thereof, seaward by sedimentary deposition of riverine sediments in the near-shore environment. Deltas and fan-delta complexes are progradational as are beaches enlarged by the combination of longshore drift and onshore movement of sand by wave action.

promotor (catalyst). A substance, when present, increasing the activity of a catalyst. The catalytic action of iron is greatly increased when the catalyst contains a small quantity of silicon.

propane. Third member of the gas family having three carbon and eight hydrogen atoms, found in most natural gas and the first product found naturally in crude; generally mixed with butane in varying proportions to make LPG, bottled gas, etc.

propane-acid process. Chemical refining process used in conjunction with propane dewaxing and propane deasphalting.

propane deasphalting and fractionation. A process separating asphalts or resins from viscous fractions modified to segregate heavy or medium neutral fractions by extraction with propane.

propane dewaxing (refin). Process separating high pour point waxy materials from lubricating oils. The products are zero or lower pour test oils with exceptionally low cloud points and wax.

propanol. A product found in petroleum having a boiling point within the gasoline boiling range.

propping agent (drill). A granular substance, as sand grains, walnut shells, or other material carried in suspension by the fracturing fluid and serving to keep the fracture open when the fracturing fluid is withdrawn.

propylene. A member of the gas family having three carbon atoms; used for making alcohol as isopropyl alcohol. A colorless, gaseous hydrocarbon of the ethylene series, $CH_3CH:CH_2$.

propylene/acetone dewaxing (refin). A process separating wax from various lube oil fractions in order to produce finished lubricants with low pour points.

propylitization. A form of hydrothermal alteration involving the introduction or formation in place of carbonates, secondary silica, chlorites, and sulfides; reaches its maximum development in fine-grained rocks in the vicinity of upper mesothermal or lower epithermal veins.

proration.   A system forced by a state or by agreement limiting the amount of oil produced from a particular well or field within a given period.

prospecting.   Activity of searching for an accumulation of natural earth materials of economic value. Prospecting may be a simple procedure as walking across an area and looking for evidence of ore or petroleum accumulations or may include use of sophisticated geophysical and geochemical techniques. Places deemed to have a high potential for yielding economically valuable quantities of ore or petroleum are called prospects.

protection casing (drill).   A string of casing set to protect a section of the hole and to permit drilling to continue to a greater depth; sometimes called "intermediate casing."

proto.   Prefix used in chemistry to denote the first or lower number of a series.

protobitumen.   First of a mixture of hydrocarbons. A partially reduced carbohydrate which, upon further reduction, yields oil. Conversely, protobitumens are partially reduced carbohydrates formed by the oxidation of oil.

proton.   An elementary nuclear particle with a positive electric charge equal numerically to the charge of the electron and a mass of 1.007594 mass units. It is one of the constituents of every nucleus.

protopetrolatum.   A material found in petroleum having a consistency of honey and contributing exclusively to the "oily" characteristic found in all lubricants produced from petroleum.

provenance.   Denotes the accumulated information pertaining to the origin of the constituent materials of a sedimentary rock; includes the geographic location and environmental conditions of the sediment source area as well as the rock types from which the sediment was derived.

pseudormorph.   Natural cast of mineral crystal composed of a substance differing from the original composition of the replaced object. A crystal or apparent crystal of some mineral having the outward form characteristic of another species replaced by substitution or by chemical alteration.

p.s.i..   Pounds per square inch.

p.s.i.a..   Pounds per square inch absolute.

p.s.i.g..   Pounds per square inch gage.

pudding grease.   A semi-fluid grease made from lubricating oil.

puking (refin).   A still or bubble tower is said to "puke" when the oil foams and rises in the vessel so high that part of the liquid is driven out of the vessel and through the vapor line.

pulling (prod).   Withdrawing sucker rods and production tubing from a pumping well to clean out or replace parts of the pump.

pulling unit (prod).   A well servicing rig used in pulling rods and taking them from the well.

pull it green (drill).   Pulling a drilling bit from the hole for replacement before it is too worn.

pump (prod).   A power consuming unit used to lift, compress, transfer or otherwise move liquids, gases and vapors by means of a pressure, vacuum or both. Diagram of the main parts of an oil pumping unit. Not all units are exactly alike, but they operate in the same general way. (See Fig. p. 277)

pump off.   Pumping so rapidly that the oil level drops below the standing valve on the pump.

purine.   A crystalline compound ($C_2H_4N_4$); parent of other compounds of the uric acid group.

push and pull pump (prod).   A reciprocating pump operated by a jack line or pull rod.

pyconometer.   An instrument used to determine the density of petroleum liquids having vapor pressure less than 600 millimeters of mercury and low viscosities.

pyrene.   A hydrocarbon, $C_{16}H_{30}$, obtained from coal tar.

pyridine.   A coal tar product in phenol oil of principal use as a denaturant for industrial alcohol.

pyrobitumen.   A naturally occurring, solid hydrocarbon complex infusible and often associated with a mineral matrix; insoluble in water, and relatively insoluble in carbon disulphide, benzol, etc. Pyrobitumens are derived from the metamorphosis of vegetable growth as peat, lignite, bituminous coal and anthracite coal, and from the metamorphosis of asphalts.

pyroclastic.   Designating detrital volcanic materials explosively erupted; formed by fragmentation as a result of volcanic action. Ejected, transported aerially and deposited upon land surfaces, in lakes, or in marine waters.

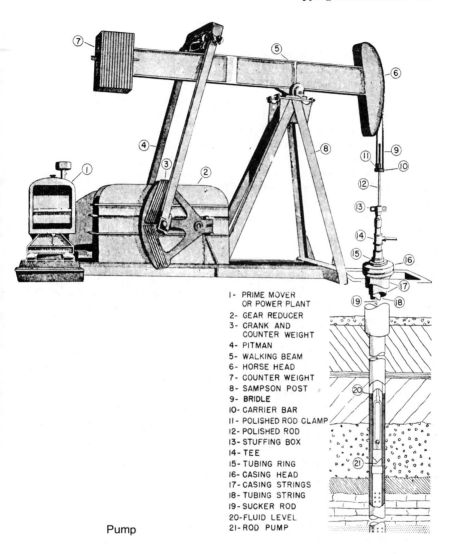

1- PRIME MOVER
   OR POWER PLANT
2- GEAR REDUCER
3- CRANK AND
   COUNTER WEIGHT
4- PITMAN
5- WALKING BEAM
6- HORSE HEAD
7- COUNTER WEIGHT
8- SAMPSON POST
9- BRIDLE
10- CARRIER BAR
11- POLISHED ROD CLAMP
12- POLISHED ROD
13- STUFFING BOX
14- TEE
15- TUBING RING
16- CASING HEAD
17- CASING STRINGS
18- TUBING STRING
19- SUCKER ROD
20- FLUID LEVEL
21- ROD PUMP

Pump

pyrogenic distillation.   Cracking distillation at high temperatures, high pressures, or both, to obtain greater yields of the lighter hydrocarbons composing gasoline.

Q

quadrangle map. A map bound by the same number of minutes of latitude as longitude. Common sizes are 7.5 minute, 15 minute and 30 minute quadrangle maps covering areas bounded by 7.5 minutes of latitude and longitude, 15 minutes of latitude and longitude and 30 minutes of latitude and longitude, respectively. Representative fraction scales of these maps are 1/24,000 for 7.5 minutes, 1/62,500 for 15 minutes, and 1/125,000 for 30 minutes. Quadrangle maps typically show topography and culture as houses, roads etc. They are used as a base to plot geology and other types of information.

quake. Earthquake.

quantum. A fundamental field particle as the photon which is the quantum of the electromagnetic field; the meson is the quantum of the nuclear field. A unit of radiant energy absorbed or emitted by an atom. Energy is radiated only in quanta and the quantum is the smallest amount of energy transmitted at any given wavelength.

quantum theory. A theory, developed in 1900 by the German physicist Max Karl Erst Ludwig Planck, states that radiation consists of discrete bundles of energy, just as matter is made up of atoms, and that the process of emission or absorption of energy by atoms or molecules is not continuous but takes place by steps. Planck called the unit of radiation the "quantum", a word derived from the Latin word for "how much." Planck's quantum theory was very helpful in explaining the behavior of atoms, electrons in atoms, and neutrons in the atoms' nuclei.

quarter section. Standard area of land under the U.S. Public Land Survey system. It is one quarter of a section and one half mile square, containing approximately 160 acres.

quartile. One means of expressing sediment partical size distribution is to state the three particle diameters for which, respectively, one quarter of the sample (by weight) is coarser than that diameter, one half of the sample (by weight) is coarser than that diameter, and three-quarters of the sample (by weight) is coarser than that diameter. Each of these units is called a quartile.

quebracho (drill). A drilling-fluid additive used extensively for thinning or dispersing to control viscosity and thixotrophy. It is a crystalline extract of the quebracho tree consisting essentially of tannic acid.

quick clay. A marine clay often of glacial origin having a modest unremolded strength and much less remolded strength. When undisturbed, quick clay will have much the same engineering properties as any clay sediment, but when disturbed either by vibrations or by cutting steep faces into it by stream erosion or human excavation, quick clay liquifies and flows with great fluidity.

quick lime.   (1) A dehydrated burned lime rock (calcium oxide) which when hydrated becomes calcium hydroxide; commonly used in the manufacture of soap employed to make calcium or lime base grease. (2) **(drill)** Calcium oxide, CaO, used in certain oil base muds to neutralize the organic acid.

quicksand.   Sand supersaturated with water and easily movable or quick; will not support a heavy object.

R

rabbit. A small plug run through a flow line by pressure to clean the line or to test it for obstructions.

rack. (1) Equipment for loading and filling tank cars at the refinery. (2) A frame built by the side of a drilling well to store drill pipe or casing horizontally.

racking pipe (drill). Placement of stands of pipe in orderly arrangement in the derrick while hoisting pipe from the well bore.

radical. Two or more atoms behaving as a single chemical unit as an atom; e.g., sulftate, phosphate, nitrate.

radicals of fixed oils. All fixed oils are combinations of alcohol radicals and fatty acid radicals. The alcohol radical occurring in the vegetable oils and many of the animal oils is glyceryl (C^3H^5, a trivalent which combines with three fatty acid radicals.

radioactive age determination. Utilization of known decay rates of radioactive isotopes to determine the age of natural earth materials. Isotopes with large half-life as uranium-238 are used to date geologically ancient materials whereas those with shorter half-life as carbon-14 are used to date geologically recent material.

radioactive decay. Disintegration of the nucleus of an unstable nuclide by the spontaneous emission of charged particles and/or photons.

radioactive half-life. Time required for 50 percent of a radioactive substance to change to a decay product.

radioactive material. A substance containing atoms that disintegrate spontaneously, throwing off parts of themselves and changing into different atoms. This disintegration of atoms is called radioactivity.

radioactive nuclides. Atoms that disintegrate by emission of corpuscular or electromagnetic radiations. The rays most commonly emitted are alpha, beta, or gamma rays. The three classes are: (1) Primary, which have a half-life exceeding 100,000,000 years, which may be alpha-emitters or beta emitters. (2) Secondary, formed in radioactive transformations starting with uranium-238 or thalium-232. (3) Induced, having geologically short half-lives and formed by induced nuclear reactions occurring in nature. All these reactions result in transmutation.

radioactive well log. Record of the radioactive characteristics of subsurface formations. The radioactive log, normally consists of two curves, a gamma ray curve and a neutron curve. The two logs may be run simultaneously and in conjunction with a collar locator in a cased or uncased hole.

radioactivity.   Number of atoms decaying per unit of time. The unit of radioactivity is the Curie, 3.7×10^{10} disintegrations per second.

radiocarbon dating.   Age calculated from the specific radioactivity of carbon-14 in the remains of a once living organism; dating is possible because carbon-14 is produced in the atmosphere by cosmic rays and is incorporated and maintained in all living things at an essentially constant ratio with the common isotope of carbon (C_{12}). After death, the carbon-14 nucleii decay exponentially with a half-life of 5,568 years and are no longer replaced as they were in living organisms. By computing the carbon-14 to carbon-12 ratio for an organism before death and comparing it with the carbon-14 to carbon-12 ratio currently existing in the remains of the organism, it is possible to compute how long ago the organism died.

radioisotope.   Any isotope of any element capable of emitting waves and/or particles spontaneously until it reaches a stable state. All elements exist as several types of isotopes, and many of these are radioactive. The term "radioactive isotope" is often shortened to "radioisotope."

radiometric dating.   Process of determining the absolute age of a material by measuring the ratio of the amount of any radioactive parent isotope contained to the amount of daughter or stable product isotopes.

raffinate (refin).   Applied in solvent refining practice to that portion of the oil remaining undissolved and not moved by the selective solvent used.

Ramsbottom carbon test.   Test determining the amount of carbon residue left by evaporating an oil under specific conditions; intended to throw some light on the relative carbon-forming propensities of oils. The results of the test must be considered in connection with other tests and the intended use of the oil.

range.   Part of the nomenclature of the U.S. Public Land Survey System. A six mile wide corridor of land running north-south, at right angles to a similar corridor called a township. Where townships and ranges intersect, they define squares of land, six miles on a side, consisting of 36 one square mile sections.

range length.   A grouping of pipe lengths. API designation of range lengths are as follows:

	Range 1	Range 2	Range 3
Casing	16–25 ft	25–34 ft	34 ft or more
Drilling pipe	18–22 in	27–30 in	38–45 in
Tubing	20–24 in	28–32 in	

rapid curing cut-back.   An asphalt cement thinned with gasoline or a naphtha-type of distillate.

rare earths.   Any of a large series of similar oxides of metals, chiefly lanthanum, cerium, praseodymium, neodymium, illinium, samarium, europium, gadolinium, terbium, gasolinium, terbium, dysprosium, holmium, yttrium, erbium, thulium, ytterbium and lutetium.

rat hole.   A hole from 30 to 35 feet deep with casing that projects above the derrick floor, into which the kelly is placed when hoisting operations are in progress.

rat tailing (drill).   Drilling of a small hole in advance of the main drilled hole for closer study of the formation.

raw distillate (fuel).   Gasoline and other fuels as they come from the condenser, before being refined to remove objectionable materials by such processes as chemical refining, solvent refining, etc.

raw lube stock.   Applied to a lubricating oil stock before it is refined by solvent or chemical refining into a finished product.

raw water white.   Refined oil cut or second cut from the crude stills before treatment or rerun. Its gravity is usually 30 to 46 degrees API.

ray pattern.   A graphic presentation of the path of sound or seismic rays in relation to depth and range.

reaction.   Chemical activity ensues when one material comes in contact and interacts with another producing new and different materials. When oxygen, for example, combines with gasoline at low temperatures, the reaction changes the gasoline to gasoline gum. When two or more substances undergo a chemical change, they are said to react with one another and the change is called the reaction.

reaction rim.   A rim of alteration products formed around an earlier crystal by reaction with a fluid.

reactivation.   Restoration of a catalyst or reagent to a chemically or physically active state.

reactor (refin).   Vessel in which all or at least the major part of a reaction or conversion takes place. (1) In catalytic cracking, the enlarged space in which hot oil is contacted and cracked with the catalyst; in isomerization, the vessel containing catalyst in which the hydrocarbon is isomerized; in alkylation plants, the vessel containing emulsified acid in which most of the alkylation occurs. (2) An assembly of fissionable atoms and control materials whose function is to release useful energy at a controllable rate.

reactor core.   In a nuclear reactor, the region containing the fissionable material.

reagent.   Any substance when added to another, results in a chemical reaction.

ream (drill).   In drilling, to enlarge a hole already drilled, to permit the entrance of casing.

reamer (drill).   Tool employed in drilling. Used for smoothing the wall of a well, enlarging the hole to the desired size, stabilizing the bit, straightening the well bore where kinks or doglegs are encountered and in drilling directionally.

reboiler (refin).   An auxilliary of a fractioning tower designed to supply additional heat to the lower portion. Liquid is usually withdrawn or pumped from the side or bottom of the tower, reheated by means of heat exchange, the vapors and residual liquid separated or together are reinforced into the tower.

receiver (refin).   Tanks into which distillates from the stills are run.

Recent.   The last 10,000 years of earth history, starting with the retreat of the last glaciation from the Northern Hemisphere.

recharge.   Percolation of meteoric water through the zone of aeration until it reaches and is added to water already present in the zone of saturation.

reconnaissance.   A preliminary geological survey usually made quickly, without attention to detail, to gain general knowledge of an area; also, an organized search for materials of economic value.

recovery acid (refin).   Sulfuric acid purified after separation from the acid sludge formed by having treated a batch of oil with the original acid. The original acid may have been used for the first time in the treatment or may have been recovered from a previous lot of acid sludge. Sulphuric acid used for treating oil and recovered for repeated use.

recovery gas-oil ratio.   Number of cubic feet of gas produced from the reservoir per barrel of oil.

recrystallization.   Enlargement or reshaping of existing mineral grains and the formation of new ones, within a solid rock. Recrystallization commonly occurs either during metamorphisis or diagenesis.

rectification (refin).   Separation of components of a liquid mixture by fractional distillation, as in a bubble tower or packed tower, under partial reflux.

recumbent fold.   A fold in which the axial plane is approximately horizontal.

recycling.   (1) In refinery operations, the redistillation of that part of the charging stock collecting in an evaporator, fractioning tower, etc. When this stock is returned to the still, the operation is termed recycling and the stock is called recycle stock. (2) In production, a process in which gas received from an oil is constantly returned to the oil sand through another well and thus completes a cycle from the oil well to the pressure well through the oil sand and back to the well.

red bed.   Red sedimentary rocks; usually terrigenous standstones and shales. Coloring of the red beds is from the trivalent, ferric state of iron. Permian and Triassic Periods of geologic history are particularly rich in red beds.

red clay.   An old term for fine grained pelagic sediments. Red clay is rich in argillaceous constituents, frequently of wind blown terrigenous origin and deficient in calcium carbonate relative to other pelagic sediments. Color is rarely true red but grades from brown to tan with lighter shades reflecting higher concentration of calcium carbonate.

redistillation (refin).   A repetition of the distillation process during the manufacturing of a product. When making lubricating oils, they are often distilled more than once. All distillations following the original distillation are referred to as redistillations or rerunning.

redox potential.   Electrical potential, positive or negative, required to drive an oxidation or reduction chemical reaction or the electrical potential generated during a spontaneous oxidation or reduction reaction. The redox potential of a natural system may be calculated by thermodynamic equation or may be measured with a standard electrode. Redox potential of a depositional system will exert a major control on the oxidation state or valence of iron and therefore the color of the sediment.

reduce.   (1) To deprive of oxygen. To add electrons to the valence shell of an atom. (2) To distill off lighter oils to obtain oils of greater gravity or viscosity.

reduced oil (refin).   (1) Oil redistilled in a vacuum or steam still from an already distilled oil, or (2) an oil made from the residue in the still after another product has been distilled from the crude.

reducing (refin).   Removal of light hydrocarbons by distillation.

reef.   (1) Rigid topographic features of the near shore marine environment. Reefs tend to parallel shore and are produced by actively growing and sediment binding organisms under high energy conditions. Reefs are, therefore, wave resistant and serve as protective barriers between the open sea on one side and lagoons, on the shoreward side. (2) Thick, lenticular masses of carbonate rock exhibiting limited lateral extent in the stratigraphic record are called reefs. Some uncertainty exists whether all stratigraphic reef structures were actually formed under the same conditions of modern day reef development.

reef talus.   Massive inclined beds of debris derived principally from a reef and deposited along the seaward margin of a living reef.

reeve the line (drill).   To string a wire rope through the sheaves of the traveling and crown block, then to the hoisting drum.

refined paraffin wax.   A hard crystalline hydrocarbon wax derived from mineral oils of the mixed-base or paraffin-base by a process of refining in which the oil is eliminated or reduced to a negligible extent.

refined solvent naphthas.   Has a specific gravity ranging between 0.850 and 0.870 and possesses other properties as described under ASTM designation.

refined tar.   A tar freed from water by evaporation of distillation continued until the residue is of desired consistency or produced by fluxing tar residuum with a tar distillate.

refined wax.   Sweated wax filtered and decolorized with fuller's earth. A grade clay filtered to improve color.

refinery gas.   Includes all gases produced at the refinery as cracking gas, many members of the gas family having free bonds. Any form of mixture of still gas gathered in a refinery from the various stills.

refining.   Manufacture of petroleum products by a process of distillation from crude petroleum in which successive components are vaporized and separately condensed. The resultant distillate is then redistilled or chemically purified to yield the finished products for marketing. Free of impurities.

reflection seismograph.   Used to pick up and record shock and/or sound waves reflected from subterranean rock strata. On land, the shock wave may be created by exploding dynamite, a thermonuclear device, dropping a large weight or a heavy duty vibrator. At sea, the shock wave may be created by an electronic transducer, electric spark, rapid release of compressed air or steam, or explosion of dynamite.

reflux (refin).   Flowing back; the reflux action in a stabilizer tower used in connection with petroleum refining; the flowing back downward of condensed fractions of the upward flowing gas. This reflux action may be accelerated by pumping back to the top of the tower a portion of the condensate taken from bottom of same. This condensate is sometimes called reflux.

reforming (refin).   A thermal or catalytic process in which light petroleum fractions have their chemical structure changed to produce gasoline components of increased octane number. The operation of heat treating or cracking a product as straight run gasoline, in order to increase its anti-knock quality.

refraction.   Process in which the direction of energy propagation is changed as the result of a change in elastic properties within the propagation medium, or a change in propagation direction as the energy passes through an interface representing a discontinuity in elastic properties between two media.

refraction shooting.   A seismic technique used to determine the depth to a rock layer conducting sound at a higher velocity than overlying layers. Frequently used in engineering geology to measure thickness of overburden on bedrock. Seismic energy travels downward from its source until it encounters a high velocity layer at which point some energy is reflected back to the earth's surface but a significant portion of the energy is bent or refracted to travel within and parallel to the high velocity layer. As the energy travels through and parallel to the high velocity layer some of the energy is continually bent or refracted back to the earth's surface.

refractometer.   An instrument for measuring the refractive index of oils as a means of identification or for determining purity.

refractories.   Materials that can resist high temperatures and large changes of temperature without melting, crumbling or cracking, and are relatively unaffected by the flow of very hot material over their surface.

regeneration (refin).   In a catalytic process, part of the system having as its primary function the revivification or reactivation of the catalyst, which is done by burning off coke deposits under carefully controlled conditions of temperature.

regional dip.   General or average dip of sedimentary strata over a large area, expressed in feet per mile, degrees, or any other convenient unit. Minor irregularities in dip as well as folds superimposed on the regional dip are ignored or discounted.

regional gravity.   Contributions to the observed gravity field due to density irregularities at great depths in the earth. Regional gravity changes little and at a slow rate.

regional metamorphism.   Large scale recrystallization of rocks without involving melting; associated with mountain building, igneous intrusion and converging crustal plates.

regolith.   Surface layer of sediment, rock waste, volcanic ash, alluvium, glacial drift, organic matter, and windblown material lying above bedrock.

regression.   Applied to bodies of water and sediments deposited during withdrawal of the sea and/or emergence of the land. (See Fig. p. 287)

RECORD OF A REGRESSION
Retreat of the sea from the land

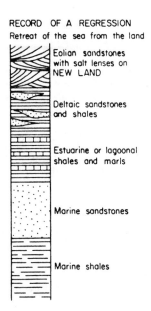

Eolian sandstones
with salt lenses on
NEW LAND

Deltaic sandstones
and shales

Estuarine or lagoonal
shales and marls

Marine sandstones

Marine shales

rejuvenated stream.   A stream whose gradient has been increased because of uplift of the land.

relative age dating.   Process of determining the chronological sequence of events in geologic history or of a series of rock units or formations. The absolute ages of events or materials are of no consequence except when determination of absolute ages is the best way to establish relative age sequence.

relict.   A deposit, structure, feature, mineral, or texture reflecting a time or environment of formation other than those of the rock or physical location in which the "relict" deposit, etc. is now found; sometimes spelled "relic."

relict sediment.   Sediment deposited under one set of environmental conditions and remaining uneffected by subsequent changes in the environment. An example would be submarine terrace sediments deposited in shallow water during low sea stands of the last glacial period now covered by more recent or post glacial continental shelf sediments.

relief, topographic.   Difference in elevation between the high and low points of a land surface.

remanent magnetism.   Permanent magnetism acquired by a rock at the time of its formation. The direction of magnetization is parallel but opposite to the earth's field at the time and place of the rock's origin.

remolding.   Disturbance of a clay sediment or soil to the end that primary structures and textures are destroyed. This is done in a laboratory in order to predict how a material will act if it is involved in a landslide.

remote sensing. Acquisition of information or data about a phenomenon, body or structure at a distance without making direct contact with object under observation; may involve study of energy which emanates from the object or it may incorporate the effect on energy generated by the observer. Some authors restrict use of the term to techniques involving aircraft or satellites.

replacement. Process whereby the substance of a rock, mineral, or an organic fragment is slowly removed by solution and material of a different composition is precipitated in its place.

repose, angle of. Slope which any given unconsolidated material will not exceed regardless of the quantity of the material added to the top of the deposit. The angle of repose is a function of grain size, grain shape, grain roundness or angularity and moisture content of the deposit.

repressuring (prod). System of increasing the flow of crude oil by forcing gas under pressure into the oil sands.

rerun oil (refin). An already distilled oil redistilled usually in a vacuum or steam still.

rerunning (refin). A general term covering the redistillation of any material and indicating that a large part of the material is distilled overhead. A rerun tower may be associated with a crude distillation unit that produces a wide cut naphtha as an overhead product. By separating the wide overhead cut fraction into a light and heavy naphtha, the rerun tower acts in effect, as an extension of the crude distillation tower.

reserves. Petroleum in the ground which is economically recoverable.

reservoir. A subsurface porous and permeable rock body in which oil and/or gas is stored. Lithologically, reservoir rocks are either limestones, dolomites or sandstones, or their combinations. A petroleum reservoir generally contains three fluids gas, oil and water separating into discrete phases because of the variance in their gravities. Gas, being the lightest, occupies the upper part of the reservoir rocks and water, the lower portion, while oil occupies the intermediate section.

reservoir gas-oil ratio. Number of cubic feet of gas per barrel of oil originating in the reservoir.

reservoir pressure. Fluids confined in the pores of the reservoir rock occur under a certain degree of pressure, generally called reservoir pressure or formation pressure. Pressure can be determined by measuring the force per unit area exerted by the fluids against the face of the reservoir rock where it has been penetrated by a well. The amount of pressure is stated in pounds per square inch (psi) or in pounds per square inch absolute (psia), or in atmospheres (multiples of 14.7 psi).

residual clay.   A clay deposit formed by the weathering of rock in place.

residual deposits.   Material formed by the decay or disintegration or weathering of rock in place.

residual soil.   Soil developed on "residual" material rather than a transported soil developed on "transported" material. An obsolete term found in older literature and occasionally occurring in modern publications. Term has found disfavor because soils are neither residual nor transported in the terms used in the definition but in their true sense the adjectives should be applied to the soil source material.

residue.   Solid matter remaining after the extraction of the oil from crude material in a standard laboratory distillation; the amount of original liquid remaining in the distilling flask when the distillation is completed. Portion of the crude remaining in the still after the products have been removed which distill at or below approximately 750°F; also referred to as residuum or bottoms.

residue gas.   Casing head gas stripped of its gasoline; generally delivered to a fuel system or returned to the oil producing reservoir for repressuring.

residuum cracking (refin).   When the residue remaining after cracking is a liquid or semi-liquid product, the process is termed residuum cracking. When the residue is coke, it is termed a non-residuum cracking process or coking process.

resin (drill).   Semi-solid or solid complex, amorphous mixtures of organic compounds having no definite melting point nor tendency to crystallize. Resins may be a component of compounded materials added to drilling fluids to impart special properties to the system, wall cake, etc.

resistivity (drill).   Electrical resistance offered to the passage of a current, expressed in ohm-meters; the reciprocal of conductivity. Fresh-water muds are usually characterized by high resistivity, salt-water muds by a low resistivity.

resistivity log.   An electric well log created by instruments designed to measure the electrical resistance of the wall rock of the well. Resistance is depended upon rock type as well as composition of rock pore fluids. (*See* resistivity method)

resistivity method.   An electrical geophysical prospecting technique used on the earth's surface or in wells. Current is introduced into the ground by current electrodes and the electrical potential generated is measured by potential electrodes. Electrode potentials when interpreted in conjunction with electrode spacings yield information about variations in ground resistance to electricity with depth in the earth or distance from the well wall. Changes in resistance or resistivity are controlled by variations in rock type and composition of rock pore fluid.

retainer (drill).   A cast-iron or magnesium, drillable tool consisting essentially of a packing assembly in the lower portion and incorporating a ball-type back pressure valve; used for closing off the annular space between tubing or drill pipe and casing to place cement or fluid through the tubing or drill pipe at any predetermined point behind the casing or liner around the shoe or into the open hole around the shoe.

retort (refining).   A still used in the retorting process producing liquids and gas from solids or semi-solids as in the retorting or distillation of oil shales and coal. A vessel or chamber in which matter is vaporized or gasified by the application of heat. A retort is distinguished from a still in that it is more often used for the treatment of a solid or semi-solid substance.

reverse circulation (drill).   Normal course of drilling fluid circulation is downward through the drill pipe and upward through the annular space surrounding the drill pipe. For special problems, normal circulation is sometimes reversed and the fluid returns to the surface through the drill pipe or tubing after being pumped down the annulus.

reverse fault.   A fault in which the hanging wall appears to have moved upward relative to the footwall; also called a thrust fault.

reversed magnetic field.   Paleomagnetic field reversed in polarity from the modern ambient field. The earth has experienced numerous polarity reversals of its magnetic field during geologic history. Evidence is preserved in rocks formed during periods of reversals by exhibiting remnent magnetism reversed in polarity from magnetism exhibited by rocks formed in modern times.

reversible reaction.   Reaction that can proceed in either direction by suitable variation in the condition of temperature, volume, pressure or of the quantities of reacting substances.

reversing thermometer.   A mercury in glass thermometer that records temperature upon being inverted and thereafter retains its reading unit until returned to the first position. Used for measuring ocean temperature at depth; usually attached to a Nansen or Niskin bottle.

revolution.   A dated term for an occurrance of mountain building of continental or worldwide extent. The term was coined before the concept of plate tectonics modified geologic thought concerning mountain building and rendered its usage unpopular.

reworked. A fragment of rock or a fossil removed by natural means from its place or origin and deposited in recognizable form in a younger deposit.

Reynolds number. A nondimensional parameter representing the ratio of the momentum forces to the viscous forces in fluid flow; named after Osborne Reynolds, an English scientist.

rheology. Study of fluid and plastic flow of substances.

rhigolene (refin). Most volatile liquid fraction obtained on the distillation of petroleum; has a boiling point at 18°C and consists largely of pentane; specific gravity is 0.60.

ria shoreline. A shoreline formed when a stream dissected coastal margin is submerged. It is highly irregular and characterized by long narrow estuaries resulting from drowned river valleys.

rich oil. A mixture of gasoline and oil in the manufacture of natural gasoline by the absorption process.

Richter Scale. Scale of earthquake magnitude devised by Charles Richter. The scale is logarithmic so that each unit on the scale represents an earthquake one order of magnitude greater than the preceeding unit. The energy of an earthquake is measured at a seismograph station. Using that value and the distance from the station to the earthquake epicenter, it is possible to compute the amount of energy released by the earthquake at its focus (magnitude).

ridge. An elongated crest of a hill or mountain. The intersection of two topographic surfaces sloping downward away from each other.

ridge and valley. *See* valley and ridge.

rift valley. (1) A long, narrow valley running along the crest of midocean ridges. The valley walls represent the edges of two oceanic plates diverging or spreading away from each other. The valley floor consists of the youngest exposed oceanic crust and is the frequent site of submarine volcanic activity. (2) A graben or a valley formed when the land between two normal faults drops down.

rift zone. A system of crustal fractures associated with plate boundaries.

rig. Structure used to drill a well, particularly an oil well; an oil derrick.

rigging down (drill). Dismantling the rig and auxiliary equipment following the completion of drilling operations.

"rigging up" (drill). Before the work of drilling can be started, the derrick has to be built, and tools and machinery installed. This operation is getting the rig ready.

right lateral fault.   A strike-slip fault where the ground opposite the observer appears to have moved to the right. The San Andreas fault is a right lateral fault.

rigidity.   Measure of the ability of a solid to resist deformation under stress. A measure of a rock's resistance to change of shape and volume.

rig irons (drill).   Various iron and steel parts used in making up power transmission and hoisting wheels and drums of the surface equipment of a cable tool outfit.

rill.   A very small rivulet or tiny stream. Valley or channel carved by a small rivulet; most often seen carved by run-off on a steep slope of poorly consolidated material unprotected by vegetation.

rill mark.   Small dendritic groves or channels, (1) on the swash zone of a beach face, formed by return of water related to ebbing tide or wave action, (2) develops where a small brook or trickling stream empties onto a mud flat.

ring and ball test.   Used to determine the melting point and softening temperature of asphalt, waxes, and paraffins. A small ring is fitted with a test sample and a small ball placed on the surface. The melting point is the temperature at which the test sample softens sufficiently to allow the ball to fall through the ring.

ring compound.   Certain hydrocarbon compounds in which the carbon atoms form a closed ring. Compounds of the ring type are known as the "aromatics" as benzol, toluene, napthalene and anthracene.

ring dike.   An igneous dike having a generally circular outline or a series of broadly arcuate dikes which as a group make up a roughly circular pattern. The dikes are vertically walled or show walls which dip steeply away from the center of the ring.

rip.   (1) A rapid seaward current moving through and disrupting the breaker zone in the near shore environment. Rips move at nearly right angles to the shore line; are a few meters to tens of meters wide and are of short, minutes to hours, duration; sometimes called rip current. (2) A zone of turbulence or water agitation usually in the marine environment and caused by the meeting and interaction of two or more currents. (3) In engineering, to excavate using hand or engine powered equipment (as a bulldozer) without benefit of blasting. A rock or deposit excavated in this way is called ripable.

ripple mark.   A series of subparallel ridges and troughs of small scale, rarely exceeding a few inches from ridge to ridge. Ridges may be asymmetrical current ripple marks or symmetrical oscillatory ripple marks. Ripple marks are created by bottom currents in streams, lakes, estuaries, lagoons or the marine environment.

They are useful in reconstructing paleoenvironments, in general and paleocurrent strengths and directions, in particular.

rip up. A sedimentary structure formed when currents lift plates or discs of finely laminated and partially consolidated sediment and transport them to a new site of deposition. The individual plates or clasts may themselves be referred to as rip-ups.

rise, continental. *See* continental rise.

riser. A vertical supply or return pipe for steam, water, gas, etc.

riser structure, drilling. An extension of the hole in an offshore drilling facility. It terminates at the rig to permit drilling operations at various water depths. Diagram of the principal designs for drilling and production riser systems.

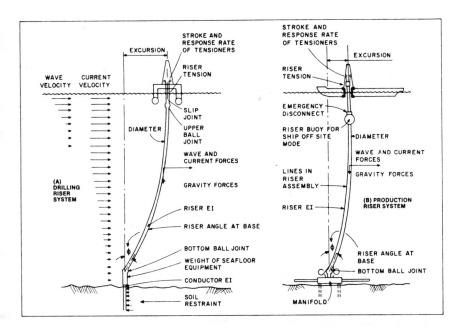

river bed. Also spelled riverbed. The bottom of a river channel. The channel may be either currently or recently occupied by a river.

riverine. Related to a river, as in "riverine sediments."

river piracy. *See* stream piracy.

river flood plain. *See* flood plain.

river terrace. *See* stream terrace.

river valley. An elongated depression in the earth's surface, carved by river or stream erosion in conjunction with mass wasting or collapse of river banks.

road log. A record of the sites of geologic interest along a road or network of roads. Sites are located by mileage from some arbitrarily established starting point; usually associated with a field trip.

roaster (refin). A heated, revolving drum used to dry materials or to oxidize materials as sulfides and carbonates, evolving SO_2 and CO_2 respectively.

rock. A rock is a naturally occurring and physically separable aggregate of minerals. Three main categories of rock exist and each is further subdivided on the terms of distinguishing characteristics as mineralogical composition and texture. The three major groups are igneous, sedimentary, and metamorphic.

rock asphalt. Sandstone or limestone mined or quarried, containing as a rule, less than 15 percent content. It is ground to a powder and used in asphalt mixtures as mastic asphalt.

rock-a-well. To bleed pressure from the casing of a well, then from the tubing, from the casing, and so on, to start the well flowing.

rock bit. A special bit designed to cut hard rock; of various designs, making use of extremely hard metals. Knurled cones are popular in the design.

rock cleavage. When shear between beds occurs, whether related to folding or faulting, a complementary system of fractures often develops. In weak beds, these fractures are often closely spaced. These closely spaced joints give the rock a capacity to split into sheets, as they impart a fracture cleavage. The plane of fracture cleavage gives a rough approximation of the axial plane of a related fold, and therefore the attitude of the intersection of the fracture cleavage plane with a bedding plane approximates the plunge of the fold.

rock creep. Slow downhill slipping of large joint blocks wherever well-jointed massive formations crop out along a slope. In this process, a joint block gradually widens the gap between itself and the parent outcrop and eventually tilts to the angle of slope of the surface. In bedded sedimentary strata or in slaty rocks, a downhill bending of the strata may be observed.

rock cycle. Concept that each of the principal rock types, igneous, sedimentary, and metamorphic, can be derived from each other by means of rock forming processes occurring at the surface or within the crust. For example, sedimentary rocks may be formed from the weathering products of igneous and metamorphic rocks while igneous magma may result from the melting of sedimentary and metamorphic rocks.

rock flour.   Fine grained rock particles resulting from glacial abrasion and subject to transportation and deposition by glacial melt-water. Streams that drain from the front of a melting glacier are charged with rock flour. Volume of this material may be so great that it may give the water a characteristically grayish-blue color similar to that of skim milk.

rock pressure.   Initial pressure of gas in a well. Usually it approximates the pressure exerted by a column of fresh water of a height equal to the depth of the gas sand below the surface.

rock slides.   A suddenly initiated and rapid movement of freshly detached portions of bedrock under the influence of the force of gravity. Bedrock failure normally is initiated along weakened zones, faults and joints and once the mass begins to move it frequently breaks up into smaller and smaller pieces.

rock-stratigraphic unit.   *See* geologic rock unit.

rock unit.   *See* geologic rock unit.

rod backoff wheel (prod).   A device used to unscrew rods when the pump is stuck or sanded up and the well has to be stripped out.

rod, elevator (prod).   A steel block with opening and latching device for the entrance of a sucker rod and provided with two long links to suspend from the elevator hook when running sucker rods into a well for pumping.

rod guide (prod).   An attachment to pump sucker rods in oil wells serving to prevent the rod from oscillating, knocking, and rubbing on the side of the tubing.

rock packing gland (prod).   A flanged ring fitting loosely upon a polished rod, a valve stem or a piston rod, forced against the packing in the stuffing box.

roller analysis (refin).   An analytical method of determining groupings of products or particle size of fine powders, as fluid cracking catalysts. Air of controlled humidity and flow rate is allowed to suspend and transport progressively larger particles upward into a filter where the fraction can be weighed.

rolly oil.   A mixture of crude oil and water in the form of an emulsion.

rope grab (drill).   A three-pronged fishing tool with barbs on inner sides of the prongs designed to recover rope or cable from a well.

rope sheaves.   Grooved pulleys for hoisting or power transmission.

rosin (refin).   Molten resin remaining in the still after the spirits of turpentine have been distilled from gum turpentine. When used as a basic resin in varnish, it is usually hardened by the addition of lime, zinc, calcium, oxide, etc. or by pro-

longed heating or both. It may be made into a rosin ester through heating with
glycerine.

rosin oil.   Obtained by distillation of rosin, varying in color from almost colorless
to dark brown. Heavy rosin oils, boiling above 300°C are produced by the destructive distillation of light rosin spirit, or pyrolin coming over the first runnings.

rotary (drill).   Turntable used to rotate pipe in a rotary rig.

rotary drilling.   A method of drilling deep holes in search of petroleum, using a
bit attached to a revolving drill pipe. At one end of the drill pipe is a shallow
squared section called the kelly held and revolved by the rotary table. Drilling
mud circulates down the inside of the kelly, the drill pipe and through holes in the
bit and back up the outside of the drill pipe to carry away the drilled materials.
Diagram of a rotary system.

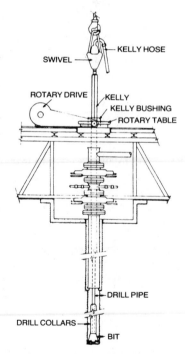

Rotary system.

rotary mud.   *See* drilling mud.

rotary table (drill).   Situated in the middle of the derrick floor, consists of a circular plate of chrome nickel steel about five feet in diameter. It revolves on a ball race and is driven by a pinion gear engaging with a crown gear. A second ball race prevents an upward movement of the table. The speed of rotation varies from about 50 to 250 revolutions per minute. The horizontal shaft running in roller bearings is driven from the drum shaft of the draw works through an intermediate shaft, by a chain engaged with the chain wheel. The rotary table causes a rotation of the kelly. The kelly fits into a hole in the table so that it may be lowered during drilling without stopping rotation.

roughneck.   A well driller's helper associated with oil wells.

roundness.   Degree to which the sharp corners and edges of a sedimentary particle have been abraded to smoothness. Measured numerically as the ratio of the average of radii of curvature of the several edges and corners to the radius of an inscribed sphere. Not to be confused with sphericity. A particle may approach a spherical shape very closely yet have many sharp corners and edges and thus not be well rounded or a platey particle may be well rounded while departing greatly from a spherical shape.

round trip.   Combined operations of going-in-the hole and coming out-of-the-hole during drilling operations.

roustabout.   An unskilled laborer performing general work in an oil field.

royalty.   Part of the value of oil, gas or minerals paid by the lessee to the lessor or to one who has acquired possession of the royalty rights, based on a certain percentage of the gross production from the property.

royalty interest.   Share of minerals, reserves, free of expense to the lessor, distinguished from working interest bearing the cost of development and operation.

rubidium-strontium age dating.   A technique measuring the age of rocks; a relatively rare variety of rubidium that has an atomic mass of 87, decays into the element strontium. Rubidium has a very slow rate of radioactive decay, consequently it can be used to measure the geologic age of very old rocks in much the same way as uranium.

rudite.   A general term for coarse grained detrital sedimentary rocks; includes breccia and conglomerate.

run.   (1) The amount of stock processed by a particular unit in a refinery in a given time. (2) Oil taken from or by a pipeline.

run a tank.   To empty oil from a tank into a pipe line.

run back (refin).   A pipe through which all or part of the condensate can be run back to a still instead of being run off. Runback lines are usually connected to take the condensate from the air-cooled condensers or heat exchangers.

run down (refin).   Distillation of an oil down to the solid residuals.

run down tanks (refin).   Tanks in which the condensation is received from the stills, agitators or other refinery equipment.

run off.   Portion of rainfall which neither sinks into the ground nor evaporates but runs off the earth's surface in stream or sheet flow.

S

sabkha. A salt flat, usually associated with an estuary or coastal lagoon which is infrequently inundated; originally applied to occurances in the Middle East.

saddle reef. An opening in the crest of a sharp fold in sedimentary rocks occupied by ore. If a thick stack of writing paper is sharply arched, openings form between the sheets at the crest of the arch. Similar ore receptacles are formed when alternating beds of competent and incompetent rocks are closely folded. When filled by ore, they resemble the cross section of a saddle, hence the name.

safety clamp (drill). A clamp used to maintain a rod string when it is desirable to take the weight of the rod string off the pumping equipment.

safety joint (drill). A fishing tool accessory placed above the tool. If the tool is engaged and the fish cannot be pulled, the safety joint will permit disengagement.

safety valve (refin). A valve on the top of a still, vapor line, or other pressure vessel set so that it will permit emission of gases when a dangerously high pressure is reached; insures against a destructive explosion within the still.

salar. A salt flat in a region of arid climate and usually associated with a salt lake or playa lake; originally applied to occurrences in the Chilean high deserts.

salinometer. An instrument for measuring the salinity of water; used widely for determining ocean water salinity and for measuring the concentration of dissolved salts in ground water. Most salinometers operate on the basis of electrical conductivity using either an electrical potential source or electromagnetic induction.

salt. (1) Any of a class of compounds derived from acids by replacement of part or all of the hydrogen by a metal or metal-like radical; $NaHSO_4$ and Na_2SO_4 are sodium salts of sulfuric acid (H_2SO_4); occurs in nature as rock salt or halite. When fatty acids contact metal, they form a salt called soap. (2) In drilling mud terminology, the term salt is applied to sodium chloride ($NaCl$). Salts are formed by the action of acids on metals, or oxides and hydroxides, directly with ammonia and in other ways.

saltation. A variety of wind or water borne sediment transport. Grains of sediment skip and hop along the sediment surface being alternately part of the suspended load and the bed load.

salt brine. Solution of a salt, as sodium chloride or table salt in water.

salt domes.   Immense columns or pillars of salt intruded from below into nearly flat lying sedimentary rocks. The intrusion occurs because the salt, which has a low specific gravity of 2.16, rises or "floats" through sediments whose specific gravity averages 2.5 to 2.6. Lying directly on the salt at the top of most domes is a caprock, consisting of anhydrite overlaid by gypsum and impure limestone. Thickness of the caprock ranges from a few feet to 1,000 feet or more.

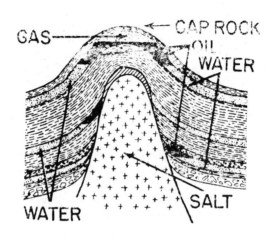

salt-water muds (drill).   A drilling fluid containing dissolved salt, brackish to saturated. These fluids may also include native solids, oil, and/or such commercial additives as clays, starch, etc.

salt water wells.   Wells flowing with salt water after the petroleum contents have become exhausted. The appearance of salt water is generally a signal that the petroleum production is nearing an end.

samples (drill).   Cuttings obtained for geological information from the drilling fluid as it emerges from the hole. They are washed, dried, and labeled as to depth. From an examination of these cuttings, the geologist determines the rock type, the formation being drilled, the formation tops, and the indications of oil and/or gas content.

sample log.   A strip of graph paper showing units of depth on which the geologist, using cores and samples, describes the rock formations penetrated by drilling.

sand.   Loose, uncemented mass of mineral grains, one-sixteenth millimeter to two millimeters in diameter, rounded or angular, produced by weathering and erosion of rocks. Quartz is the most common constituent of sand but other minerals frequently found in sand are calcite or gypsum.

sand dune.   *See* dune

sand flat.   A flat depositional surface in coastal estuaries or lagoons. Devoid of vegetation and subject to regular tidal inundation. Similar to mud flat but differs only in terms of the coarser grain size of sand on the flat.

sand line (drill).   A wire rope used on well-service rigs to operate a swab or bailer. It is usually nine-sixteenth inch in diameter and several thousand feet long.

sand reel (drill).   A drum, operated by chain or a friction drive from the bandwheel, used to raise or lower the sand pump or bailer, for bailing and swabbing operations.

sandstone.   A consolidated sedimentary rock composed of sand grains, compressed or cemented together. The size range and composition of the constituents are the same as for sand; also called arenite.

sand trap (prod).   A tubular connection placed below the working barrel in a well pump for the purpose of receiving oil and liquids but designed to exclude sand.

sanfrac.   Injection of heated oil and graded sand into formations under high pressure to increase permeability.

saponification number.   A means of determining both free and combined fatty acids in fixed oils, fats or compounded oils in terms of the number of milligrams of potassium hydroxide consumed by one gram of the test sample under test conditions.

saponifier.   A substance as an alkali changing a fat into a soap. Any compound, as caustic alkali used to convert fats, esters, or fatty acids into soap.

saturate.   To completely fill to capacity; fully penetrated or impregnated.

saturated compounds.   Compounds to which no other elements can be added for the reason there are no free valence to which hydrogen atoms or their equivalent can be added.

saturated hydrocarbons.   Hydrocarbons of a molecular structure that all adjacent carbon atoms are connected by not more than one valence or bond. Each valence not taken up by adjacent carbons atoms connects with or is satisfied by a hydrogen atom. These compounds cannot take on another product or the atoms of another element without giving up an equivalent amount of hydrogen.

saturated solution.   A solution is saturated if it contains at a given temperature as much of a solute as it can retain. At 68°F it takes 126.5 lb/bbl salt to saturate 1bbl of fresh water.

saturated steam. Steam or water vapor carrying all of the water that can be evaporated at a given temperature and pressure.

saturated vapor pressure. Vapor pressure when vapor and liquid are in equilibrium.

Saybolt chromometer. An instrument used to classify the color of oils by comparing them with colored discs viewed through a predetermined height of the test sample.

Saybolt Furol Viscosity. A viscosity test similar in nature to the Saybolt Universal viscosity test, appropriate to testing high viscosity oils.

scale wax. Paraffin derived by a process of sweating the greater part of the oil from slack wax. In contains up to six percent oil.

scarp. An escarpment or cliff.

schist. Foliated metamorphic rock consisting of thin crystalline mineral plates or rods lying parallel to one another, generally very crumpled.

schistosity. Irregular, roughly parallel cleavage induced in some metamorphic rocks by the orientation of platey or rod-like mineral grains.

schlieren. Elongated segrations of light or dark minerals in igneous rocks, presumably caused by magma or lava flow prior to immobilization by crystallization.

Schlumberger. Refers to electric well logging. It is derived from the name of the French scientist who first developed the method. One of the leading companies in this field of operation bears this name. Around drilling rigs, it is pronounced "slumberjay."

scoria. A frothy volcanic rock containing a large volume of gas bubbles or vesicles; similar to pumic but with larger holes.

scour. Downward and sideward erosion of rock or sediment deposits by sediment ladened wave and/or current action.

screen analysis (drill). Determination of the relative percentages of substances, as in the suspended solids of a drilling fluid, passing through or retained on a sequence of screens of decreasing mesh size. Analysis may be by wet or dry methods.

screen pipe (drill). Perforated pipe having a straining or filtering device, usually closely wound coils of wire wrapped around the pipe, for the purpose of admitting well fluids while excluding sand.

scrubbers (refin).   A unit cleaning natural or artificial gas by removing undesirable substances as sulfur compounds, ammonia, carbon, etc. Hydrogen sulfide is removed by washing the gas in a scrubber containing soda ash or lime, or filtering the gas through a mixture of ferrous and calcium hydroxide made porous with wood shavings.

scrubbing (refin).   Purification of a gas or liquid by washing it in a tower or agitator.

sea floor spreading.   *See* spreading.

sealing bed.   Impermeable bed restricting the movement of hydrocarbons out of a reservoir bed; used in conjunction with either stratigraphic or structure traps.

seal off (drill).   Penetration of a drilling fluid into a potentially productive formation disabling the formation from producing.

seamounts.   Dotting the ocean floors, drowned isolated steep-sloped peaks standing at least 1,000 meters above the surrounding ocean floor, their crests covered by varying depths of water which may exceed 1,000 meters. Seamounts are of volcanic origin; most are peaked but some are flat topped.

secondary mineral.   A mineral formed later than, and often from, the substance of earlier mineral deposits as by weathering or by groundwater action.

secondary recovery (gas injection).   Used under a variety of conditions in the early stages of development of an oil field. Gas may be injected at the top of the reservoir into the gas cap to reduce the normal decline in reservoir pressure. This is known as pressure maintenance. Gas is also injected at later stage in the life of a well.

secondary recovery (water flooding).   Water under pressure injected into a reservoir through special input wells located at selected points in the structure. The effect is to displace the oil towards the producing wells.

secondary waves (s).   Earthquake waves that vibrate at right angles to their direction of travel. These shear waves move about two-thirds as fast as primary waves and can pass only through solids.

sectile.   Capable of being cut with a knife without breaking off into pieces.

section.   Sequence of strata found in a particular area; exposure of a succession of rocks or graphic representation of a succession.

secunda oil.   A solar oil obtained by shale oil distillation, having a high gravity and boiling point.

sediment.   Particulate organic and inorganic matter accumulating in a loose un-consolidated form. In the singular the word, usually applied to material in suspen-sion in water or recently deposited from suspension. In the plural, applied to all kinds of deposits from the water of streams, lakes or seas, and in a more general sense, to deposits laid down by wind and ice. Deposits that have been consolidat-ed are generally called sedimentary rocks.

sediment and water.   A test on crude and fuel oils to measure the volume and in-soluble sediments in the oil.

sediment basin.   A sea floor depression in which sediments are deposited. The de-posits are usually thickest in the center and thinner toward the edges.

sedimentary facies.   Aggregate or set of factors as composition, texture, struc-ture, color, fossil content observed in a sedimentary rock reflecting the environ-mental conditions of deposition. Hence, a change in environmental conditions of deposition results in a rock "facies change." The total aspect of a sedimentary rock reflecting conditions of origin. The term is more often applied to areal changes in conditions of deposition rather than temperal changes, although the latter is by no means excluded.

sedimentary rock.   Sedimentary rock is formed by the following steps: (1) Weath-ering of parent rock. (2)Transportation of the weathered products as sediment. (3)Deposition of the sediment, and (4) Transformation of the sediment to rock by lithification or diagenesis. Sedimentary rocks may be detrital, consisting of bro-ken fragments of pre-existing rocks or they may be crystalline, consisting of dis-solved sediment precipitated out of solution in water. Sedimentary rocks combin-ing attributes of the two are not uncommon, as in detrital grains held together or cemented by precipitated mineral crystals. Limestones and dolomites do not fit readily into a clastic-crystalline system of classification because although many were formed originally from broken shell fragments (bioclastic) they have ex-perienced profound recrystallization during lithification or diagenesis.

sedimentation.   Process of breakup and separation of particles from the parent rock, their transportation, deposition, and consolidation into another rock. Some authors restrict the meaning to the deposition of sediment.

sediment load.   *See* load.

seeps.   An oil spring, out of which issues anywhere from a few drops to several barrels of oil daily. Almost invariably, seeps are at topographically low spots where water has also accumulated. The lighter oil rises to the top of the water and covers it with an irridescent "rainbow film."

seismic basement.   *See* basement, seismic.

seismic exploration.   Elastic properties of earth materials vary widely. Seismic methods of geophysical exploration are based on detecting and recording the variation of elastic properties. Differences in the elastic coefficients of different layers give rise to reflections and refractions of seismic waves, which are treated in the same manner as the comparable phenomena of geometrical optics. The instruments are designed to measure and record the arrival, period, and velocity of such waves in earth materials. Inferences may be made from these data regarding attitude, nature, distribution and structure of subsurface materials. Diagram shows the principles of seismic reflection surveying. An explosion at shot point number one creates shock waves reflected by subsurface formations to seismometers and recorded in the truck.

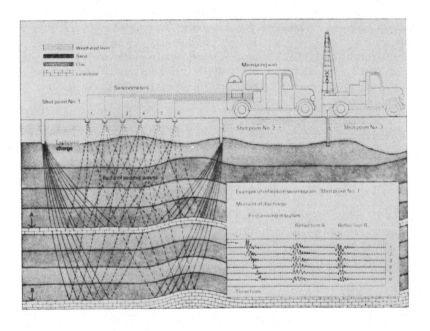

seismic reflection.   An event or pulse recorded on a seismogram and identifiable by its unique or nearly unique amplitude, period or wave form. Created when the induced energy travels through the earth, reflects off a seismic or elastic discontinuity, and returns to the earth's surface to be detected by a seismometer.

seismogram.   A graphic record of earthquake waves against time. Wave length amplitude and period may be measured directly. Distance and direction to seismic energy source or to reflection/refraction horizon may be calculated.

seismograph.   A device for detecting vibrations in the earth. One variety is used to record naturally occurring earthquakes while another is used in prospecting for probable oil-bearing structures. When prospecting, vibrations are created in several ways: discharge of explosives, weight drop, and at sea, by electrical discharge, sound producing "pinger" or steam and air blast. The nature and velocity of vibrations recorded by a seismograph are controlled by the general composition and structure of the section of earth through which the vibrations pass. A seismograph consists of the following components: seismometer, amplifier/filter and recorder. Seismometers transfer earth vibrations into electrical signals by electromagnetic induction or variable capacitance. The principle of operation is to hold a mass still and allow the earth to move under it. This is accomplished by use of pendulum or spring suspension system. Amplification and filtration is accomplished either mechanically or electronically, the former being rarely used in regions subject to strong earthquakes, while the latter is used universally for recording earthquakes or exploration. The seismogram is recorded in several ways: light beam on photo sensitive paper, ink pen on paper, hot pen on heat sensitive paper, electrostatically or on magnetic tape. Magnetic tape is most widely used in seismic prospecting whereas the others are used in earthquake seismology. Diagram illustrates the essential parts of a horizontal seismograph of the Milne-Shaw type, used for recording naturally occurring earthquakes.

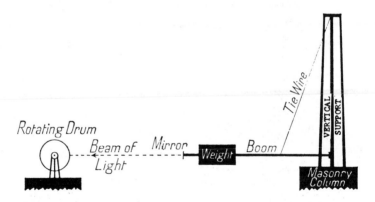

seismology. (1) Scientific study of earthquakes. (2) Use of earthquakes and artifically generated seismic signals to study the structure, composition and physical state of the earth's interior.

selective cracking process (refin). A process by which recycled stock is heated in a separate still and not in the same tubes in which the original or virgin charging stock was heated. The same fractioning equipment may or may not be used.

selective evaporation (refin). Partial evaporation of certain components in a compound fluid by reason of differences in their boiling points.

selective polymerization process (refin). A process in which the charging stock is carefully selected and processing given precise and exacting control; used to produce special products as aviation fuels.

selective solvent (refin). A solvent which will, at certain temperatures and ratios, selectively dissolve more of one component of a mixture than of another and thereby permit partial separation.

semi-drying oil. Possesses drying properties between drying and non-drying oils as soy bean, cotton and cotton seed oils.

semi-refined wax. Commercial grade wax not fully refined but meeting certain requirements as to color and oil content.

SeNo. Steam-emulsion number indicates the time it takes an oil to separate from an emulsion under specified test conditions.

sensitizer (refin). In the catalytic process, a sensitizer is a solution other than the catalyst whose presence facilitates the start of the catalytic action.

separation. Disassociation in space of once continuous linear or planar features by faulting. Data is insufficient to establish whether the disruption measures actual slip or a component of slip.

separator. (1) Machine for separating, with the aid of water or air, materials of different specific gravity. (2) An apparatus for separating the oil mechanically carried over by the vapor in distillation. (3) A closed steel vessel or tank of special design having interior baffles and automatic regulating valves used to separate gas in large volumes from oil as they flow from a well.

separator, centrifugal. Machine for separating liquids of different gravities by virtue of differences in centrifugal forces developed at high speeds or rotation.

septaria. Concretions crossed by networks of cracks in which minerals have been deposited from solution. Most septaria are claystones. Calcite is the most common mineral filling cracks; it may be clear and crystalline or amorphous and ironstained or both in a single septarium.

sequestration (drill). Formation of stable calcium, magnesium, iron complex by treating water or mud with certain complex phosphate.

set casing. To install a steel pipe or casing in a well bore. An accompanying operation is the cementing of the casing in place by surrounding it with a wall of cement extending for all or part of the depth of the well.

set grease soap. An inexpensive soap made cold by saponifying slack lime with resin oil; used to make the cold set type of grease, as axle grease.

setting point. Temperature at which liquids congeal or set and become solid or semi-solid. In many petroleum oils, determined by the amount of paraffin contained in the oil which becomes solid when the temperature is reduced below a certain point.

settled production. A loose term used to describe oil fields that produce at nearly the same rate from day to day.

settler. A separator. A tub, pan, vat or tank in which the partial separation of a liquid from its impurities can be effected by settling.

settling velocity. Terminal velocity attained by a sediment particle when falling through quiet water or air. Settling velocity varies according to particle size, particle shape and difference between the specific gravity of a particle and the settling medium. Mineral grains of like shape and specific gravity may be sorted according to grain size by means of their settling velocities.

shake out. To spin a sample of oil at high speed to determine its basic sediment and water content.

shale. Composed of compacted or cemented beds of mud or clay, usually thin-bedded, showing frequent changes in the fineness of materials. Shales that contain sand are arenaceous; those containing iron are ferruginous; and those containing large amounts of organic matter are carbonaceous. Carbonaceous shales usually are black and sometimes grade into beds of coal. Thin bedded limestones or sandstones often are referred to as shaly, many of them grading into shale.

shale naphtha. Naphtha from shale oil generally consists of 60 percent to 70 percent olefines and other hydrocarbons acted on by fuming nitric acid. Obtained at different specific gravities and boiling points according to requirements; for instance 0.660 for gasoline; 0.690 for motor spirits; and 0.72 to 0.75 for lighting purposes and as solvents.

shale, oil. Compact rock of sedimentary origin, with an ash content of more than 33 percent, containing organic matter yielding oil when destructively distilled but not appreciably when extracted with ordinary solvents for petroleum.

shale shaker. A vibrating sleeve removing cuttings from the circulating fluid stream in rotary drilling operations.

shard. A curved, blade- or needle-like fragment of volcanic glass.

shearing stress. Internal force acting along a plane between adjacent parts of a body, when two equal forces, parallel to the plane act on each part in opposite directions. The shear resists the tendency of one part to slide over the other part.

shear zone. A layer or slab-like portion of a rockmass traversed by closely spaced surfaces along which shearing has taken place. Many faults are characterized by closely spaced fractures among which movements have been distributed. Fracture or shear zones are suggestive evidence of faults. The shear zones of some faults are silicified to varying degrees or intruded by a network of quartz veins which fill the fractures. Many ancient faults have been completely sealed and healed by mineral fillings and placements.

sheave (drill). A grooved pulley. The grooved wheel or pulley of a pulley block.

sheeted complex. A rock mass formed of multiple intrusive dikes. The sheeted complex is presumed to form below the eruptive centers at a spreading oceanic ridge.

sheet flow. Continuous thin layer of water flowing downhill along the ground surface as a result of rainfall. Flow is continuous and essentially laminar, particularly when the ground surface is smooth and unbroken. Ground surface irregularities will disrupt sheet flow, instigating formation of rills thereby. Rills, once formed, lead to channels and gulleys which further disrupt sheet flow.

sheeting. Joints essentially parallel to the ground surface; more closely spaced near the surface and become progressively farther apart with depth. Particularly well developed in granite rocks, but sometimes in other massive rocks as well. Sheeting develops when rock masses of deep seated origin are exposed by erosion and expand upward as a result of reduced lithostatic pressure.

shell. (1) The crust of the earth or some other continuous layer beneath the crust. (2) A layer of hard rock. (3) In the structure of an atom, the shell is an energy level in which the negatively charged electrons are distributed. The positive charge is in the atomic nucleus. (4) Faunal exoskeleton; calcareous, siliceous or chitinous.

shell still (refin). A still where oil is charged into a closed cylindrical shell and the heat required for distillation is applied to the outside or the bottom from a firebox.

shield. Precambrian nuclear mass of a continent, around which and to some extent on which, the younger sedimentary rocks have been deposited. The term was originally applied to the shield shaped Precambrian area of Canada, but it is now used for the primitive areas of other continents, regardless of shape. Large areas of ancient rocks that have remained relatively stable through most of geologic time.

shield volcano. A gently sloping volcanic cone built chiefly of overlapping and interfingering basaltic lava flows, as in Hawaii.

shingle beach. A narrow rocky beach almost totally devoid of sand-size or finer material. Rocks called shingles are frequently elongated and flattened, and are steeply imbricated.

shoe (drill).   A heavy steel ring with the end beveled on the inside and then hardened; used as a shoe in the bottom end of a string of casing.

shoe string sand.   A very narrow, highly elongated occurrence of sandstone usually representing a buried and preserved sand bar or stream channel. Such deposits hold the potential for being excellent hydrocarbon reservoirs but are difficult to find by geophysical or geological prospecting techniques.

shooting.   Exploding nitroglycerin or other high explosives in a hole to shatter the rock and increase the flow of oil. In seismographic work, refers to the discharge of explosives to create vibrations in the earth's crust.

short residuum.   A type of residual oil from distillation operation, the reduction being made so as to take off "neutral oils" with the distillate.

shot.   A charge of high explosives, usually nitroglycerine, deposited in a well to shatter the sand and to expedite the recovery of oil. This method of formation stimulation has been completely replaced by fracturing and/or acid treatment.

shot depth.   Distance below surface in a shot hole at which an explosive charge is detonated in seismic prospecting; also used to designate the water depth at which an explosive charge is detonated in marine seismic prospecting.

shot hole.   A shallow hole drilled for purposes of detonating explosive charges below ground surface in seismic prospecting.

show.   An appearance of oil or gas in the cuttings, samples, cores, etc. of a drilling well.

shut down time (drill).   A rate provision usually contained in a drilling contract specifying the compensation to the independent drilling contractor when drilling operations have been suspended at the request of the operator.

shut in.   To close valves on a well so that it stops producing.

shut in pressure.   Pressure noted at the well head when the well is completely shut in.

sial.   A contraction of the term "silica-alumina" applied to the whole assemblage of relatively light weight, high standing continental rocks including granite, granodiorite, and quartz diorite. The contrasting term is "sima", a contaction of silica-magnesia designating the heavier deepseated rocks of the earth.

sidehill bit (drill).   A bit, eccentrically dressed, used to cut a hole larger in diameter than the diameter of the bit.

side door elevator (drill).   An elevator with a hinged part at one side opening to permit it to receive casing or tubing.

side irons (drill).   Iron or steel parts making up the bearings for the walking beam and the supports for these bearings. These are butted to the sides of the top end of the Sampson post.

side stream (refin).   A liquid stream taken from any one of the plates of a bubble tower. In cracking coils, side streams are used as a source of heating oil or gas oil or to keep too much liquid from going to the bottom of the bubble tower.

side tracking (drill).   Drilling past a broken drill or casing permanently lodged in the hole. This operation is usually accomplished by the use of a special tool known as a whipstock.

side wall coring (side wall sampling).   Geological samples of the formation constituting the wall of the well bore. A gun shoots from six to thirty hollow cylindrical bullets into the formation exposed in the well bore. Each bullet is fired separately by electrical control and the bullets with their rock cores are retrieved by wires attached to the gun. Cores are three-quarters of an inch in diameter and may be up to two inches in length.

sieve analysis.   A technique used to study the grain size distribution in a sediment or disaggregated sedimentary rock. A set of wire or plastic mesh seives are nested with the coarsest sieve on top of progressively finer mesh sieves. The sample to be analyzed is placed on the top seive and the whole nest shaken by hand or mechanically until the sample has been sorted according to the intermediate dimension of the constituent grains. The number of grains held on each sieve is measured indirectly by aggregate weight.

signal to noise ratio.   Ratio of signal amplitude to the amplitude of the background. Term is frequently used when interpreting seismograms. Optimizing the signal to noise ratio usually allow more accurate interpretations of the data. The ratio may be improved by (1) modifications of the equipment to minimize instrumental noise, (2) electronic filtration to eliminate as many naturally occurring and spurious signals as possible and, (3) mixing outputs from several sources so that signals constructively interfere and enhance, noise destructively interferes and diminishes. Signal is defined as useful or potentially useful data or information such as a seismic energy pulse reflected off of a reflecting horizon.

silica gel.   A manufactured amorphous silica which is exceedingly porous and with the property of selectively removing and holding certain chemical compounds.

silica sand.   Sand high in quartz or silicon dioxide.

silicates.   Silicate group contains the most important of the rock forming minerals. Silicates are combinations of silicon and oxygen with metallic or basic elements.

siliceous oolite.   Rounded grains of chalcedony deposited around sand grains, forming balls like those of true oolite. There is commonly a matrix of amorphous chalcedony between the grains.

siliceous ooze.   A fine grained pelagic sediment containing more than 30 percent siliceous skeletal remains of pelagic organisms.

siliceous sinter.   Geyserite; hydrated varieties of silica deposited from waters of hot springs and geysers.

silicification.   A special type of fossil replacement in which the original material is replaced by silica.

silicified.   A fossil preserved by replacement of organic matter by silica.

silicone.   A member of the family of polymeric materials characterized by a recurring chemical group that contains silicon and oxygen atoms as links in the main chain. These components are derived from silica or sand and methyl chloride; one of their important properties is resistance to heat.

sill.   A tabular shaped concordant igneous intrusion. Sheet of igneous rock intruded between strata while still in the molten state. Some sills lift the overlying rocks, wedging their way into the spaces so made, others digest some of the strata between which they are injected. Sills may be connected by dikes which follow oblique joints cutting across several beds, others are connected by dikes to laccoliths or batholiths. Most sills consist of dark rocks as basalt and andesite. Sills may be hundreds of feet in thickness, miles wide and long, but most are of more modest proportions. They may be horizontal, inclined or vertical in attitude but are always parallel to the strata into which they were forced.

silled basin.   A submarine basin separated from the open ocean by a narrow submerged ridge. Deep water in the silled basin may be stagnant and anaerobic.

silt.   A sediment composed of particles generally ranging from 1/256 to 1/16 millimeter in diameter. These particles may consist of any kind of mineral or rock fragment though quartz and feldspar generally predominate. Carbonate, sulphates and organic matter may also be present. The fine particles are produced mainly by abrasion, grinding, and impact, though some are small grains of resistant materials or minerals which have survived physical or chemical weathering of older rocks.

siltstone.   Rock composed of silt and lacking the fissility of shale.

sima.   Lower layer of the earth's crust, lying under ocean basins and under the sial layer of the continents. The term is derived from the abbreviations for silicon (Si) and magnesium (Ma). Major subdivision of the earth's crust characterized by relatively high specific gravity of approximately 3.3, composed mainly of siliceous (Si), magnesium (Mg) and aluminum (Al) materials.

single (drill).   A joint of drill pipe.

single effect evaporation.   An evaporator in which complete evaporation takes place in one vessel or by means of a single heating unit.

sinker bar (drill).   A bar added to the drilling tools to give the required weight for effective operation of jars.

sinkhole.   A depression in the surface occurring where rocks as salt, gypsum or limestone have been dissolved and the roof of the solution cavern has collapsed; a vertical hole dissolved into limestone by ground water, usually connected with an underground channel.

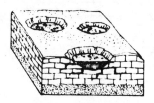

Sinkholes

sixes (drill).   A stand of drill pipe made by forming six lengths of drill pipe.

skid (drill).   Moving a rig from one location to another usually on tracks where little dismantling is required.

skimming (refin).   Removal by distillation of all the lighter fractions of the oil including kerosene and often a part of the gas oil.

skimming plant (refin).   An oil refinery designed to remove and finish only lighter constituents from crude oil, as gasoline and kerosene. In such a plant, the portion of the crude remaining after the above products are removed, is uaually sold as fuel oil.

slack lime.   Quicklime which has absorbed water, chemically and referred to as hydrated lime in grease making. It is the same as calcium hydrate.

slack wax.   Soft crude wax obtained from the pressing of paraffin distillate or wax oil.

slate.   A homogeneous, fine-grained rock splitting into thin or thick sheets with relatively smooth surfaces. Cleavage is often parallel to bedding, but at many places it intersects the bedding at high angles. Slates range in color from gray, through red, green, and purple, to black. Slates are usually named from the predominant color or most conspicuous mineral. There is no sharp boundary between shales and slates or between slates and phyllites.

slickenside.   A smooth, striated, polished rock surface in a fault zone where two masses of rock slide over one another. A fracture surface along which movement has taken place, may locally display parallel striations or grooves called slickensides. The direction of movement is indicated by the trend of the striae, and the direction of relative displacement can be determined from many slickensided and polished fault surfaces, by passing the hand over the surface to find the rough and smooth direction. Diagram of a slickensided surface showing relative movement. Steplike surface is exaggerated in diagram.

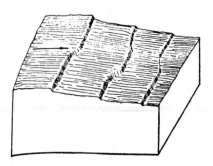

Sligh oxidation test (lube oil).   Apparatus used for this test is a water bath, heated and circulated during the test. The test sample, in a flask containing oxygen, is submerged in the bath and heated for several hours at test temperatures. The solids and insolubles forming during the test are weighed and reported as the Sligh oxidation number.

slim hole drilling.   Drilling operations in which the hole size is smaller than the conventional hole diameter for a given depth; enables the operator to run a smaller casing, thereby decreasing the cost of completion.

slip.   Actual displacement or movement along a fault plane; usually includes direction as well as sense of movement.

slips (drill).   Two or more pieces of steel of wedge-shape design and circular in form with exterior surface roughened, tooth-like; used for fishing tools, tubing hangers and other devices that are attached to either the interior or exterior of casing or tubing, slips up and down on a wedge shaped surface, thus the name. Wedge-shaped collars made in two sections and designed to hold the string of casing while the adjoining top section is being removed or added during the drilling operations. Rotary slips fit around drill pipe and wedge against the master bushing to support pipe. Power slips are pneumatically or hydraulically activated devices operated by the driller at his station and which dispense with the manual handling of slips when making a connection. Packers and other down-hole equipment are secured in position by means of slips that are caused to engage the pipe by action performed at the surface.

slip velocity (drill). Difference between the annular velocity of the fluid and the rate at which a cutting is removed from the hole.

slop (refin). A term rather loosely used to denote odds and ends of oil produced at various places in the plant, which must be rerun or further processed in order to get in suitable condition for use. When there is no other suitable use oil usually goes into pressure still charging stock, or to coke stills.

sloughing (drill). Partial or complete collapse of the walls of a hole resulting from incompetent, unconsolidated formations, high angle of repose and wetting along internal bedding planes.

sludge. A residue accumulating during chemical refining.

sludge acid (refin). Sulfuric acid or a tarry substance from the bottom of oil refining tanks consisting of the impurities from the oil mixed with the strong sulfuric acid used as a refinery acid.

sludge coking plant (refin). A plant for the recovery of sulfuric acid from dry acid sludge.

slug the pipe (drill). A procedure before pulling the drill pipe whereby a small quantity of heavy mud is pumped into the top section to cause an unbalanced column. As the pipe is pulled, the heavier column of fluid in the drill pipe will fall, thus keeping the inside of the drill pipe dry at the surface when the connection is unscrewed.

slump. Downward and outward movement of rock or unconsolidated material traveling as a unit or a series of units. Slump usually occurs where the original slope has been sharply steepened either artificially or naturally. The material reacts to the pull of gravity as if it were an elastic solid, and large blocks of the slump move downward and outward along curved planes. The upper surface of each block is tilted backward as it moves. (See Fig. p. 316).

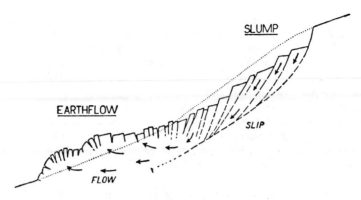

slurry (drill).   A plastic mixture of portland cement and water pumped into the well to harden, after which it supports the casing and provides a seal in the well bore to prevent migration of underground fluids.

slush pit (drill).   Pit adjacent to a drilling rig into which the mud and cuttings are discharged. The mud pit in which rotary drill cuttings are separated from the mud stream, mud treated by means of additives, and temporarily stored before being pumped back into the well. Modern rotary drilling rigs are generally provided three or more pits, usually fabricated steel tanks fitted with built-in pipe, valves and mud agitators.

smoke point.   Maximum height at which a burning oil will burn without producing smoke in a standard lamp.

smoke test.   A test made on kerosene showing the highest point at which the flame can be turned before it will smoke.

snake (drill).   To pull out.

snatch block (drill).   A block that can be opened up for putting a line over the roller or sheave.

snubbing (drill).   A method of running tubes in a high pressure well while oil is flowing through the casing.

soaker (soaking drum).   A heavy drum, usually set up vertically and employed in connection with cracking coils in order to furnish a necessary delay in cracking time. Oil slowly passes through the drum under elevated pressure and temperature after it emerges from the cracking coils allowing time for the decomposition reaction to proceed as far as possible.

soap.   Base of lubricating greases, forming the thickening and retaining medium for lubricating oil compounds. The soap is formed by the process of "saponification," the addition of fatty acids to metals, metallic salts or alkalis.

soap, aluminum. A soda soap is first made, since it is not possible to saponify aluminum and a fat directly, and then a solution of alum and aluminum soap, an insoluble compound which is then combined with mineral oil to form oil pulp. After heating and driving off the water, the mixture becomes very gelatinous and stringy.

soap, calcium or lime. In making cup-grease, lime or calcium hydroxide is used as the alkaline base which when combining with fats forms a soap, insoluble in water. These soaps are soft when cold as they exist in the form of an emulsion.

soap, metallic. Soaps formed by the combination of fatty acids with metals or salts of metals. The soaps are generally stearates of the metals and are greasy with fair lubricating value.

soap, soda. Caustic soda, sodium hydrate is very extensively used as a grease soap alkali. Soda soaps are soluble in water and are generally firecooked to expel the great amount of water contained in the soap at the beginning of manufacture. The water would reduce the melting point if allowed to remain.

soda. A popular name for commercial sodium salts as caustic soda, sodium hydrate, sodium carbonate, etc. The word "soda" is a corruption of sodium.

soda ash. A common name for crude sodium carbonate made from common salt. When causticized with quick lime, it is changed to caustic soda used to neutralize acids during the chemical refining process.

soda grease. A grease made by dissolving soda soaps in hot mineral oils. It is similar to fiber or sponge grease.

sodium bicarbonate . $NaHCO_3$, A material used extensively for treating cement contamination and occasionally other calcium contamination in drilling fluids. It is the half-neutralized sodium salt of carbonic acid.

sodium carbonate. Na_2CO_3, A material used extensively for treating out various types of calcium contamination. It is commonly called "soda ash." When sodium carbonate is added to a fluid, it increases the pH of the fluid by hydrolysis. Sodium carbonate can be added to salt water to increase the density of the fluid phase.

sodium chloride. Commonly known as table salt, sea salt, rock salt, halite or plain salt. A compound of sodium with chlorine used in refining and the manufacture of grease soaps. $NaCl$.

sodium hydroxide hydrate.   Used in petroleum refining and grease making, strongly alkaline and saponifys readily with fatty acids to form soap; absorbs moisture and carbon dioxide gas.

sodium plumbite.   Solution consisting of a mixture of sodium hydroxide, lead oxide and distilled water; used in making the doctor test and in connection with some sweating processes.

sodium polyacrylate.   A synthetic high-molecular weight polymer of acrylonitrile used primarily as a fluid-loss control agent.

sodium silicate muds.   Special class of inhibited chemical muds using as their bases sodium silicate, salt, water, and clay.

soda catalytic process (refin).   Process which activates water by means of a catalytic or acitive element contained in a small hermetically sealed metal cylinder introduced into the water system.

soft rock.   Applies generally to sedimentary rock in opposition to igneous and metamorphic rocks called "hard rock."

soil.   Product of weathering of the bed rock before removal by erosion. The upper portion of the regolith, capable of supporting plant life.

soil laterization.   Process of forming a laterite. *See* laterite.

soil podsolization.   Process of forming a podsol. *See* podsol.

soil mechanics.   Physical and engineering methods of investigation applied to the analysis of soils to determine strength, stability, compaction, and other physical properties.

soil profile.   A vertical section of the soil from the surface downward through all of its horizons into the parent materials. The layering of successively more mature weathering products over successively less mature ones.

sol (drill).   A general term for colloidal dispersions, distinguished from true solutions.

sole marks.   Cracks, tracks, grooves or other ridges and depressions preserved on the underside of a bed of sandstone or siltstone which has been deposited gently on a soft sedimentary bed. The marks are molds of structures created on the top of the underlying soft bed by currents or other directional processes.

solex oil process.   A propane solvent process used to separate fixed oils and fats into parts or fractions in much the same manner that fractional distillation separates petroleum into parts or fractions.

solid carbon.   Black carbon in a pure state and free from hydrocarbons; combined hydrogen, oil or tarry matter.

solid residue (refin).   Dry coke remaining after complete distillation.

solidified gasoline.   Gasoline converted into a jelly or stearic acid, the acid having previously undergone treatment with hydrochloric acid, at a high temperature in the gasoline. This solution is then mixed with an alcohol solution of sodium hydroxide.

solidified kerosene.   Kerosene jellified by treatment with stearic acid.

solifluction.   (From Latin *solum*, "soil" and *fluere* "to flow"), refers to the downslope movement of debris under saturated conditions in high latitudes where the soil is strongly affected by alternate freezing and thawing. Solifluction is most pronounced in areas where the ground freezes to great depths, but even moderately deep seasonal freezing promotes solifluction. In the Arctic and subarctic regions, and above the timberline in temperate zones, slow downslope movement called solifluction is intermediate between creep and debris flow. Frost and other weathering agents produce abundant fine rock fragments. When this debris is saturated with water, which freezes and melts, it spreads slowly downslope and along the valley floors.

soluble.   Substance dissolved by a prescribed solvent. Salt, for example, is water soluble and water is the prescribed solvent. By the same rule, salt is gasoline insoluble as it will not be dissolved by gasoline.

soluble bitumens.   Bituminous matter in asphalt subject to solution in petroleum ether, carbondisulphide, acetone, etc.

solute.   Substance that dissolves in a solvent to produce a solution.

solution.   A mixture of two or more components that form a homogeneous single phase, as solids dissolved in liquid, liquid in liquid, gas in liquid.

solution gas.   Gas produced in the form of a solution in a formation liquid. Reduction in pressure permits gas to separate or break out.

solution process (refin).   Sweating process removing mercaptans from gasoline by washing with a caustic solution containing organic compounds capable of increasing the solubility of mercaptans.

solvation. Chemical combination of molecules of solvent with solute ions or molecules.

solvent. A substance, usually a liquid, capable of absorbing another liquid, gas or solid to form a homogeneous mixture. The two most important properties of solvents are volatility and solvent power. Volatility, or the tendency of a liquid to change to vapor is indicated by a standard distillation test and measurement of the rate of evaporation.

solvent, aliphatic. Usually has a "gasoline smell," but through processing can be made odorless. The deodorized solvents are usually straight run naphthas in the 300° to 400°F boiling range specifically treated by hydrofining, while the odorless type are synthetic hydrocarbons obtained as a by-product in the manufacture of aviation alkylate.

solvent, aromatic. Aromatic solvents as benzene, toluene, and xylene have a much greater ability to dissolve resinous materials than aliphatic solvent. Catalytic reforming permits production of fractions very rich in aromatics, converting naphthene hydrocarbons into aromatic hydrocarbons. The mixture of aromatics produced by the reforming process is separated and purified to produce aromatic solvents of any desired purity.

solvent dewaxing (Ketone dewaxing). Process carried out by mixing waxy oil with one to four times its volume of ketone and heating the mixture until the oil is in solution. The solution is then chilled at a slow, controlled rate.

solvent, hydrocarbon. Composed of hydrocarbons of the paraffin, naphthene and aromatic families. These hydrocarbons do not normally react within the materials they contact, nor do they decompose under moderate heat or in the presence of water, therefore their chemical stability; Non-corrosive to metals.

solvent naphtha. Coal-tar distillation products consisting essentially of the xylenes and possibly their higher homologues.

solvent refined oils. Oils, solvent treated during the refining process and said to be solvent refined. Most high quality lubricating oils are solvent refined.

solvent refining. Use of a solvent as a means of selective extraction of compounds in petroleum products, principally, sulfur, asphalt, wax and low viscosity index compounds; consists of mixing the petroleum stock with a suitable solvent which preferentially dissolves undesirable constituents, separating the resulting two layers and recovering the solvent from the raffinate or the purified fraction and from the extraction by distillation.

solvent refining (Edeleunu process). A process using sulfur dioxide (SO^2) as the solvent. When used to solvent refine kerosene, it removes aromatic and other unsaturated hydrocarbons, and sulfur compounds.

sonic log. A geophysical well logging technique recording the velocity of sound in the strata punctured by a well as a function of depth below earth surface; also called acoustic or velocity log.

sorption. A general term used for three processes; adsorption, absorption, and persorption.

soundhead (seis). In marine seismology, an enclosure containing the sound transmitter and the receiving hydrophone. When towed behind a vessel it may be referred to as a fish.

sound pressure level (seis). Twenty times the logarithm to the base ten of the ratio of the pressure of sound to the reference pressure in decibels at a specific point. The reference pressure is explicitly stated.

soup. In the treatment of oil with sulfuric acid, the oil left after the sludge has settled out and before neutralization is often called "soup" oil.

source rock (petroleum). A rock unit or sequence of strata responsible for the generation of oil and/or gas from organic remains. The term usually implies that oil and gas have been generated in sufficient quantity to be economically important. Generally the oil and gas have long ago migrated away or been expelled from the source bed.

source rock (sedimentology). Rocks weathered to produce sediment of which a sedimentary rock or deposit is composed.

sour dirt. A dirt impregnated with sulphates resulting from the escape of SO_2 or H_2S. Because sulphur is often associated with petroleum in the salt domes of the Gulf Coast field, sour dirt, also known as copper dirt, is considered by many as indication of oil in that region.

sour gas. Gas containing objectionable amounts of sulfuretted hydrogen and other sulfur compounds.

sour oil. The word sour, as used in the petroleum industry has a special technical meaning. When used in a general way it refers to odor; any gas or crude oil having a disagreeable, unpleasant odor is said to be sour. Usually the bad odor is due to some sulfur compound, especially hydrogen sulfide.

spacing. Distance between wells, producing from the same pool; usually expressed in terms of acres, e.g., 20 acre spacing.

spacing clamp (drill).   A clamp used to hold the rod string up in pumping position while the well is in the final stages of being put back on the pump.

sparker (geoph).   A marine seismic prospecting technique in which a bubble is created by a high potential arc or spark. Sudden collapse of the bubble generates a seismic pulse.

spatter cones.   Small structures built around unimportant or immature vents. Explosions throw lumps of sticky lava a few feet into the air, and falling, they build cones or domes. Some spatter cones from on top of lava flows, where gas gathers in pockets and makes tiny explosions. A small cone or cones that form in lava fields on the sides of or away from the main vent.

spear (drill).   Fishing tool used in recovery of ropes and other materials lost in a well.

specific gravity.   Specific gravity and API gravity are expressions of the density or weight of a unit volume of material. The specific gravity is the ratio of the weight of a unit volume of the substance in question to the weight of a unit volume of water at standard temperature. Unless otherwise stated both specific gravity and API gravity refer to those constants at 60°F. The ratio of the weight of a volume of a body to the weight of an equal volume of some standard substance. In the case of liquids and solids, the standard is water; and in some gases, the standard is hydrogen or air.

specific heat.   Number of calories required to raise gallon of a substance one degree Centigrade.

specific volume.   Volume per unit mass of a substance; the reciprocal of density.

spheroidal weathering.   Formation of large rounded boulders by weathering along sets of intersecting joints within rock masses. This phenomenon is common where granitic intrusive rocks are exposed to weathering in arid or semiarid regions. Production of rounded, residual boulders by chemical weathering of rock along fractures.

spider.   A steel block having a tapered opening therein to permit the passage of pipe through it when run into or pulled from a well. Its purpose is to hold pipe in suspension in the well when the slips are placed in the tapered opening and in contact with the pipe.

spike test (grease).   Designed to determine the viscosity of a grease by passing a spike through the grease under test conditions.

SP log.   Spontaneous-potential log. A well logging technique which measures the naturally occurring electrical potentials created by chemical differences across sedimentary contacts and by flow of fluids through the restricted interstices of sedimentary strata. It is most frequently used to measure changes in porosity and permeability of strata with depth in the well.

sponge grease. A soda-base grease differing from smooth, buttery, soda-base greases in that it is more fibrous and sponge-like in structure.

spreading. Describes the movement of the sea floor along a divergent plate boundary. The sea floor spreads at right angles to the midocean ridge marking the divergent plate boundary. The ocean floor moves in opposite directions on either side of the midocean ridge as new simatic (possibly mantle) material rises from within the earth to fill the void left by the diverging plates.

spudder. A spudding bit or a small drilling rig used while spudding.

spudding (drill). Initial step in drilling or boring regardless of the method used. Spudding with a cable tool rig is accomplished when a spudding bit is actuated by a jerk line from the wrist pin of the crank.

spudding bit (drill). A broad, dull drilling tool for working in earth down to rock.

spudding in (drill). See *spudding.*

spudding shoe (drill). A device used for attaching the jerk line to the drilling cable for the purpose of the spudding operation.

spud mud (drill). Fluid used when drilling starts at the surface, often as thick bentonite-lime slurry.

squeeze (drill). A procedure whereby slurries of cement, mud, gunk plug, etc. are forced into the formation by pumping into the hole while maintaining a back pressure, usually by closing the rams.

stabbing (drill). Inserting the threaded end of a joint of pipe into the collar of the joint already in the hole and rotating slowly to engage the threads properly prior to screwing up.

stabbing board (drill). A temporary platform erected in the derrick from twenty to forty feet above the derrick floor. The derrickman or other crew members work on the board while casing is being run in a well.

stabilized. A well is considered "stabilized" when, in the case of a flowing well, the rate of production through a given size of choke remains constant or, in the case of a pumping well, when the fluid column within the well remains constant in height.

stabilizer (drill).   A tool placed near the bit in the drilling assembly to change the deviation angle in a well by controlling the location of the contact point between the hole and drill collars.

stabilizer tower (refin).   A tower where gases are separated from cracked and natural gasoline. Gasoline from a stabilizer is often referred to as stabilizer bottoms or stabilized gasoline.

stable emulsion.   An emulsion, the components not separating readily during long standing or storage.

stable isotope.   A non-radioactive isotope of an element.

stack.   An isolated, steep-sided rock mass standing as a small island in front of a cliff line or off the end of a promotory along a coast.

stacked seismic data.   A process of interpreting seismic data which involves mixing the output of several detectors in order to enhance the signal to noise ratio.

stacking a rig (drill).   Storing a drilling rig upon completion of a job when the rig is to be withdrawn from operation for a period of time.

stalactite.   An icicle-like formation of calcium carbonate precipitating when percolating ground water encounters the roof of a cave and evaporates. It hangs down from the roof of limestone caverns.

stalagmite.   Cone-shape posts of dripstone growing upward from the floor of a cave. Stalactites and stalagmites often meet forming a pillar from floor to roof. A raised deposit on the floor of a cavern formed by calcium carbonate precipitating from dripping ground water.

stand (drill).   A stand of pipe with tubing or drill stem, may be two or more joints forming a portion of the string; as the pipe is taken from the well it is disconnected by stands and the stands are set upright in the corner of the derrick.

standard colors.   Oils are commonly graduated by color, although the color has little significance except to the refiner. There are a number of standard color comparison systems of which the Lovibond, Saybolt and Union are the most used. In all cases, comparisons of the colors are made with standard colored glasses or standard solutions of potassium bichromate.

standard colors, kerosene.   Colors are all slight variations of white, making very close comparison necessary. Among the most common trade name colors for these oils are standard white, prime white, superfine or water white, the latter being the lightest grade.

standpipe (drill).   A rig pipe, part of the drilling fluid circulation system, extending up into the derrick to a height suitable for attaching the rotary hose.

stands (drill).   Connected joints of pipe racked in the derrick when making a trip. On a drilling rig, the usual stand is approximately 90 feet long, comprised of three lengths of drill pipe screwed together.

stand, tubing (drill).   To support tubing in the derrick when it is out of the well rather than laying it on a rack.

starch (drill).   A group of carbohydrates occurring in many plant cells. Starch is specially processed, pregelatinized, for use in muds to reduce filtration rate and occasionally to increase the viscosity.

static head (prod).   Distance from the surface of the ground to the fluid level in the well when not being pumped.

steam distillation (refin).   Steam introduced into a still during petroleum distillation to lower the boiling point of the oils being distilled in order to minimize cracking.

steam refined.   As applied to lubricating oils, particularly cylinder oils, refers to products distilled with heat aided by steam and which have not been filtered.

steam rig.   A drilling rig which has a battery of portable boilers as a source of power.

steam still (refin).   A still in which steam provides the major portion of the heat. The use of steam makes it possible to distill a product from a charging stock at a lower temperature than would otherwise be possible except by vacuum distillation.

stearate.   (1) A metal salt formed by the union of the metal and stearic acid. (2) In drilling, salt of stearic acid, a saturated, 18-carbon fatty acid. Certain compounds as aluminum stearate, calcium stearate, zinc stearate, have been used in drilling fluids for one or more of the following purposes: defoamer, lubrication, air drilling in which a small amount of water is encountered.

stearic acid.   One of the free fatty acids occurring naturally in animal and vegetable fats as tallow, suet, palm oil, castor oils. It has a corrosive effect on many metals forming the greasy salts known as "stearates," used as soaps in the manufacture of greases. Stearic acid is obtained by breaking down stearin and then neutralizing the acids free by decomposition.

stearin.   A white, odorless, tasteless grease, obtained from animal fats, used in soap and candle making and in compounding lubricating oils. It is technically known as tristearine and glyceryl-stearic ester. Chemically it is an "ester" produced by the combination of stearic acid and glyceryl radicals.

step-out time (geoph). Difference in arrival time of a seismic event with variations in distance between seismometers and shot point. In the case of a symmetrical array of seismometers on either side of the shot point, a horizontal reflection horizon is represented by a sequence of recordings of the reflection event concave in the direction of increasing travel time on the seismogram.

step-out well. A well drilled adjacent to or near a proven well to ascertain the limits of the reservoir.

stereochemistry. Study of the spatial arrangement of the atoms making up a moloclue or compound.

sterols. A group of alcohols related to the terpenes. They are of high molecular weight and are found in nature with fatty acids.

still (refin). A device for evaporating liquids by the application of heat, the vapor afterwards being condensed to a liquid state by means of a condenser. In petroleum refining, the still is a sort of boiler for heating crude oils or raw distillate and vaporizing them at different temperatures so that various hydrocarbons can be separated and collected in the order of their boiling points. The still may consist of a cylinder boilerlike tank or of oil-filled tubes heated on their outer surface by the fire or steam.

still, batch. Any still in which a charge of oil is completely distilled or "run down" before replacing or refining it with fresh stock.

still, blowing. A still in which the asphalt residuals are oxidized and thickened by blowing.

still coke. Residue left in the still on distilling crude shale oil to dryness.

still, continuous. A still in which distillation proceeds continously to the intermittent action of the batch still. A number of stills are connected together with each of the stills vaporizing a certain band of fractions. The oil is pumped from one still to the next without intermission. The first still takes off the lightest fractions, the next receives the oil from the first and distills off the next heavier series, and so on, through the apparatus until the oil is finally reduced to residual.

still, cracking. High pressure, high temperature still for cracking oils.

still gas. Mixture of extremely low-boiling point hydrocarbons produced during the distillation of crude petroleum. These light vaporous products are formed generally by the "cracking" which occurs.

still, tar.   Large stills in which the heavier portions of crude petroleum are distilled by intense firing, including the heavy lubricants, tar, etc. Process continued until nothing is left in the still but solid coke.

still, tower.   Crude oil stills preventing accumulation of coke and excessive cracking.

still, vacuum.   A still working under a partial vacuum as in the distillation of heavy lubricating oils.

stirrup (drill).   A loop over the end of the walking beam and attached to the pitman. A hanger.

stock.   A large mass of intrusive rock that worked its way close to the surface before solidifying. Most stocks are round to oval in cross section and are high in proportion to their width. Their sides are steep and irregular, and they cut across strata instead of bending them upward. Many stocks consist of diorite, gabbro or granite. The principal difference between a stock and a batholith is size. Batholiths exceed 40 square miles of surface exposure while stocks do not.

stockworks.   An interlacing network of small, tabular intrusive igneous rocks. A mass of rock irregularly fractured along which mineralization has spread.

Stoddard solvent.   A petroleum distillate of low flammability used for dry cleaning purposes. According to ASTM specifications, ranges between a minimum flash point of 100°F and a maximum end point of 410°F and possesses other properties as described under ASTM designations.

stoichiometry.   Aspect of chemistry concerned with the exact proportions of constituents of chemical compounds. Calculations involving balancing atomic weights and valences between products and reactants during a reaction.

stone.   General term for rock fragments and rocks ranging from pebbles and gravels to boulders or large rock masses.

stony meteorite.   A meteorite consisting principally of olivine and pyroxene and resembling a terrestrial ultramafic rock.

stop-cocking.   A method for inducing oil to flow from a well through a string tubing. A stop-cock placed in the tubing near the connection to the flow-line is alternately closed and opened to allow an accumulation of oil and gas pressure in the well. The length of time required for each accumulation may be determined by trial.

stove pipe (drill). A large diameter casing made of riveted rolled steel or iron sheets. Frequently used for the first string of casing, inserted in a drilled well.

straight chain hydrocarbons. Members of the chainlike, open-end structure, as in normal octane with its eight carbons and eighteen hydrogens in the shape of a straight chain with open ends.

straight run distillation. A continuous distillation of the products of petroleum in the order of their boiling points without cracking or compounding with other products.

straight run gasoline. Gasoline produced by the process of distillation with still pressure at or near atmospheric as in batch distillation.

straight run pitch. A pitch distilled at the desired consistency in the initial process of distillation and without subsequent fluxing or thinning. Pitch produced in the initial distillation of crude by simply reducing the residue in the same still to the desired consistency with the aid of steam.

straight run solvents. Includes products in the gasoline and kerosene range possessing marked solvent action or power, produced by straight run fractional distillation and frequently called solvent naphtha.

strain. Deformation resulting from a stress. Strain is measured by the ratio of a deformed dimension to the same dimension prior to deformation.

strata. Plural of stratum.

stratification. Characteristic structural feature of sedimentary rock produced by deposition of sediments in beds, layers, strata, laminae, lenses, wedges, and other essentially tabular units. Stratification stems from many causes, differences of texture, hardness, cohension or cementation, color, mineralogical or lithological composition and internal structure.

stratified drift. Deposits made by glacial melt water. The two requisites for stratified drift deposits are a supply of till carried and sorted by melt water, and a reduction in velocity of the transporting melt water current.

stratified rock. Rock deposited in distinct parallel beds. The distinction between the beds owing to differences in texture, mineralogy, structure or color. The individual bed is a stratum and the dividing plane is called a bedding plane.

stratigraphy. Systematic treatment of composition, sequence, correlation and formation of stratified rock in the earth's crust.

stratigraphic trap. A situation where a reservoir bed is confined by a sealing bed to the extent that petroleum and natural gas are accumulated or trapped. The configuration develops because of sedimentalogic or stratigraphic factors and not by structural deformation, as in a channel deposit or shoe-string sand overlain by shale beds; often referred to as a "strat-trap."

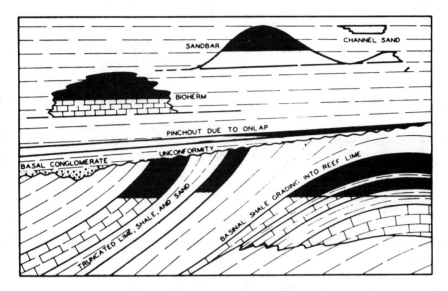

strato volcano.   Volcano built by both violent eruptions of pyroclastic material and more quiescent flows. The shape of the strato volcano is intermediate between shield volcanoes and pyroclastic cones. A large volcanic cone built of alternating layers of lava and pyroclastic debris, therefore a stratified cone.

stratum.   A single tabular layer of sedimentary material exhibiting homogeneous or gradational lithology. It is separated from adjacent strata or cross-strata by surfaces of erosion, non-deposition, or abrupt changes in character; synonymous with bed or lamination, frequently used in the plural form, strata.

straw oil.   A light oil used as an absorbent for natural gasoline, for rubbing and polishing, and as a high grade fuel oil. It belongs to the gas oil division of distillates and has a very low olefin content.

streak.   Color of the fine powder of a mineral obtained by scratching or by rubbing against a "hard white surace," usually a piece of unglazed tile or ceramic.

stream day (refin).   An operating day in a refinery unit, as distinguished from a calendar day, bearing a prorated allowance for regular "downtime" in the unit.

stream, effluent.   A stream fed by groundwater seepage. Most perennial streams are effluent. (See Fig. p. 330).

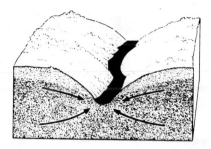

stream, ephemeral.   Streams flowing only during and immediately after a rain.

stream, influent.   Streams that lose water to the groundwater reservoir. Streams which supply water to groundwater reservoir in arid regions. Discharge of an influent stream decreases down stream until the stream ceases to flow.

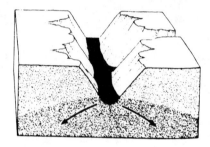

stream, intermittent.   A stream carrying water only during wet seasons when frequent rains and some base flow sustains it.

stream, piracy.   Circumstance which arises when a headward eroding stream intersects a portion of the drainage of another stream system and diverts the other stream's run-off into its own system.

stream terrace.   Flat surface, erosional or depositional, located in a stream valley. Stream terraces are elevated above current stream channel and are remnants of stream activity at a higher base level or gradient.

stream valley.   *See* river valley.

streaming potential (drill).   Electrokinetic portion of the spontaneous potential (SP) electric-log curve which can be significantly influenced by the characteristics of the filtrate and mud cake of the drilling fluid used to drill the well.

stretch (drill).   Distance a string of pipe will stretch before the drill leaves the bottom of the hole.

striae.   Minute grooves or scratches in fault surfaces or in rocks over which glacial ice has been moved.

striations.   (1) Fine parallel lines common on plagioclase feldspars. (2) Scratches on rocks and bedrock due to glacier movement; also striae.

strike.   Azimuth or bearing of the line of intersection of a horizontal plane with an inclined bed, structure or surface; the direction or bearing of a horizontal line in the plane of an inclined structural plane. Strike direction is perpendicular to the direction of the dip.

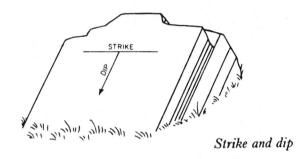

Strike and dip

string (drill).   A series of well drilling tools arranged for lowering into the hole, including the drilling bit, auger, stem, jars, sinker bar and rope socket.

string shot.   An explosive method to back off pipe utilizing primacord. Primacord is an instantaneous, textile covered fuse with a core of very high explosives. It is used to create an explosive jar inside stuck pipe so it may be backed off at the joint where the explosion takes place.

stringing up (drill).   Act of threading the drilling line through the sheaves of the traveling block and crown block. One end of the line is secured to the hoisting drum and the other is anchored to the derrick substructure.

strip a well (prod).   To pull rods and tubing from a well at the same time. Tubing must be "stripped" over the rods a joint at a time.

stripper.   A well producing a very small amount of oil.

stripping (refin).   Substantially complete removal of the more volatile components from a cut in order to raise the flash point of kerosene, gas oil, or lubricating oil, or to remove heavy slop cut from asphalt to improve its stain test.

stromatolite.   General term for those blue-green algae and other micro-organisms living in shallow water and which by their metabolic processes cause calcium carbonate to be deposited; also refers to the irregularly shaped laminated structures, domed to colloform, in limestone which are attributed to micro-organism instigated precipitation of calcium carbonate. Stromatolites are found in rocks of pre-Cambrian to recent age.

structural relief. Vertical distance between the highest and lowest points of a stratigraphic unit involved in a structure as between the crest of an anticline and the trough of an adjacent syncline.

structure. (1) (**petrology**) Features or inter-relationships in rocks of a larger scale than texture, as foliation and cross lamination (2) (**geomorphology**) A general term relating to the terrain underlying a landscape. (3) (**structural geology**) Attitude of beds, surfaces and bodies of rock and the geometry of changes in these factors as well as their interrelationships. Usually related to tectonic deformation, faults and folds. (4) (**minerology**) Arrangement of a mineral's constituent atoms. (5) (**petroleum geology**) Subsurface folds, faults, unconformities or other arrangements of rock exhibiting a potential for containing petroleum or natural gas.

structure contour. A contour used to express the irregularities induced on a once flat, horizontal surface by tectonic deformation. These are drawn on a structure contour map which portrays the subsurface configuration of faults, folds, domes and basins as they have developed, tectonically on a stratigraphic horizon.

structure trap (or structural trap). A pocket or enclosure created by tectonic deformation and capable of containing or trapping petroleum or natural gas.

stuffing box (prod). A means of preventing leakage around a polish rod or valve stem, the point at which the rod stem passes through the partition being provided with a small annular cavity well packed with some fibrous material clinging tightly to the mounted rod stem.

stylolites. Striated, or polished, columnar and variously shaped projections of rock that form an interlocking and interpenetrating series along partings in limestones. Projections vary in length from a fraction of an inch to a foot or more and are equally variable in width. Stylolites result from different amounts of solution along a bedding plane or crevace, their formation caused by the increased pressures at the points of contact on the two rock surfaces.

styrene ($C_6H_5CHCH_2$.) A constituent of petroleum and coal-tar. The gummy sediments in gasoline are usually ascribed to oxidation of such substances as styrene and indene. Uncracked petroleum is likely to yield gas containing indene and styrene which may polymerize to produce gum.

sub-bottom profiling (geoph). A seismic prospecting technique used at sea. Term originally developed when acoustic techniques for recording water depth were modified to cause emitted energy to penetrate ocean bottom sediments and thereby yield sub-bottom information. Now used in a variety of ways including general reference to all ocean-going seismic profiling and specific reference to high resolution (frequency = 3.5 to 7.0 KHz) procedures.

subcrop map.   A paleogeologic map. A geologic map drawn as it would have looked at some time in the past. A geologic map drawn on an unconformity or old erosion surface. All rocks of a younger age than the surface being mapped are discounted or hypothetically removed.

subduction.   Plunging of the margin of an oceanic plate under the margin of another plate (oceanic or continental). The usual state of affairs when converging plates interact if at least one of the plates is oceanic. The region from where the oceanic plate begins to bend downward to where it is consumed at depth is called the subduction zone, coinciding with the Benioff zone of earthquake foci. Subduction is usually accompanied by volcanic arcs and oceanic trenches.

sublimate.   A solid condensing directly from a gas without first passing through a liquid state.

sublimation.   Vaporization of a solid directly, without first passing through liquid phase.

submarine fan.   *See* fan, deep sea.

submergence.   A change in relative sea level so that either the sea rises or the land sinks, or both, and once dry land, is covered by water.

subsidence.   An essentially downward movement of the surface. Subsidence is due largely to plastic outflow of underlying strata, to compaction of the underlying material, or to collapse.

sucker rod (prod).   A section of steel rod inside the production tubing transmitting power from the pump jack on the surface of the well to a pump at or near the liquid level in the well. Opens and closes valves in the pump so that on a downstroke, oil that surrounds the pump flows into the pump and on the upstroke is lifted up the tubing.

suitcase sand.   A formation found to be nonproductive in oil and gas. Derived from the fact that operations are suspended and the crews pack their suitcases and move to another job.

sulfonated oils.   A fatty oil or fat chemically refined with sulfuric acid, followed by neutralization.

sulfonic acid.   An acid obtained by the addition of sulfur trioxide to a hydrocarbon, usually by treatment of an oil with strong acid.

sulfur. Often occurs at or near the crater rims of active extinct volcanoes where it has been derived from the gases given off in fumaroles; also formed by the reduction of sulfates as gypsum in the vicinity of oil seeps.

sulfuric acid treating (refin). Sulfuric acid partly removes sulfur and nitrogen compounds, precipitates asphaltic and resinous material and removes olefinic and unstable compounds. Sulfuric acid treating improves the color, odor and stability of cracked naphthas, kerosene, lubricating oils and waxes.

supercooled (refin). Applied to a gas cooled below the dew point but where vapor has failed to begin to condense due to the lack of nuclei on which liquid droplets can form.

super-fine kerosene. A standard color kerosene corresponding to water white.

super-fractionation (refin). A process of fractional distillation separating and collecting gases as butane, propane, etc.

superposition, law of. A basic law of geology, first proposed in 1669, by Nicholaus Steno, stating that in an undisturbed sequence of sedimentary rock the lowest beds are the oldest and each successive overlying bed is younger.

supersaturation (drill). If a solution contains a higher concentration of a solute in a solvent than would normally correspond to its solubility at a given temperature, this constitutes supersaturation. This is an unstable condition, as the excess solute separates when the solution is seeded by introducing a crystal of the solute.

surface pipe (drill). First string of casing set in a well generally used to shut off and protect shallow, fresh-water sands from contamination by deeper saline waters.

surface (L) waves. Earthquake waves that travel along the earth's surface. These waves move more slowly than either primary or secondary waves but have the greatest amplitudes and are therefore the waves "felt" during an earthquake.

surface tension. Generally, the force acting within the interface between a liquid and its own vapor tends to maintain the area of the surface at a minimum and is expressed in dynes per centimeter. Since the surface tension of a liquid is approximately equal to the interfacial tension between the liquid and air, it is common practice to refer to values measured against air as surface tension, and to use the term "interfacial tension" for measurements at an interface between two liquids, or a liquid and a solid.

surfacant (drill). A material tending to concentrate at an interface; used in drilling fluids to control the degree of emulsification, aggregation, dispersion, interfacial tension, foaming, defoaming, wetting, etc.

surge (refin).   An upheaval of liquid in a vessel frequently causing a carry-over of liquid through the vapor lines.

surge drums (refin).   A vessel or accumulation serving as a reservoir for liquid being pumped through a line, thereby overcoming fluctuations in the rate of flow caused by the pump.

suspended load.   Sediment particles kept in suspension by turbulence of the current of water or air which is transporting them.

swab.   A device fitting the inside of tubing closely so that it may be pulled through the tubing to clean it or lift fluid from it.

swash.   Thin, one to five centimeters sheet of water representing the final uprush of a breaking wave. Furthest advance of the swash is indicated by a swash mark at the point where the water flow reverses or all water sinks into the beach sand.

swage.   A tool used for straightening out collapsed casing.

sweat.   To remove oil from paraffin by a process of heating in shallow pans. The oil being lighter in weight than the wax separates when heated and floats off. This is done after the excess oil has been removed.

sweated scale wax.   A white, moisture-free wax coming from the sweaters in a semirefined conditions, finally filtered or rerun through a sweater to yield a completely refined commercial product.

sweet crude oil.   A crude oil having so little sulfur requiring no special treatment for the removal of sulfur compounds.

sweetening.   Process by which petroleum products are improved in odor and color by oxidizing the sulphur products and saturated compounds. The most generally employed method is agitation with sodium lumite or aluminum chloride, cuprous oxide, and sodium hypocholorite.

swivel (drill).   Provides the connection between the kelly and mud pump for the mud fluid to be circulated to the bit. The swivel is so designed as to allow the drilling string to rotate freely on roller bearings, and oil or other seals prevent mud or extraneous matter from reaching the bearings. The goose neck of the swivel is attached by flexible hose to the mud system.

symmetrical fold.   A fold in which the axial plane is essentially vertical and the limbs dip in opposite directions by the same amount.

synclinal.   Referring to strata so bent that fold is concave upward.

synclinal axis.   Central line of a syncline, toward which the beds dip from both sides.

syncline.   A fold in rocks in which the strata dip inward from both sides toward the axis. A concave upward surface. The opposite of anticline.

synclinorium.   A broad regional synclinal trough in which minor folds are superimposed. Schematic section through an anticlinorium bordered on each side by a synclinorium. Beds destroyed by erosion are indicated by the dotted lines.

T

tackiness agent. An additive used to impart adhesive properties to an oil or grease.

tagged atoms. Radioactive atoms or isotopes used for various types of research.

Tagliabue closed tester. An apparatus used for the determination of the flash point of mobile liquids flashing below 175°F with the exception of fuel oils.

tail (refin). Portion of an oil vaporizing near the end of the distillation; the heavy end.

tail house (refin). A building where the condensors are observed through a look box, samples are taken for testing, and from where products are diverted to the proper storage tanks or to other parts of the refinery for further distillation or processing.

tail out (prod). To lay sucker rods on a rack as they are pulled from a well when laying down rods.

tally. To measure and record lengths of pipe or tubing.

talus. A slope built up by an accumulation by mass wasting of rock fragments at the foot of a cliff or ridge. Talus has also come to be used as a name for the rock debris itself.

talus creep. Slow, down-slope movement of moderately coarse, irregular blocks of a talus or scree. Such blocks fall from receding cliffs where they are loosened by various agents of weathering. The type of creep is found wherever a steep talus slope exists and its rate of movement is determined by the climate conditions of the region.

tank baloons. Air and vapor tight flexible receivers connected to the breather pipe of gasoline storage tanks to receive and prevent the loss of gasoline vapors. As the tank cools, the vapors return to the tank from the balloon.

tank farm. Land on which a number of storage tanks are located, generally crude oil storage tanks for the producer, refiner, or pipe line.

tank wagon price. In general practice, the price charged upon delivery at the retailer's place of business or at the service station by the tank truck.

tar.   Heavy viscous product obtained when distilling organic materials as wood, coal, peat, etc. Although a tar-like product is obtained from petroleum, the term tar does not properly apply to such a product secured from this source.

tar, bituminous.   A dark colored liquid or semi-fluid seldom soluble in water, but soluble in ordinary solvents as carbon disulphide, benzol, petroleum ether, etc. On distillation produces pitch.

tar, coal.   Hydrocarbon distillates, mostly unsaturated ring compounds produced by the destructive distillation of coal.

tar distillates.   Distillates produced from a tar still or from a still in which are charged the tar bottoms from continuous crude stills, pressure stills, or cracking stills.

tar heat exchanger (refin).   Heat exchanger in which the heat of the tar bottom being drawn from the still, is transmitted to the charge within the still.

tar oils.   Generally coal tar distillates separated into commercial fractions, as crude, naphtha, light oil, crude carbolic, creosote oil.

tar sand.   A sandstone once containing petroleum but from which the lighter, more volatile fractions have escaped leaving a residue of asphalt in the pore spaces.

tar stills (refin).   Large oil stills in which the heaviest portions of crude petroleum or residues from other stills are distilled by intense firing until nothing but coke remains in the bottom of the still.

T.D.   Abbreviation for total depth.

tectogenes.   Deficiency of gravity measured over the sites of mountain ranges led to an earlier belief in roots of mountains, or crumpled and thickened portions of the outer earth shell that had been compressed and forced down into the sima. This concept of "sialic roots" was later extended to cover similar gravity deficiencies associated with island arcs and the name "tectogene" was coined for the elongated zone of crustal thickening and downfolding thought to be associated with mountain building. Plate tectonics has increased understanding of the relationship between mountain roots and orogeny. Tectogene is little used at present because of its strong association with invalid aspects of the geosynclinal theory of mountain building preceding plate tectonics.

tectonics.   Study of origin and development of the broad structural features of the upper portion of the earth's crust; occassionally used as a synonym for mountain building.

teleseism. An earthquake recorded by a seismograph at a great distance. By international convention, this distance is required to be over 1,000 kilometers from the epicenter.

telescoping derrick (drill). A portable derrick or mast as one capable of being erected as a unit, usually by means of wire line hoisting tackle, but sometimes by hydraulic pistons. Generally the upper section of a portable mast is "nested", telescoped inside the lower section of the structure.

telluric currents. Natural electric currents that flow in the earth. Methods have been developed for using these currents to make resistivity surveys.

temperature, absolute. Temperature measured from absolute zero ($-273.16°C$); thus the absolute temperature is plus $273.16°$ in degrees centigrade.

temperature survey (drill). An operation to determine temperatures at various depths in the well bore. Survey is used in instances where there is doubt as to the height of cement behind the casing, to find the location of water influx into the welbore, and for other reasons.

temper screw (drill). A long screw encased in a steel frame for connecting a wire line to a walking beam and lowering a string of tools as the hole is made.

tender. A barge anchored alongside an offshore drilling platform; usually contains living quarters, storage space, and mud system.

tension, interfacial. Tension existing between two contacted surfaces, notably between the surface of an oil film and the surface of the bearing metal to which it adheres; theory of lubricating advanced by Southcombe and Wells.

tension, surface. Forces of attraction, tension existing between the outermost atoms or molecules in the surface film of an exposed liquid, resulting in a greater cohesive force existing at the surface than within the body of the liquid.

tephra. A collective term for all pyroclastic materials ejected during a volcanic eruption; includes volcanic dust, ash, cinders, lapilli, scoria, pumice, bombs, and blocks.

terminal moraine. A rugged ridge or belt of unconsolidated till marking the outermost margin of a glacier.

terpinene. An unsaturated hydrocarbon, $C_{10}H_{16}$, an essential oil found in many plants and in oil of turpentine.

terrace. A level-topped surface bordered by a steep escarpment; may be composed of alluvium or of solid rock. A level and narrow plain usually with a steep front, bordering a river, lake or sea.

terrain correction (geoph). A correction applied to observed values obtained in gravity surveys in order to remove the effect of variations in topography in the vicinity of the site of observation.

terrigenous deposit. Material derived and deposited above sea level as volcanic ash and river flood plain sediments.

terrigenous sediments. Sedimentary deposits consisting of debris derived from weathering and erosion of land areas; may be deposited on land or in the marine environment.

test (paleontology). External shell, hard covering or external support structure of a fossil invertebrate.

test (drilling). Procedure for measuring the potential petroleum productivity of a horizon or sequence of strata in an oil well as in a drill stem test or wire line test.

test well. Exploratory well drilled to test the potential for petroleum production in an unproven area; a wildcat well.

Tethys. A geologically ancient sea resembling the modern day Mediterranean Sea. It was present from Permian to early Tertiary between the super continents of Laurasia and Gondwanaland or as in incursion into Pangea.

tetracosane ($C_{24}H_{50}$) A very heavy solid, saturated paraffin hydro-carbon; melting point 50°F; specific gravity 0.690 at 60°F.

tetradecane. ($C_{14}H_{30}$) A heavy liquid saturated paraffin hydrocarbon.

tetraethyl lead. A compound added to a fuel for spark ignition engines increasing the octane number and anti-knock quality. A colorless, oily liquid, boiling at 392°F, specific gravity 1.62. Manufactured by reacting ethyl chloride with an alloy of metallic sodium and lead at moderate temperatures and pressures.

tetravalent. An atom exhibiting a valence of four or having the power to combine with four atoms of hydrogen.

texture. Characteristic physical appearance of a rock, involving the size, shape, and arrangement of its constituent grains.

therm. Taken from Greek word therme meaning heat. A prefix used to construct compound terms. A general term for any of several quantitative units of heat as calorie or BTU.

thermal cracking (refin).   Distillation under pressure whereby hydrocarbons of larger molecular size are broken down or "cracked" into mixtures of smaller molecules. It is a process for converting heavy petroleum products as gas oil, wax distillate, and fuel oil into lighter materials of greater value. It breaks down the molecular structure of hydrocarbons by the application of heat without employing catalysts.

thermal decomposition.   Chemical breakdown of a compound or substance at elevated temperature. Simple substances or constituent elements are produced.

thermal gradient.   Change in temperature with distance as applied to thermal metamorphism. When applied to change in temperature with depth in the earth, the term geothermal gradient is preferred. The average geothermal gradient is approximately 25° centigrade per kilometer.

thermal reforming (refin).   Cracking converts heavier oils into gasoline; reforming converts or reforms gasoline into higher octane gasoline. The equipment for thermal reforming is essentially the same as for thermal cracking but higher temperatures are used. The products of thermal reforming are gases, gasoline, and residual oil or tar.

thermal unit, British (Btu).   Heat required to raise the temperature of one pound of water, one degree Fahrenheit.

thermal unit, calorie.   "Great calorie" is the heat required to raise one kilogram of water, one degree Celsius or from 0°C to 1°C. The small calorie is the heat required to raise the temperature of one gram of water by the same amount.

thermic.   Pertaining to heat.

thermie.   Unit of energy; the quantity of heat required to raise the temperature of one metric ton (1,000 kilograms) of water through 1°C.

thermometer, high reading.   A special type of thermometer designed for high temperatures, generally made by introducing nitrogen gas, under pressure, above the mercury column, thus raising the boiling point of the mercury. Mercury boils at a comparatively low temperature and measurements are impossible unless the boiling is suppressed.

thermometer, low reading.   Thermometer for reading low temperatures below freezing point using alcohol, ether, or some other fluid that has a lower freezing point than mercury.

thermo remnent magnetism.   Magnetism acquired by an igneous rock as it cools past the curie temperature of its constituent magnetic minerals.

thickened oils.   Mineral oils thickened by dissolving in them small amounts of vulcanized rubber or aluminum soap; oils are used for special lubricating purposes.

thief (prod).   A small cylindrical vessel designed so that it can take a sample of liquids from any depth desired. It is lowered into a tank from the top and the sample is taken by pulling a little wire opening and closing it. Used by pipe line gaugers to determine the percentage of water or basic sediment in the oil before taking it into the pipe line system.

thinner (drill).   Any of various organic agents: tannins, lignins, lignosulfonates, and inorganic agents: pyrophosphates, tetraphosphates, that are added to a drilling fluid to reduce the viscosity and/or thixotropic properties.

thin section.   A thin (ca. 0.03 mm) slice of a rock, suitable for viewing in transmitted light under a microscope. A relatively thick slice is made with a diamond bladed rock saw and then ground down to appropriate thickness on a grinding wheel or lap.

thio.   A combining form indicating the name of a compound in which a sulphur atom occupies a place ordinarily filled by an oxygen atom.

thixotropy   Ability of a fluid to develop gel strength with time. Property of a fluid causing it to build up a rigid or semi-rigid gel structure if allowed to stand at rest. It can be returned to a fluid state by mechanical agitation. The change is reversible. Most drilling fluids, or muds exhibit this property owing to the presence of the clay mineral, betonite, or its equivalent.

thorium.   A chemical element which is a source of fuel for atomic furnaces. Thorium atoms do not split but can be converted inside an atomic furnace into uranium atoms that do split. A radioactive element, symbol Th; atomic number 90.

three phase separator.   Production vessel capable of separating gas, oil and water, and discharging in three separate streams.

three point method.   Technique for determining the attitude (dip and strike) of a surface or bed when the elevation or depth below a datum of the surface or bed is known at three localities. The procedure is trigonometric and involves formula or graphic manipulations of the data. A basic assumption is that the surface or bed is flat and not deformed.

thribble.   A stand of drill pipe made up of three joints, each about 30 feet in length.

thribble board (drill).   Working platform of the derrick man, located at a height in the derrick equal to three lengths of pipe joined together.

throw. Vertical displacement of a particular stratum by a fault, usually measured in feet, yards, or meters. The vertical component of the net slip of a fault.

throwing the chain (drill). Act of flipping the spinning chain up from a tool joint box, to place several wraps around the tool joint pin after it is stabbed into the box. The stand or joint of drill pipe is turned by pulling on the spinning chain from the cathead on the draw works.

thrust fault. A fault along which the hanging wall appears to have been raised relative to the footwall. Generally characterized by a low angle of inclination with reference to the horizontal; commonly called a reverse fault. Diagram of a reverse or thrust fault.

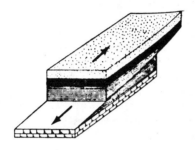

thumping (geoph). A source of seismsic energy produced by dropping a heavy weight; used when detonating explosives might be damaging to surface structures, aquifers or other features; popular in the 1960's but used less frequently in the 1970's.

tight. Used in reference to strata lacking in porosity; impervious.

tight hole. A well from which the information obtained is restricted and passed only to those who may be authorized to receive it.

till. Unsorted and unstratified drift or miscellaneous mixture of rock fragments deposited by melting glaciers. Material consists of angular or partly rounded pieces, many of which are marked by lines or striae made when other rocks are rubbed against them. They range in size, from fine clay to boulders, many feet in diameter.

tidal flat. A flat, essentially horizontal land surface that is covered by water at high tide and exposed at low tide. Surface may be vegetated or barren. Barren tidal flats may be called mudflats or sandflats depending on the grain size of the sediment.

tie line (geop). A seismic profile or traverse crossing a series of parallel or subparallel seismic profiles or traverses. The purpose is to assure proper three dimensional control at depth and to allow accurate computation of the dip and strike of beds.

tiff.   Originally a colloquial term for the mineral barite ($BaSO_4$) in Southeast Missouri. The term has come to wider use in the field of well drilling because barite is used as a drilling fluid additive increasing specific gravity, thereby improving the ability of the fluid to lift well cuttings.

tillite.   (1) Indurated till. The term is reserved for pre-Pleistocene tills that have been indurated or consolidated by normal post-depositional diagenetic processes. (2) Cemented till, recognized by an almost complete lack of sorting, bedding, and generally by the presence of striae on at least a few of the largest pebbles and boulders.

time break (geoph).   A distinctive mark on an exploration seismogram indicating the exact time of detonation of the explosive energy source.

time unit.   *See* geological time unit.

tintometer.   A colormeter used for determining the color of oils by comparison with arbitrary standards.

titration.   An analytical procedure for determining the concentration of a substance in a solution. Involves the addition of a liquid reagent to a known amount of another liquid until a change of color or effect takes place; also referred to as volumetric analysis.

toluene.   An aromatic hydrocarbon somewhat similar to benzene but of higher boiling point produced in coking of coal and also by petroleum processing. A solvent secured from coal tar which like benzene is ring-shaped and found sparingly in crude oil; also known as toluol.

tongs (drill).   Large wrenches used for turning to make up or break out drill pipe, casing, tubing and other pipe; variously called casing tongs, rotary tongs, etc., according to their designated use. Power tongs are pneumatically or hydraulically operated tools that serve to spin the pipe up tight, and in some instances, to apply final makeup torque.

tongue.   A subdivision of a formation which passes into a similar type of rock and dies out. A formation that extends laterally into another formation or between other formations.

tool joint (drill).   A heavy, special alloy steel coupling element for drill pipe. Tool joints have coarse, tapered threads and seating shoulders designed to sustain the weight of the drill stem, to withstand the strain of frequent coupling and uncoupling and to provide a leakproof seal.

tool pusher.   An individual who supervises the crew of roughnecks at an oil derrick during drilling operations. A foreman in charge of a string of drilling tools.

topographic high.   A general term indicating a higher elevation relative to surroundings; opposed to topographic low indicating a lower elevation.

topography.   (1) Practice of surveying the physical features of a portion of the earth's surface and the art of delineating them on maps. (2) General shape or configuration of a portion of the earth's surface.

topped crude oil (refin).   Residue remaining in a topping or skimming plant after distilling off the tops constituting the major part of the more volatile components of the crude oil.

topping (refin).   Practice of distilling only the light gasoline and burning oil fractions from the crude petroleum and using the heavier oil for fuel; also known as "stripping" or "skimming."

topset bed.   Layer of sediment constituting the upper most surface of a delta; usually nearly horizontal and covers or truncates the edges of inclined forset beds.

top water.   Water entering an oil well from a sand above the production sand.

torque.   A measure of the force or effort applied to a shaft causing it to rotate. On a rotary rig, this applies especially to the rotation of the drill stem in its action against the bore of the hole. Torque reduction can usually be accomplished by the addition of various drilling fluid additives.

torque gauge (drill).   On a drilling rig, the torque gauge gives a measure of the twisting force in the drill stem and enables the driller to keep this at a safe point below the breaking strength of the drill pipe.

total depth (T.D.).   Greatest depth reached by a well bore.

total magnetic intensity.   Vector resultant of the intensity of the horizontal and vertical components of the earth's magnetic field at a specified point.

tour.   Shift of a drilling crew or other oil field workers. Usually pronounced as if it were spelled tower. The day tour starts at seven or eight a.m., the evening tour starts at three or four p.m. and the morning tour starts at eleven p.m. or midnight, sometimes referred to as the graveyard shift or tour.

tower (refin).   An apparatus for increasing the degree of separation obtained during the distillation of oil in a still. Towers may be divided into two general classes, those which secure separation by fractionization and those which take only advantage of partial condensation. Towers of the first class are used when accurate work is necessary, as in the production of naphtha and gasoline to meet rigid distillation specifications. Towers operating by partial condensation are used to divide roughly the vapors from a still into several liquid portions.

township. A subdivision of land under the U.S. Public Land Survey system. On maps, it is a horizontal (parallel to lattitude lines) tier of land, six miles wide. Successive township tiers are numbered sequentially north and south of an arbitrarily designated latitude line, called a Base Line. T3N would be the third tier north of the base line. North-south trending tiers of a similar nature are called Ranges. The square of land delineated by the intersection of a Township tier and a Range tier is of itself called a township. Townships are approximately 36 square miles in area and subdivided into 36 sections.

trace fossils. Preserved tracks, trails, and burrows of ancient organisms.

transform fault. *See* fault, transform.

transgression. Sea invades a previously emergent region. The sediment is laid down by the incursion in transgressive deposits. Conversely, when a region is abandoned by the sea it is called a regression.

RECORD OF A TRANSGRESSION
Advance of the sea on the land

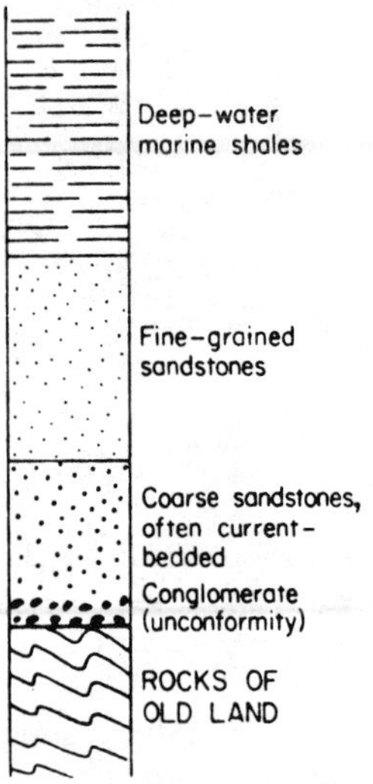

Deep-water marine shales

Fine-grained sandstones

Coarse sandstones, often current-bedded
Conglomerate (unconformity)
ROCKS OF OLD LAND

transmutation. Conversion of an atom or an element into an atom of a different element either by radioactive decay or as a result of bombardment by high energy radiation.

transported soils.   Soils derived from regolith or previous soils which have been carried to their present positions from their place of origin; made up largely of material that is not weathered or that is only partly weathered. Their present positions are due to some agent of transportation, as running water, wind, moving ice sheets, or gravity. Transported soils vary in texture, from fine silts to coarse gravels; vary in chemical composition.

transverse dune.   A dune formed in areas of scanty vegetation in which sand has been moved in a ridge at right angles to the wind. It exhibits the gentle windward slope and the steep leeward slope characteristic of other dunes.

trap.   Any geological barrier to oil migration as a fault, dome, anticline or pinch out of the reservoir that may localize the accumulation of an oil pool. *See* stratigraphic trap and structural trap. Types of structural traps.

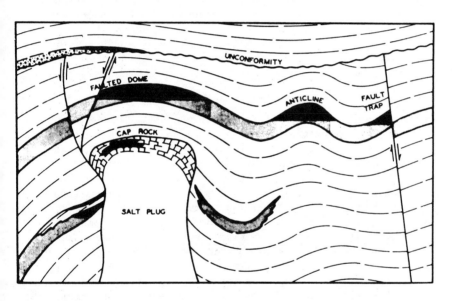

trap, gas.   A closed steel tank provided to receive a mixture of gas and liquid, usually oil, to separate the gas from the oil or other liquid and automatically discharge the gas and liquid into separate deliver pipe lines.

trap, stratographic.   *See* stratigraphic trap.

traveling block (drill).   Multiple sheaves hoisting block for raising and lowering drill pipe and casing.

traverse.   Geologic survey designed to gather data sufficient to prepare both geologic maps and sections. In order to gather this data, field traverses are made according to some systematic plan. Three principal systems of traversing are (1) boundary tracing, (2) cross-structural traversing, and (3) multiple outcrop mapping.

traverse (barometer-compass). Geologic surveying technique used in regions of moderate to high relief. The compass is used to locate positions along the traverse and the barometer to measure elevations above sea level. Use of the two instruments, in conjunction with each other, provides three dimensional control for the construction of maps and profiles.

traverse (boundary tracing). Location of margins of rock types is critical in geologic mapping. By tracing the boundaries or contacts between rock types, a map deliniating the geologic units is compiled.

traverse, (chain and compass). Chain and compass method provides a means for obtaining horizontal control of the location of sites of interest along a traverse. The compass indicates direction of traverse from one site to another and the chain measures the distances between sites. Chain and compass mapping is most frequently used for surveying mines, mine prospects, or detailing large exposures of particular interest.

traverse (cross structural). In the areas of strongly folded sediments and metamorphics, the most information can be gathered in the least time by making traverses roughly at right angles to the prevailing structural trends. If the traverses are closely spaced, the contacts between formations can be closely drawn.

traverse (multiple outcrop mapping). If sufficient time is available or great detail sought, all exposures within the map are studied. The results therefore are as complete as possible without subsurface exploration and if the latter is necessary, the locations where supplementary information is needed, become apparent. Location of outcrops is established by pace and compass, chain and compass, or barometer and compass methods depending on accuracy desired and on terrain.

traverse (pace and compass). In the pace and compass method, horizontal distances from known points are determined by pacing along compass courses. Essentially the same as chain and compass traverse with pacing replacing the chain.

travertine. *See* tufa.

treater (refin). Equipment for removal of mercaptans, hydrogen sulfide, absorbed water, salt, and other undesirable constituents from gas and liquid.

treating (refin). Purifying petroleum intermediates by agitation with chemical, or by physical absorbents, in a specially constructed apparatus known as an agitator or washer.

treating, acid (refin).   Acid treating of light petroleum products is carried out in lead-lined agitators or in continuous treaters. The process consists of mixing the petroleum product and the acid, then allowing the mixture to separate into a naphtha layer and an acid layer. The impurities remain in the acid layer which is removed.

treating, bauxite (refin).   A vaporized petroleum fraction is passed through beds of bauxite. The bauxite acts catalytically to convert many different sulphur compounds, in particular mercaptans into hydrogen sulphide. The hydrogen sulphide is subsequently removed by lye treatment.

trellis drainage.   A drainage pattern paralleling deeply eroded folded strata; pattern resembles a garden trellis. A structurally controlled drainage pattern that forms where the underlying bedrock comprises a series of strata of alternating high and low resistance to erosion.

trench.   A narrow elongated depression in the sea floor usually associated with convergent plate boundaries and subduction. Trenches usually have steep walls and are filled to varying degrees with sediment depending largely on their geographic location. Trenches in close proximity to continents usually contain great thickness of terrigenous or hemipelagic sediments. Open ocean trenches contain lesser amounts of sediment owing to their greater distance of separation from abundant sediment sources.

tricone bit.   A type of rock bit in which each of three toothed, conical cutters is mounted on friction reducing bearings and is forced into the formation by the weight and rotary motion of the drill stem. The bit body is fitted with nozzles, jets, through which the drilling fluid is discharged.

trip.   To pull or run a string of rods of tubing from or into a well.

trip spear (drill).   A fishing tool for recovering lost casing. If the casing is found immovable the hold is broken by operating the trip release and the spear is then withdrawn.

tritium.   Hydrogen isotope having one proton and two neutrons in the nucleus with mass of 3.

truncation.   Erosional or fault surface cutting across bedding or other structures.

tube bundle (refin).   Group of fixed parallel tubes as used in a heat exchanger. The tube bundle includes the tube headers with the tubes, the baffles, and spacer rods.

tube sheets (refin).   Flat plates in a heat exchanger with holes for the necessary number of tubes and onto which the tubes are inserted.

tubing block (drill).   Sheave block used to travel and hoist tubing by means of a wire line hoist.

tubing catcher (drill).   Device attached to a string of casing to prevent the tubing slipping back into the well when it is being pulled.

tubing head (drill).   Similar to casing head, connecting the tubing to the flow lines and control equipment; also supports the tubing and seals off the space between casing and tubing.

tubing spider (drill).   Device used to prevent tubing from falling into the hole while a joint of pipe is being unscrewed and racked.

tubular goods.   Term covering all classes of pipe, casing and tubing used in drilling or operating oil or gas wells.

tufa.   Porous or "spongy" limestone formed by deposition of calcium carbonate in desert lakes, in shallow bays, around hot springs; consolidated tufa is often called travertine.

tuff.   Fine grained pyroclastic or volcanic rock of light weight, chalky or dense.

tundra.   An undulating treeless plain characteristic of arctic regions, having a black muck soil and a permanently frozen subsoil.

turbidite.   A deposit of sediments or a sedimentary rock. Sediments transported by and deposited from a turbidity current. Turbidites exhibit moderate sorting and graded bedding.

turbidity current.   A subaqueous, usually ocean or lake bottom, flow or current. The current has a higher density than surrounding water owing to a sediment load maintained in suspension by turbulence within the current or flow. The high density current moves downslope under the influence of gravity. Turbidity currents may be initiated in a number of ways including submarine slumping, discharge of river water heavy laden with sediment, over-supply of sediment to continental shelf or slope and tectonic movement. Turbidity currents lose energy as the turbid water is diluted by surrounding clear water, resulting in deposition of turbidites. Opinions differ concerning the efficiency of turbidity currents to carve submarine canyons.

turbo-drill (drill).   System of drilling similar to the rotary method except the bit is rotated by means of a fluid turbine actuated by the drilling mud. The mud turbine is usually placed in the drilling string just above the bit.

turbulent flow.   Fluid in which individual fluid molecules do not move parallel to each other nor to channel walls. A confused or heterogeneous flow is most apparent in river rapids. Nearly all earth surface water flows with some degree of turbulence.

turbulent velocity.   Velocity of a stream above which the water must move by turbulent flow. Care should be taken not to confuse this term with "velocity of turbulence" which is occasionally used in reference to the individual water molecule velocity in turbulent flow.

turnaround (refin).   Time necessary to clean and make minor repairs in refinery equipment after a normal run. It is the elapsed time between drawing the fires or shutting the unit down and putting the unit onstream again.

turpentine.   Water white volatile liquid secured from distilling pine trees and other forms of vegatation.

Twaddel Hydrometer.   Hydrometer scale divided into degrees, used ordinarily for liquids lighter than water. To convert Twaddel degrees into specific gravity, multiply the Twaddel degrees by five, add 1,000 and divide the result by 1,000.

twist off (drill).   To twist a joint of drill pipe in two by excessive force applied by the rotary table. Many failures which result in parting of the drill pipe in the well bore are erroneously referred to by this term.

Tyler standard grade scale.   Scale for sizing particles based on the square root of two, used as specifications for sieve mesh. Alternate class limits closely approximate the class limits on the Udden grade scale and the intermediate limits on the geometric mean of the Udden scale values, 0.50, 0.71, 1.00, 1.41, 2.00.

type, locality.   Place where a rock formation was first named and described or a type specimen of a fossil collected.

type specimen.   Individual fossil chosen as the most authentic representative of a species, primarily for the purpose of stabilizing nomenclature.

U

U-233. Symbol for one type (isotope) of uranium that splits making it unstable for atomic fuel. U-233 is not found in natural uranium minerals, but must be manufactured from thorium in an atomic furnace. U stands for uranium, and 233 identifies the number of protons and neutrons in the nucleus (central core) of each atom.

U-235. Symbol of one type (isotope) or uranium which is radioactive. U-235 is a naturally occurring atomic fuel found mixed with other isotopes of uranium. U-235 makes up only 0.7 of one per cent of the total. The mixture called natural uranium can be used to operate an atomic furnace, although only a small percentage of U-235 atoms split.

U-238. Symbol for the type (isotope) of uranium that makes up more than 90 per cent of natural uranium. U-238 is radioactive but with too long a half life (decays too slowly) to be useful as an atomic fuel. U-238 can be transformed inside an atomic reactor into plutonium which is an effective atomic fuel.

Ubbelohde melting point (grease). Temperature during heating when the first drop separates for a test sample supported around the bulb of a thermometer in a glass cup.

Ubbelohde viscosimeter. An instrument incorporating a suspended level apparatus which determines to a high degree of accuracy the viscosity of liquids.

ultimate analysis. The extreme chemical analysis of an organic (petroleum) compound showing the proportions of the elements or the classification of the atoms. Thus, the ultimate analysis will give the carbon, hydrogen, oxygen and nitrogen content of the compound and not the proportions of elementary compounds such as methane, decane, etc.

ultra basic. Igneous rock containing less silica (SiO_2) than a basic rock and consisting essentially of ferromagnesian silicates. The percent silica allowed in ultrabasic rocks is not universally agreed upon but is generally considered to be less than 45 per cent.

ultramafic rocks. Igneous rocks consisting chiefly or entirely of mafic minerals. Most, but not all, ultramafic rocks are ultrabasic. An ultramafic rock consisting entirely of hornblende is not ultrabasic because the silica content is higher than that allowed for ultrabasic rocks.

unconformity. A plane of erosion or nondeposition between two groups of rocks. Major breaks in sedimentation. Three types of unconformity are recognized: disconformity, angular unconformity, and nonconformity. A disconformity is an unconformity in a sequence of essentially parallel beds; beds which should be present representing the time interval of their deposition are lacking either because of nondeposition or because they were eroded before succeeding beds were deposited. An angular unconformity is an erosion surface truncating folded or tilted strata and upon which a sequence of strata were deposited parallel to the surface. A nonconformity is created when sedimentary strata are deposited on a surface eroded on older plutonic (igneous or metamorphic) rocks. Block diagram A depicts an angular unconformity; diagram B, a disconformity; and diagram C, a nonconformity. The topographic forms are examples of many possible land forms which might develop.

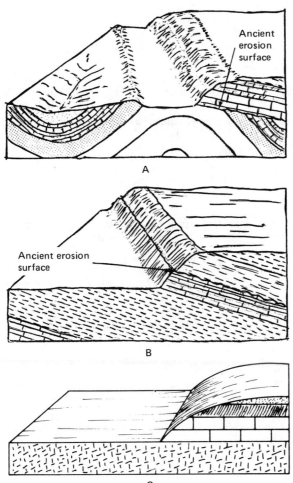

Ancient
erosion
surface

A

Ancient erosion
surface

B

C

unconsolidated sand (drill). Loose, caving sand frequently causing trouble when drilling for oil.

unctiousness. Term or name for the oiliness or slipperiness as found in lubricating oils.

underclay. Layer of clay under a bed of coal; represents the soil in which the plants that formed the coal once grew.

undercutting. Removal of material at the bottom of river bank, steep slope, or cliff usually resulting in the upper portions of the bank, slope or cliff hanging over the lower portions; accomplished by stream, wave or wind erosion and less frequently by alpine glaciers.

underground water. Water occupying the pore spaces in the regolith and solid rock below the surface of the ground; also called ground water and subsurface water.

underlie (stratig). Rocks found in position directly under other rocks. Frequently, but not exclusively, applied to sequence of sedimentary strata where strata occupying a lower position are said to underlie those layers immediately above.

under-ream (drill). To enlarge a drill hole below the casing.

under-reamer (drill). A reamer with expanding cutters used for enlarging the bore of the hole below the drive shoe.

undisturbed. Sample taken so gently and carefully that the sampling process has in no way distorted, disrupted or otherwise affected the sample.

uniformitarian principle. Idea that geological processes operate at essentially the same rates and in the same ways now as they did in the past; sometimes stated as "The present is a key to the past."

unloading. Reduction of pressure on underlying rocks by erosion or denudation of overlying material.

unsaturated. Having the property of taking on additional products as the halogen elements without giving up hydrogen, as the olefin and acetylene series, etc.

unsaturated carbon. Organic compounds having carbon atoms without hydrogens attached. Compounds containing carbon bonds, as found in ethene, propene, butene, and all of those ending in "ene" are called unsaturated hydrocarbons.

unsaturated hydrocarbons. Hydrocarbons of such molecular structure that at least two adjacent carbon atoms are connected by two or three valences or bonds. Each valence not taken up by an adjacent atom is satisfied by a hydrogen atom.

unstable. Easily disintegrated. Radioactive materials are unstable, their atoms break up into different kinds of atoms; not in equilibrium.

unsulfonated residue (UR). Percentage of oil insoluble in sulfuric acid under test conditions. The higher the unsulfonated residue, the more paraffin in the oil. Usually applied to spray oils where the content of certain unsaturated or reactive fracions are significant.

uphole shooting (geoph). A procedure used in seismic prospecting to measure the velocity of sound in various subsurface strata or formations. Explosive charges are placed at various depths in a shot hole and detonated sequentially from bottom to top. The time of transit for a shock wave from the site of each explosion to the earth's surface is recorded and used to calculate variations in the velocity of sound with depth in the rock penetrated by the shot hole.

uphole time (geoph). Length of time for a seismic pulse to travel from an explosion detonated at some depth in a shot hole to the earth's surface. The pulse is recorded by a seismometer placed on the ground immediately adjacent to the bore hole.

uplands. Elevated areas between stream channels.

uplift. General process of raising the land surface, usually by tectonic processes.

upset. To forge the ends of tubular products in such a way that the wall is given extra thickness and strength near the end.

upset end joint (drill). Pipe in which the metal at the ends is thickened for a short distance, usually to such an extent that the threading operation leaves a great thickness of metal below the roots of the thread as in the main body of the tube imparting strength for use in oil drilling operations.

upthrow. Block or mass of rock on that side of a fault displaced relatively upward. Term normally refers merely to a relative and not absolute displacement.

upwarp. A broad area uplifted gently by tectonic forces.

uranium. Heaviest naturally occurring atoms; its two principle isotopes, U^{235} and U^{238} differ greatly in their nuclear behavior, the first fissioning readily with slow neutrons, the second only with very fast neutrons. Each decay, ultimately to a characteristic isotope of lead.

uranium-lead radioactive clock.   Uranium provides an extremely accurate means of measuring geological time. Uranium is continuously giving off rays some of which are actually particles of the nucleus. As these particles are emitted, the uranium atom disintegrates at a constant and known rate into lighter elements. The uranium continues to break down until it forms atoms of lead. With the formation of lead, the radioactivity stops. The original amount of uranium in a rock can be calculated from the amount of uranium and radiogenic lead in the rock today. Combining this information with the known rate of decay of uranium, allows scientists to calculate the age of formation of the rock. This radioactive age dating technique was first developed and used in the late 1940s.

V

vacuum. Space containing little or no air or molecules of any kind; a void.

vacuum distillation (refin). A method of distillation carried on at a low pressure area or partial vacuum in a still, usually associated with steam distillation to reduce to a minimum decomposition the lubricating oil being distilled. Such a still is called a vacuum still and the residue remaining in the still is frequently referred to as short residuum.

vacuum gauge. A gauge for indicating the degree of vacuum, usually expressed in terms of inches of mercury or in millimeters of mercury for laboratory determinations. This can be in the form of an ordinary direct reading spring gauge or a simple mercury column or "U" tube.

vacuum gas oil (refin). An overhead product used as a catalytic cracking stock or, after suitable treatment, may be used as a very light lubricating oil.

vacuum still (refin). A still in which a partial vacuum is created by means of a pump drawing off the vapor over the distillate flowing out of the rundown line. The pump also serves to remove all air from the still and prevents oxidation.

vadose water. Ground water in the zone of aeration above the water table.

valence. Valence is a number representing the combining power of an atom, as the number of electrons lost, gained, or shared by an atom in a compound. It is also a measure of the number of hydrogen atoms with which an atom will combine or replace, as an oxygen atom combines with two hydrogens, hence has a valence of 2.

valley. A relatively shallow, wide depression with gentle slopes, the bottom grading continuously downward. The term is used for features that do not have canyon-like characteristics in any significant part of their extent.

valley-and-ridge topography. A ground surface exhibiting a succession of parallel or sub-parallel ridges and valleys. Such topography develops when erosion by running water operates on a bedrock of folded sedimentary strata. Resistant or hard beds erode to ridges whereas softer strata erode to valleys; typical of the valley and ridge region of the Appalachian Mountains; also called ridge-and-valley topography.

valve. One of two usually articulated shells of a bivalved invertebrate organism. A single clam shell, separated from its counterpart would be an example.

vapor.   Any substance existing in the gaseous states at a temperature lower than that of its critical point; that is, a gas cool enough to be liquified if sufficient pressure were applied to it. If any vapor is cooled sufficiently at constant pressure, it ultimately reaches a state of saturation that further removal of heat is accompanied by condensation to the liquid phase.

vapor density.   Quantity of liquid while in vapor form in any given volume and determined by weight, as the weight of a given volume of vapor compared with the same volume of air or hydrogen free of vapor.

vapor phase cracking process (refin).   A process in which the charging stock is heated to higher temperatures than in liquid phase cracking and taking place in a matter of only seconds. Generally kerosene or a light gas oil are used as charging stock.

vapor phase system (refin).   A catalytic process passing superheated oil vapors through a bed of solid catalyst, for the production of anti-knock gasolines.

vapor pressures.   Pressure exerted by molecules of a given vapor. For a pure, confined vapor, it is the vapor's pressure on the walls of its containing vessel; for a vapor mixed with other vapors or gases, it is that vapor's contribution to the total pressure, its partial pressure. That part of the pressure in an enclosed space due to the vapor of the substance contained in that space. A measure of the tendency of the substance to evaporate.

varves.   Regular layers or alternations of materials in sedimentary deposits caused by annual seasonal influences. Each varve is usually interpreted to represent the deposition during a year and consists ordinarily of a lower part deposited in summer and an upper, fine-grained part, deposited in winter. Care must be taken when interpreting varves because documented occurrences of varve-like layers have been observed associated with single storms or flood. Varves of silt and clay-like material occur abundantly in lake sediments associated with glaciers and in certain marine shales and slates. Diagram showing delta and varved sediments built up year by year by a sub-glacial stream entering a lake.

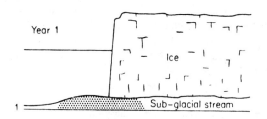

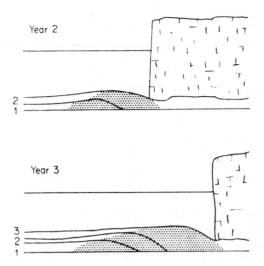

vein. A fissure or elongated opening in a rock mass filled with mineral matter formed later than that of the rock surrounding it; ore or other material filling a fissure in a rock. Diagram of a cross section of a vein filling a fault fissure.

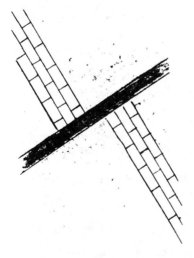

velocity, critical. Velocity at the transitional point between laminar and turbulent types of fluid flow. This point occurs in the transitional range of Reynolds numbers of approximately 2,000 to 3,000.

velocity discontinuity. An abrupt change of the rate of propagation of seismic waves within the earth as at an interface.

velocity hydrophone.   A hydrophone in which the electric output substantially corresponds to the instantaneous particle velocity in the impressed sound waves.

veneer.   A thin layer of sediment covering a rock surface.

ventifact.   A pebble, cobble, or boulder with its shape or surface modified by wind-driven sand.

venturi meter.   Instrument for measuring the velocity or indirectly the volume of flowing fluids. Consists of two parts, the tube through which the fluid flows and a set of indicators or a recorder showing the pressure, rate of flow, or quantity discharged through the tube.

venturi tube.   Short tube of smaller diameter in the middle than at the ends. When a fluid flows through such a tube, the pressure decreases as the diameter becomes smaller, the amount of the decrease being proportional to the speed of flow and the amount of restriction.

verifier (drill).   Tool used in deep drilling for detaching and bringing to the surface portions of the wall of the drill hole, at any desired depth.

vertical exaggeration.   (1) A purposely increased scale in the vertical dimension over the horizontal dimension of a topographic or geologic cross-section in order to emphasize vertical irregularities or variations. (2) The difference between the vertical and horizontal scales on the records of marine seismic profiles, caused by the fact that the horizontal velocity of the ship is far less than the vertical velocity of the seismic or sound pulses; results in exaggerated thickness of strata and exaggerated steepness of dips. These exaggerations must be corrected before the record can be interpreted accurately.

vertical intensity.   Magnetic intensity of the vertical components of the earth's magnetic field, reckoned positive if downward, negative if upward.

vesicle.   Small cavity in a fine grained or glassy igneous rock, formed by a bubble of gas during the solidification of lava. Openings in lavas formed by expanding gas. Expansion of gases take place with reduction in pressure as lavas come up from depth to the earth's surface giving lava a spongy or vesicular appearance. If the bubbles are very small, producing a glassfroth, the structure is pumaceous; if the bubbles are larger, the structure is scoriaceous. Pumice is very light and will float until its cavities are filled by water; scoria, though also light in weight, will rarely float.

virgin stock.   Oil derived directly from crude oil containing no cracked materials.

viscosity.   A measure of a liquid's resistance to flow, such resistance being brought about by the internal friction resulting from the combined effects of molecular or atomic cohesion and adhesion. The viscosity of petroleum products commonly expressed in terms of the time required for a specified volume of the liquid to flow through an orifice of specific size.

viscosity, absolute.   Force moving one square centimeter of a plane surface with a speed of one centimeter per second relative to another parallel plane surface from which it is separated by a film of the liquid one millimeter thick.

viscosity breaking (refin).   Sometimes a reduced crude is obtained that is too viscous for use as a heavy fuel oil. Such a residual oil can be subjected to a mild thermal cracking which will reduce its viscosity, lower its boiling range and pour points, and in addition produce gas oils and gasoline. The cracking process is called viscosity breaking or visbreaking.

viscosity index.   A series of numbers ranging from zero to one hundred which indicates the rate of change of viscosity with temperature. A viscosity index of one hundred indicates an oil that does not tend to become viscous at low temperatures.

viscous.   Adhesive or sticky, having a ropy or glutinous consistency.

vitreous.   (1) Having the luster of broken glass. (2) of amorphous atomic structure.

vitrified rocks.   Extreme example of cataclastic metamorphism are rocks transformed into vitreous material, which under microscope appears as a noncrystalline glassy substance but which x-rays may reveal to be only cryptocrystalline.

volatile.   Easily evaporated; a term applied to a liquid as gasoline having a high vapor pressure.

volatile matter.   Normally gaseous products, except moisture, given off by a substance, as gas breaking out of live crude oil that has been added to a mud. In distillation of drilling fluids, the volatile matter is the water, oil, gas, etc. vaporized, leaving behind the total solids consisting of both dissolved and suspended solids.

volatility.   Extent to which oils vaporize. The ease with which a liquid is converted into a vaporous state. Inversely proportional to the amount of heat required to vaporize the liquid.

volatility test.   A fractional distillation test made by standard methods in which the "initial boiling point" and the "end point" are noted.

volcanic arc.   An arcuate string of volcanic islands associated with and parallel to a subduction zone along convergent plate boundaries; often found closely adjacent to a deep ocean trough as the Aleutian Islands.

volcanic ash.   Uncemented pyroclastic material consisting of fragments mostly under four millimeters in diameter. Coarse ash is one-fourth to four millimeters in grain size; fine ash is below one-fourth millimeter.

volcanic blocks.   Essentially, accessory or incidental volcanic ejecta, usually angular and larger than 32 millimeters in diameter, erupted in a solid state.

volcanic bombs.   Fragments of lava up to several feet long thrown out of a volcano in a liquid, semifluid, or plastic state and solidified in flight or soon after landing.

volcanic breccia.   A rock composed predominantly of angular volcanic fragments greater than two millimeters in size set in a subordinate matrix or composed of fragments other than volcanic set in a volcanic matrix.

volcanic cone.   Conical hill or mountain with a crater or cup-shaped hollow at the summit constructed of ash, scoria, lava, and other volcanic materials discharged through the summit crater.

volcanic dust.   Very fine particles of lava shot out of volcanoes. They are carried great distances by wind and may remain in the air for a long time.

volcanic eruption.   Forceful extrusion of molten lava onto the earth's surface with accompanying gases and pyroclastic materials. A volcanic cone usually develops around the point of eruption. The type of eruption and associated volcanic cone vary with the composition of the lava. Diagrams of the chief type of volcanic eruptions.

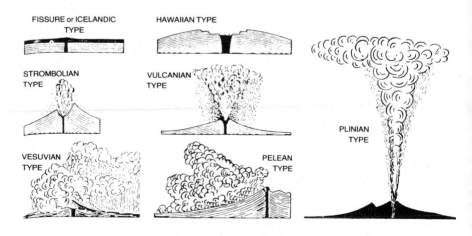

volcanic glass.   Natural glass produced when lava cools too rapidly to permit crystallization, as obsidian, pitchstone, pumice and occasionally scoria.

volcanic neck.   A rock plug formed in the lava conduit or neck of a volcano when magma slowly cools and solidifies there.

volcanism.   Process of lava generation and eruption.

volcano.   A conical hill or mountain formed around an opening in the earth's surface through which hot rock fragments, gases and lavas are ejected. As the solid materials accumulate around the conduit, they build up a cone, increasing in size until a huge volcanic mountain is formed. A cone so constructed is called a volcano.

vug.   In petroleum geology, an opening in rock usually three or four millimeters in diameter or larger, a potential or actual site for the accumulation of petroleum or natural gas.

W

wacke. A poorly sorted to unsorted or dirty sandstone consisting of angular fragments of rocks and minerals surrounded by a clay matrix; sometimes used as a shortened form of graywacke, but such usage is confusing; may also be called microbreccia.

waiting on cement (WOC) (drill). After the casing has been cemented, it is necessary to suspend operations and allow time for the cement to set or harden in the well bore. The time during which operations are suspended, is designated as waiting on cement.

walking beam. An oscillating bar or beam pivoted at the center and free to rock up and down. In oil derricks rigged for cabletool drilling, the walking beam carries the string of drilling tools suspended from one end and is connected to a crank at the other. Rotation of the crank causes the tools to lift and drop and thus to drill the hole by concussion.

walking out. A field mapping technique involving walking along an outcrop to determine if two relatively widely spread exposures are stratigraphically equivalent.

wall cake (drill). Solid material deposited along the wall of the hole resulting from filtration of the fluid part of the mud into the formation.

wall hook. A device used in fishing for a drill pipe. If the upper end of the lost pipe is leaning against the side of the well bore, the wall hook centers it in the hole so that it may be recovered with an overshot, running in on the fishing string above the wall hook.

wall hook packer (drill). A packer supporting itself against the walls of the casing.

warp. Deformed slightly from the original shape. A broad regional flexure of the earth's crust.

warping. A gentle bend of the crust of the earth not resulting in the formation of pronounced folds or faults.

wash (refin). Signifies cleaning or purifying a product by agitating it with a liquid cleaning agent; also means to purify a gas by passing it through a liquid dissolving and carrying away impurities.

wash down spear (drill). A fishing tool used in rotary drilling equipped with a bit for penetrating casings over the fishing tool and fitted with slips to obtain a bull-dog hold.

washout (drill).   An excessive well bore enlargement caused by solvent and erosional action of the drilling fluid; also a fluid-cut opening caused by leaking fluid.

washover pipe (drill).   An accessory used in fishing operations to go over the outside of tubing or drill pipe to clean out the annular space and permit recovery.

wash pipe (drill).   Small diameter pipe run inside perforated pipe to conduct water to the lower end of the screen pipe when washing out wells.

wash plug (drill).   A wooden plug run inside the perforated pipe to make a seating for the wash pipe.

water block.   Reduction of the permeability of a formation caused by the invasion of water into the pores or capillaries.

water box (refin).   A hollow iron box case cooled by a constant stream of water; for cooling any equipment or portion of a furnace.

water cushion (drill).   A water load pumped into drill pipe during a drill stem test to retard fill-up and prevent collapse of pipe under sudden pressure change.

water drive.   Oil and gas driven by a body of water to a well from which they can

Water drive.

be produced. There are two principle kinds of water drives, one utilizes natural water pressure, often called edgewater encroachment and the other method induces migration of hydrocarbons by introduction of water into the reservoir bed through an existing well or through a well specially drilled for that purpose called an injection well. The most efficient propulsive force in driving oil into a well is natural water drive where the pressure of the edgewater forces the light oil ahead and upward until all of the recoverable oil has been flushed out of the reservoir into the producing wells.

water flooding (prod).   A secondary recovery method by which water is forced down a hole specially drilled in the center of an oil field thereby forcing crude oil to migrate toward surrounding oil wells. Borax is frequently added to the water as a tracer, detected by spectrographic analysis of the crude oil recovered.

water flush system (drill).   A system of cable drilling in which the water is forced to the bottom of the hole through the drilling rods.

water gap.   A pass or steep walled valley cut all the way through a mountain ridge. The floor of the water gap is occupied by the stream responsible for eroding the gap. In most cases, the stream was in existence prior to uplift of the ridge and was able to erode at a rate which kept pace with the rate of uplift.

water gas.   A combustible gas used for industrial heating and as a gas engine fuel. It burns with a blue flame unless enriched with the vapors of petroleum oil. Water gas is the result of passing steam through deep beds of incandescent coal where it decomposes and increases the yield of carbon monoxide and methane, both combustible gases.

water gas tars.   Tars produced by cracking oil vapors at high temperatures in the manufacture of carbureted water gas.

water-in-oil emulsion.   A mixture of tiny drops of water surrounded by oil.

water level.   In an oil well, the oil-water interface in a reservoir. Pores are saturated with water below the water level and with economically recoverable quantities of petroleum or natural gas above the water level.

water separator (refin).   An apparatus used to separate mechanically the water carried over the light petroleum vapors in a distillation process.

water shed.   Entire area drained by a stream and its tributaries.

water soluble grease.   Grease wholly or partly soluble in water. A small quantity of soda base grease and a few drops of water mixed in the palm of the hand produces a soapy lather with a distinct soapy odor.

water soluble oils. Oils having the property of forming permanent emulsions or almost clear solutions with water.

water string (prod). A string of casing used to shut off all water above the oil sand.

water white distillate. A kerosene cut or refined oil cut coming from the crude oil stills before the distillate is treated or rerun.

wave cut terrace. A leveled rock bench produced by the retreat of a sea cliff through wave erosion; also called wave platform, shore platform, plain of marine abrasion, and wave cut platform.

wax. A term used loosely for any group of substances similar to beeswax in appearance and character. In general, waxes are distinguished by their composition or esters of the higher alcohols and by their freedom from fatty acids. Mineral waxes include ozokerites and paraffin, both hydrocarbons.

wax distillate. A neutral oil distillate containing a high percentage of crystallizable paraffin wax obtained in the distillation of paraffin-base or mixed-base crude and on reducing of neutral lubricating stocks. It is a primary base for paraffin wax and neutral lubricating oils.

wax oil. Immediate base from which paraffin and neutral oils are manufactured.

wax paraffin. A wax occurring principally in residues of paraffin-base petroleum, although it also occurs in limited quantities in mixed-base crudes and asphalt-base oils. It is removed by pressing, sweating, and filtering from the wax distillate taken from the tar stills. When thoroughly refined, it is a pure white wax having an unctuous surface.

wax press (refin). Mechanical unit consisting of a series of canvas covered plates designed to remove certain forms of paraffin from chilled lubricating oil prediluted with naphtha to prevent the mass from congealing. The paraffin or wax collects between the plates from where it is removed.

wax tailing (refin). A heavy distillate obtained during the final stages of distilling certain mixed-base oils down to coke. The product contains anthracene and chrysene produced by cracking. It is used for weatherproofing and waterproofing compounds.

weak acid. An acid with little ionization and yielding but a few hydrogen ions in aqueous solutions, as acetic acid.

weathered gas. Gas, obtained from the top of the stabilizer tower during the absorption process, weathered and frequently added to gasoline in small quantities. The gas is first weathered, exposed to the weather to evaporate the light volatile unstable gases before it is added to the gasoline.

weathering. A process of alteration of rocks near the earth's surface chiefly by atmospheric gases O_2 and CO_2, water, and water solutions. It is partly a mechanical breakdown and partly a chemical change into rock fragments, new minerals, and solutions.

weathering, chemical. The alteration of rocks by means of mineralogical or chemical changes induced by constituents of the atmosphere. The active ingredients in rock weathering are oxygen, carbon dioxide, water vapor, and acids principally H_2CO_3 carbonic acid dissolved in the moisture falling as precipitation, and carried into the rock as a certain amount of moisture penetrates the earth's surface. Decaying vegetation and organic wastes contribute organic acids and carbon dioxide to the water penetrating the ground. The chief processes of chemical weathering are oxidation, hydrolysis, hydration, carbonation and solution.

weathering correction (geoph). Velocity correction applied to seismic data owing to the reduced velocity of seismic wave propagation in weathered rock.

weathering, mechanical. Breakdown of a rock mass into smaller particles without chemical alteration. The chief types of mechanical weathering are **(1)** frost wedging, resulting from the freezing and expanding of water in joints breaking the rock mass into individual blocks or fragments; **(2)** root wedging, resulting from the growth of tiny roots which wedge joints apart from the pressure of their expansion and, **(3)** spheroidal weathering, resulting from expansion of the rock mass as it chemically weathers. Extreme changes in temperature related to forest and brush fires may physically weather rock by thermal expansion and dehydration.

weathering oils. Exposure of crude oils to the weather causes a loss due to the evaporation of the volatile constituents and oxidation of the heavier residuals with a tendency toward developing asphaltic contents. Weathering lubricating oils bleaches them and decomposes certain of the hydrocarbons so that there is a decided change in the gravity and viscosity.

weathering velocity (geoph). Velocity of propagation of seismic waves through weathered rock, obtained by uphole shooting and used to apply the weathering correction.

wedge out. Disappearance of a thinning, lensing, or truncated layer as its thickness reduces to zero.

weeping core. Applied to a core cut from a formation brought to the surface covered with drops of fluid that have the general appearance of tears. Such a formation is usually low in production of oil.

weight (drill).   In mud terminology, refers to the density of a drilling fluid. This is normally expressed in either pound gallon, pound cubic foot, psi hydrostatic pressure per 1,000/ft of depth.

weight of oils.   Weight of an oil can be found by multiplying the weight of an equal volume of water by the specific gravity of an oil at the same temperature.

Weiland oxidation test (lubricating oil).   A heated oil bath with a glass tube containing the test sample, air is circulated through the sample and the change in the color of the sample is used to report the results in terms of Weiland oxidation number, often called the Weiland sludge number.

well.   In the petroleum industry, a hole or boring sunk into the ground with the objective of (1) bringing petroleum and natural gas to the surface, (2) injecting water into the ground to increase petroleum production and, (3) detonating explosives during seismic prospecting, (4) obtaining subsurface samples and information.

well log.   A record of variations with depth in a well of various geological features and parameters; included are resistivity, spontaneous potential, radioactivity, acoustic velocity, gravity and lithology. Recording the log is called "well logging."

well record.   An abbreviated statement of all available information on a particular well on a day-to-day basis from spudding to cessation of production. The most widely used record is a "scout ticket."

well shooting.   A technique for measuring changes in acoustic velocity with depth in the earth. A string of geophones is suspended in a well and a small explosive charge is detonated at the well head. Occasionally a single geophone is lowered to various depths in the well while a series of small explosions are detonated at the well head. A reversal of procedure from uphole shooting. Well shooting is preferred where it is desirable to avoid explosive damage to the well.

well site.   Position selected to drill a well.

well ties.   Wells existing or specially drilled yielding information to aid interpretation or test the accuracy of a seismic profile or map.

Wentworth grade scale.   A logarithmic grade scale for size classification of sediment particles, starting at one millimeter and using the ratio of one-half in one direction and two in the other, providing diameter limits to the size classes of one, one-half, one-fourth, etc., and one, two, four, etc. Adopted by Wentworth from Uden's scale with slight modification of grade terms and limits.

Wesphal balance.   A delicate scale used to determine the specific gravity of liquids and solids. A balance in which the buoyance of a float is balanced by sliding weights; used for determining the specific gravity of liquids, minerals, fragments, etc.

wet density.   Density of a sediment or sedimentary rock when the pores are saturated with water, as opposed to dry density measured when the pores are empty or dry. In the oil industry the presence of petroleum rather than water may be considered when determining wet density.

wet gas.   Natural gas containing condensible gasoline recovered by the compression, refrigeration, or absorption process to make natural gasoline. Gas partly saturated with liquid hydrocarbons; gas that carries a lot of liquid with it.

wet hole (drill).   Hole in which drilling is retarded owing to a heavy influx of water.

wet natural gas.   Natural gas containing condensible gasoline.

wetting agent.   A substance, as soap added to a liquid reducing surface tensions and increasing the spreading of the liquid on a surface or the penetration of the liquid into a material.

whipstock.   A long steel casting using an inclined plane to cause the bit to deflect from the original borehole at a slight angle. Whipstocks are used in controlled directional drilling for (1) straightening crooked boreholes, (2) sidetracking in order to avoid unretrieved fish, (3) drilling toward a reservoir located elsewhere then directly beneath the well head, and (4) drilling several wells from one well head.

white mineral oil.   An odorless, tasteless, and colorless oil either a liquefied petroleum or a distilled and refined mineral oil.

wildcat.   A well in an unproved territory. An exploratory well.

wildcatter.   Person drilling wells in the hope of finding oil in territory not known to be an oil field.

wire line.   Any rope made out of twisted strands or wires of steel; used in well drilling to hoist drilling pipe, casing and operate a cable tool; synonymous with steel cable.

working interest.   Portion of oil production out of which operating and development costs are paid.

working pressure.   Pressure to which a particular piece of equipment is subjected during normal operations.

workover. To clean out or otherwise work on a well in order to increase or restore production. Examples of work-over operations are deepening, plugging back, pulling and resetting liners, squeeze cementing, shooting, fracturing (fraccing) and acidizing.

X

xanthate. Any salt of xanthic acid. Xanthates are widely used in oil-floatation cycles which may or may not utilize petroleum oils.

xenolith. A strange rock broken from the wall surrounding a magma chamber and frozen in the intrusion as it solidified. A fragment of country rock incorporated in congealed lava.

xylene. An aromatic hydrocarbon produced in the destructive distillation of coal to coke and in the petroleum refining processes. Of importance as a solvent in manufacture of paint; a coal tar, distillate resembling benzol and toluol, but of higher boiling and slower evaporating range.

xylol. Any of the metameric dimethylbenzenes.

Y

yellow wax. A viscous, semi-solid not very volatile obtained from still residuum, resulting from coal tar distillation; contains some anthracene and other hydrocarbons of complex structure.

yield. Percentage of specification material obtained in distilling, extracting, etc.

yield curves (refin). If a property is not additive, properties of various ranges or fractions of material can be determined experimentally by blending fractions together and plotting the property value so obtained as a function of the yield or amount of blended material.

yield point. Unit stress at which the deformation first increases markedly without any increase in the applied load.

Z

zigzag ridges. Ridges looping back and forth, formed by resistant layers in deeply eroded series of alternating plunging anticlines and synclines.

zones. (1) A subdivision of stratified rock based primarily on fossil content; may be named after the fossil or fossils it contains. No fixed thickness or lithology is implied by the term "zone." (2) Applied to reservoirs to describe an interval which has one or more distinguishing characteristics, as lithology, porosity, saturation, etc.

zone of aeration. A zone immediately below the surface of the ground and above the water table. Openings are partially filled with air, and partially with water trapped by molecular attraction. Subdivided into (1) belt of soil moisture, (2) intermediate belt, and (3) capillary fringe.. Some of the water that moves down from the surface is caught by rock and earth materials and is checked on its downward progress. The water itself is called suspended or vadose water.

zone of capillarity. An area overlying the zone of saturation and containing capillary voids, some or all of which are filled with water held above the zone of saturation by molecular attraction acting against gravity.

zone of oxidation. Upper zone of a mineral deposit that has become oxidized.

zone of saturation. Part of the ground within which all openings are filled with water. Its upper surface is the water table, extending as far down within the earth as connected openings can exist. The zone of saturation lies immediately below the zone of aeration.

zone of weathering. Uppermost portion of the earth's crust extending to various depths below the exposed surface, created by the destructive action of weathering both physical and chemical.